T0203089

Undergraduate Lecture Notes in Physics

Undergraduate Lecture Notes in Physics (ULNP) publishes authoritative texts covering topics throughout pure and applied physics. Each title in the series is suitable as a basis for undergraduate instruction, typically containing practice problems, worked examples, chapter summaries, and suggestions for further reading.

ULNP titles must provide at least one of the following:

- An exceptionally clear and concise treatment of a standard undergraduate subject.
- A solid undergraduate-level introduction to a graduate, advanced, or non-standard subject.
- A novel perspective or an unusual approach to teaching a subject.

ULNP especially encourages new, original, and idiosyncratic approaches to physics teaching at the undergraduate level.

The purpose of ULNP is to provide intriguing, absorbing books that will continue to be the reader's preferred reference throughout their academic career.

More information about this series at http://www.springer.com/series/8917

Kerry Kuehn

A Student's Guide Through the Great Physics Texts

Volume IV: Heat, Atoms and Quanta

 Springer

Kerry Kuehn
Milwaukee
Wisconsin
USA

ISSN 2192-4791   ISSN 2192-4805 (electronic)
Undergraduate Lecture Notes in Physics
ISBN 978-3-319-79364-1   ISBN 978-3-319-21828-1 (eBook)
DOI 10.1007/978-3-319-21828-1

Springer International Publishing AG Switzerland is part of Springer Science+Business Media
(www.springer.com)

For Cindy.

Preface

What is the Nature of This Book?

This four-volume book grew from a four-semester general physics curriculum which I developed and taught for the past decade to undergraduate students at Wisconsin Lutheran College in Milwaukee. The curriculum is designed to encourage a critical and circumspect approach to natural science while at the same time providing a suitable foundation for advanced coursework in physics. This is accomplished by holding before the student some of the best thinking about nature that has been committed to writing. The scientific texts found herein are considered classics precisely because they address timeless questions in a particularly honest and convincing manner. This does not mean that everything they say is true—in fact many classic scientific texts contradict one another—but it is by the careful reading, analysis and discussion of the most reputable observations and opinions that one may begin to discern truth from error.

Who is This Book For?

Like fine wine, the classic texts in any discipline can be enjoyed by both the novice and the connoisseur. For example, Sophocles' tragic play *Antigone* can be appreciated by the young student who is drawn to the story of the heroine who braves the righteous wrath of King Creon by choosing to illegally bury the corpse of her slain brother, and also by the seasoned scholar who carefully evaluates the relationship between justice, divine law and the state. Likewise, Galileo's *Dialogues Concerning Two New Sciences* can be enjoyed by the young student who seeks a clear geometrical description of the speed of falling bodies, and also by the seasoned scholar who is amused by Galileo's wit and sarcasm, or who finds in his *Dialogues* the progressive Aristotelianism of certain late medieval scholastics.[1]

---

[1] See Wallace, W. A., The Problem of Causality in Galileo's Science, *The Review of Metaphysics*, *36*(3), 607–632, 1983.

Having said this, I believe that this book is particularly suitable for the following audiences. First, it could serve as the primary textbook in an introductory discussion-based physics course at the university level. It was designed to appeal to a broad constituency of students at small liberal arts colleges which often lack the resources to offer the separate and specialized introductory physics courses found at many state-funded universities (*e.g. Physics for poets, Physics for engineers, Physics for health-care-professionals, Physics of sports, etc.*). Indeed, at my institution it is common to have history and fine arts students sitting in the course alongside biology and physics majors. Advanced high-school or home-school students will find in this book a physics curriculum that emphasizes reading comprehension, and which can serve as a bridge into college-level work. It might also be adopted as a supplementary text for an advanced placement course in physics, astronomy or the history and philosophy of science. Many practicing physicists, especially those at the beginning of their scientific careers, may not have taken the opportunity to carefully study some of the foundational texts of physics and astronomy. Perhaps this is because they have (quite understandably) focused their attention on acquiring a strong technical proficiency in a narrow subfield. Such individuals will find herein a structured review of such foundational texts. This book will also likely appeal to humanists, social scientists and motivated lay-readers who seek a thematically-organized anthology of texts which offer insight into the historical development and cultural significance of contemporary scientific theories. Finally, and most importantly, this book is designed for the benefit of the teaching professor. Early in my career as a faculty member, I was afforded considerable freedom to develop a physics curriculum at my institution which would sustain my interest for the foreseeable future—perhaps until retirement. Indeed, reading and re-reading the classic texts assembled herein has provided me countless hours of enjoyment, reflection and inspiration.

How is This Book Unique?

Here I will offer a mild critique of textbooks typically employed in introductory university physics courses. While what follows is admittedly a bit of a caricature, I believe it to be a quite plausible one. I do this in order to highlight the unique features and emphases of the present book. In many university-level physics textbooks, the chapter format follows a standard recipe. First, accepted scientific laws are presented in the form of one or more mathematical equations. This is followed by a few example problems so the student can learn how to plug numbers into the aforementioned equations and how to avoid common conceptual or computational errors. Finally, the student is presented with contemporary applications which illustrate the relevance of these equations for various industrial or diagnostic technologies.

While this method often succeeds in preparing students to pass certain standardized tests or to solve fairly straightforward technical problems, it is lacking in important respects. First, it is quite bland. Although memorizing formulas and learning how to perform numerical calculations is certainly crucial for acquiring a working knowledge of physical theories, it is often the more general questions about the assumptions and the methods of science that students find particularly stimulating and enticing. For instance, in his famous *Mathematical Principles of Natural Philosophy*, Newton enumerates four general rules for doing philosophy. Now the reader may certainly choose to reject Newton's rules, but Newton himself suggests that they are necessary for the subsequent development of his universal theory of gravitation. Is he correct? For instance, if one rejects Rules III and IV—which articulate the principle of induction—then in what sense can his theory of gravity be considered universal? Questions like "is Newton's theory of gravity correct?" and "how do you know?" can appeal to the innate sense of inquisitiveness and wonder that attracted many students to the study of natural science in the first place. Moreover, in seeking a solution to these questions, the student must typically acquire a deeper understanding of the technical aspects of the theory. In this way, broadly posed questions can serve as a motivation and a guide to obtaining a detailed understanding of physical theories.

Second, and perhaps more importantly, the method employed by most standard textbooks does not prepare the student to become a practicing scientist precisely because it tends to mask the way science is actually done. The science is presented as an accomplished fact; the prescribed questions revolve largely around technological applications of accepted laws. On the contrary, by carefully studying the foundational texts themselves the student is exposed to the polemical debates, the technical difficulties and the creative inspirations which accompanied the development of scientific theories. For example, when studying the motion of falling bodies in Galileo's *Dialogues*, the student must consider alternative explanations of the observed phenomena; must understand the strengths and weaknesses of competing theories; and must ultimately accept—or reject—Galileo's proposal on the basis of evidence and reason. Through this process the student gains a deeper understanding of Galileo's ideas, their significance, and their limitations.

Moreover, when studying the foundational texts, the student is obliged to thoughtfully address issues of language and terminology—issues which simply do not arise when learning from standard textbooks. In fact, when scientific theories are being developed the scientists themselves are usually struggling to define terms which capture the essential features of their discoveries. For example, Oersted coined a term which is translated as "electric conflict" to describe the effect that an electrical current has on a nearby magnetic compass needle. He was attempting to distinguish between the properties of stationary and moving charges, but he lacked the modern concept of the magnetic field which was later introduced by Faraday. When students encounter a familiar term such as "magnetic field," they typically accept it as settled terminology, and thereby presume that they understand the phenomenon by virtue of recognizing and memorizing the canonical term. But when they encounter an unfamiliar term such as "electric conflict," as part of the scientific argument from which it derives and wherein it is situated, they are tutored into the original argument and are thus obliged to think scientifically, along with the great scientist. In other words, when reading the foundational texts, the student is led into *doing* science and not merely into memorizing and applying nomenclature.

Generally speaking, this book draws upon two things that we have in common: (i) a shared conversation recorded in the foundational scientific texts, and (ii) an innate faculty of reason. The careful reading and analysis of the foundational texts is extremely valuable in learning how to think clearly and accurately about natural science. It encourages the student to carefully distinguish between observation and speculation, and finally, between truth and falsehood. The ability to do this is essential when considering the practical and even philosophical implications of various scientific theories. Indeed, one of the central aims of this book is to help the student grow not only as a potential scientist, but as an educated person. More specifically, it will help the student develop important intellectual virtues (*i.e.* good habits), which will serve him or her in any vocation, whether in the marketplace, in the family, or in society.

How is This Book Organized?

This book is divided into four separate volumes; volumes I and II were concurrently published in the autumn of 2014, and volumes III and IV are due to be published approximately a year later. Within each volume, the readings are centered on a particular theme and proceed chronologically. For example, Volume I is entitled *The Heavens and the Earth*. It provides an introduction to astronomy and cosmology beginning with the geocentrism of Aristotle's *On the Heavens* and Ptolemy's *Almagest*, proceeding through heliocentrism advanced in Copernicus' *Revolutions of the Heavenly Spheres* and Kepler's *Epitome of Copernican Astronomy*, and arriving finally at big bang cosmology with Lemaître's *The Primeval Atom*. Volume II, *Space, Time and Motion*, provides a careful look at the science of motion and rest. Here, students engage in a detailed analysis of significant portions of Galileo's *Dialogues Concerning Two New Sciences*, Pascal's *Treatise on the Equilibrium of Fluids and the Weight of the Mass of Air*, Newton's *Mathematical Principles of Natural Philosophy* and Einstein's *Relativity*.

Volume III traces the theoretical and experimental development of the electromagnetic theory of light using texts by William Gilbert, Benjamin Franklin, Charles Coulomb, André Marie Ampère, Christiaan Huygens, James Clerk Maxwell, Heinrich Hertz, Albert Michelson, and others. Volume IV provides an exploration of modern physics, focusing on the mechanical theory of heat, radio-activity and the development of modern quantum theory. Selections are taken from works by Joseph Fourier, William Thomson, Rudolph Clausius, Joseph Thomson, James Clerk Maxwell, Ernest Rutherford, Max Planck, James Chadwick, Niels Bohr, Erwin Schrödinger and Werner Heisenberg.

While the four volumes of the book are arranged around distinct themes, the readings themselves are not strictly constrained in this way. For example, in his *Treatise on Light*, Huygens is primarily interested in demonstrating that light can be best understood as a wave propagating through an aethereal medium comprised of tiny, hard elastic particles. In so doing, he spends some time discussing the speed of light measurements performed earlier by Ole Rømer. These measurements, in turn, relied upon an understanding of the motion of the moons of Jupiter which had recently been reported by Galileo in his *Sidereal Messenger*. So here, in this *Treatise on Light*, we find references to a variety of inter-related topics. Huygens does not artificially restrict his discussion to a narrow topic—nor does Galileo, or Newton or the other great thinkers. Instead, the reader will find in this book recurring concepts and problems which cut across different themes and which are naturally addressed in a historical context with increasing levels of sophistication and care. Science is a conversation which stretches backwards in time to antiquity.

How Might This Book be Used?

This book is designed for college classrooms, small-group discussions and individual study. Each of the four volumes of the book contains roughly thirty chapters, providing more than enough material for a one-semester undergraduate-level physics course; this is the context in which this book was originally implemented. In such a setting, one or two 50-min classroom sessions should be devoted to analyzing and discussing each chapter. This assumes that the student has read the assigned text before coming to class. When teaching such a course, I typically improvise—leaving out a chapter here or there (in the interest of time) and occasionally adding a reading selection from another source that would be particularly interesting or appropriate.

Each chapter of each volume has five main components. First, at the beginning of each chapter, I include a short introduction to the reading. If this is the first encounter with a particular author, the introduction includes a biographical sketch of the author and some historical context. The introduction will often contain a summary of some important concepts from the previous chapter and will conclude with a few provocative questions to sharpen the reader's attention while reading the upcoming text.

Next comes the reading selection. There are two basic criteria which I used for selecting each text: it must be *significant* in the development of physical theory, and it must be *appropriate* for beginning undergraduate students. Balancing these criteria was very difficult. Over the past decade, I have continually refined the selections so that they might comprise the most critical contribution of each scientist, while at the same time not overwhelming the students by virtue of their length, language or complexity. The readings are not easy, so the student should not feel overwhelmed if he or she does not grasp everything on the first (or second, or third...) reading. Nobody does. Rather—like classic literature—these texts must be "grown into" (so to speak) by returning to them time and again.

I have found that the most effective way to help students successfully engage foundational texts is to carefully prepare questions which help them identify and understand key concepts. So as the third component of each chapter, I have prepared a study guide in the form of a set of questions which can be used to direct either classroom discussion or individual reading. After the source texts themselves, the study guide is perhaps the most important component of each chapter, so I will spend a bit more time here explaining it.

The study guide typically consists of a few general discussion questions about key topics contained in the text. Each of these general questions is followed by several sub-questions which aid the student by focusing his or her attention on the author's definitions, methods, analysis and conclusions. For example, when students are reading a selection from Albert Michelson's book *Light Waves and their Uses*, I will often initiate classroom discussion with a general question such as "Is it possible to measure the absolute speed of the earth?" This question gets students thinking about the issues addressed in the text in a broad and intuitive way. If the students get stuck, or the discussion falters, I will then prompt them with more detailed follow-up questions such as: "What is meant by the term absolute speed?" "How, exactly, did Michelson attempt to measure the absolute speed of the earth?" "What technical difficulties did Michelson encounter while doing his experiments?" "To what conclusion(s) was Michelson led by his results?" and finally "Are Michelson's conclusions then justified?" After answering such simpler questions, the students are usually more confident and better prepared to address the general question which was initially posed.

In the classroom, I always emphasize that it is critical for participants to carefully read the assigned selections before engaging in discussion. This will help them to make relevant comments and to cite textual evidence to support or contradict assertions made during the course of the discussion. In this way, many assertions will be revealed as problematic—in which case they may then be refined or rejected altogether. Incidentally, this is precisely the method used by scientists themselves in order to discover and evaluate competing ideas or theories. During our discussion, students are encouraged to speak with complete freedom; I stipulate only one classroom rule: any comment or question must be stated publicly so that all others can hear and respond. Many students are initially apprehensive about engaging in public discourse, especially about science. If this becomes a problem, I like to emphasize that students do not need to make an elaborate point in order to engage in classroom discussion. Often, a short question will suffice. For example, the student might say "I am unclear what the author means by the term *inertia*. Can someone please clarify?" Starting like this, I have found that students soon join gamely in classroom discussion.

Fourth, I have prepared a set of exercises which test the student's understanding of the text and his or her ability to apply key concepts in unfamiliar situations. Some of these are accompanied by a brief explanation of related concepts or formulas. Most of them are numerical exercises, but some are provocative essay prompts. In addition, some of the chapters contain suggested laboratory exercises, a few of which are in fact field exercises which require several days (or even months) of observations. For example, in Chap. 3 of Volume I, there is an astronomy field exercise which involves charting the progression of a planet through the zodiac over the

course of a few months. So if this book is being used in a semester-long college
or university setting, the instructor may wish to skim through the exercises at the
end of each chapter so he or she can identify and assign the longer ones as ongoing
exercises early in the semester.

Finally, I have included at the end of each chapter a list of vocabulary words
which are drawn from the text and with which the student should become
acquainted. Expanding his or her vocabulary will aid the student not only in their
comprehension of subsequent texts, but also on many standardized college and
university admissions exams.

What Mathematics Preparation is Required?

It is sometime said that mathematics is the "language of science." This sentiment
appropriately inspires and encourages the serious study of mathematics. Of course if
it were taken literally then many seminal works in physics—and much of biology—
would have to be considered either unintelligible or unscientific, since they contain
little or no mathematics. Moreover, if mathematics is the *only* language of science,
then physics instructors should be stunned whenever students are enlightened by
verbal explanations which lack mathematical form. To be sure, mathematics offers
a refined and sophisticated language for describing observed phenomena, but many
of our most significant observations about nature may be expressed using everyday
images, terms and concepts: heavy and light, hot and cold, strong and weak, straight
and curved, same and different, before and after, cause and effect, form and function,
one and many. So it should come as no surprise that, when studying physics *via*
the reading and analysis of foundational texts, one enjoys a considerable degree of
flexibility in terms of the mathematical rigor required.

For instance, Faraday's *Experimental Researches in Electricity* are almost
entirely devoid of mathematics. Rather, they consist of detailed qualitative descriptions of his observations, such as the relationship between the relative motion
of magnets and conductors on the one hand, and the direction and intensity of
induced electrical currents on the other hand. So when studying Faraday's work,
it is quite natural for the student to aim for a conceptual, as opposed to a quantitative, understanding of electromagnetic induction. Alternatively, the student can
certainly attempt to connect Faraday's qualitative descriptions with the mathematical methods which are often used today to describe electromagnetic induction (*i.e.*
vector calculus and differential equations). The former method has the advantage
of demonstrating the conceptual framework in which the science was actually conceived and developed; the latter method has the advantage of allowing the student to
make a more seamless transition to upper-level undergraduate or graduate courses
which typically employ sophisticated mathematical methods.

In this book, I approach the issue of mathematical proficiency in the following
manner. Each reading selection is followed by both study questions and homework
exercises. In the study questions, I do not attempt to force anachronistic concepts or

methods into the student's understanding of the text. They are designed to encourage the student to approach the text in the same spirit as the author, insofar as this is possible. In the homework exercises, on the other hand, I often ask the student to employ mathematical methods which go beyond those included in the reading selection itself. For example, one homework exercise associated with a selection from Hertz's book *Electric Waves* requires the student to prove that two counterpropagating waves superimpose to form a standing wave. Although Hertz casually mentions that a standing wave is formed in this way, the problem itself requires that the student use trigonometric identities which are not described in Hertz's text. In cases such as this, a note in the text suggests the mathematical methods which are required. I have found this to work quite well, especially in light of the easy access which today's students have to excellent print and online mathematical resources.

Generally speaking, there is an increasing level of mathematical sophistication required as the student progresses through the curriculum. In Volume I students need little more than a basic understanding of geometry. Euclidean geometry is sufficient in understanding Ptolemy's epicyclic theory of planetary motion and Galileo's calculation of the altitude of lunar mountains. The student will be introduced to some basic ideas of non-Euclidean geometry toward the end of Volume I when studying modern cosmology through the works of Einstein, Hubble and Lemaître, but this is not pushed too hard. In Volume II students will make extensive use of geometrical methods and proofs, especially when analyzing Galileo's work on projectile motion and the application of Newton's laws of motion. Although Newton develops his theory of gravity in the *Principia* using geometrical proofs, the homework problems often require the student to make connections with the methods of calculus. The selections on Einstein's special theory of relativity demand only the use of algebra and geometry. In Volume III, mathematical methods will, for the most part, be limited to geometry and algebra. More sophisticated mathematical methods will be required, however, in solving some of the problems dealing with Maxwell's electromagnetic theory of light. This is because Maxwell's equations are most succinctly presented using vector calculus and differential equations. Finally, in Volume IV, the student will be aided by a working knowledge of calculus, as well as some familiarity with the use of differential equations.

It is my feeling that in a general physics course, such as the one being presented in this book, the extensive use of advanced mathematical methods (beyond geometry, algebra and elementary calculus) is not absolutely necessary. Students who plan to major in physics or engineering will presumably learn more advanced mathematical methods (*e.g.* vector calculus and differential equations) in their collateral mathematics courses, and they will learn to apply these methods in upper-division (junior and senior-level) physics courses. Students who do not plan to major in physics will typically not appreciate the extensive use of such advanced mathematical methods. And it will tend to obscure, rather than clarify, important physical concepts. In any case, I have attempted to provide guidance for the instructor, or for the self-directed student, so that he or she can incorporate an appropriate level of mathematical rigor.

Figures, Formulas, and Footnotes

One of the difficulties in assembling readings from different sources and publishers into an anthology such as this is how to deal with footnotes, references, formulas and other issues of annotation. For example, for any given text selection, there may be footnotes supplied by the author, the translator and the anthologist. So I have appended a [K.K.] marking to indicate when the footnote is my own; I have not included this marking when there is no danger of confusion, for example in my footnotes appearing in the introduction, study questions and homework exercises of each chapter.

For the sake of clarity and consistency, I have added (or sometimes changed the) numbering for figures appearing in the texts. For example, Fig. 16.3 is the third figure in Chap. 16 of Volume I; this is not necessarily how Kepler or his translator numbered this figure when it appeared in an earlier publication of his *Epitome Astronomae Copernicanae*. For ease of reference, I have also added (or sometimes changed the) numbering of equations appearing in the texts. For example, Eqs. 31.1 and 31.2 are the equations of the Lorentz and Galilei transformations appearing in the reading in Chap. 31 of Volume II, extracted from Einstein's book *Relativity*. This is not necessarily how Einstein numbered them.

In several cases, the translator or editor has included references to page numbers in a previous publication. For example, the translators of Galileo's *Dialogues* have indicated, within their 1914 English translation, the locations of page breaks in the Italian text published in 1638. A similar situation occurs with Faith Wallis's 1999 translation of Bede's *The Reckoning of Time*. For consistency, I have rendered such page numbering in bold type surrounded by slashes. So **/50/** refers to page 50 in some earlier "canonical" publication.

Acknowledgements

I suppose that it is common for a teacher to eventually mull over the idea of compiling his or her thoughts on teaching into a coherent and transmittable form. Committing this curriculum to writing was particularly difficult because I am keenly aware how my own thinking about teaching physics has changed significantly since my first days in front of the classroom—and how it is quite likely to continue to evolve. So this book should be understood as a snapshot, so to speak, of how I am teaching my courses at the time of writing. I would like to add, however, that I believe the evolution of my teaching has reflected a maturing in thought, rather than a mere drifting in opinion. After all, the classic texts themselves are formative: how can a person, whether student or teacher, not become better informed when learning from the best thinkers?

This being said, I would like to offer my apologies to those students who suffered through the birth pains, as it were, of the curriculum presented in this book. The countless corrections and suggestions that they offered are greatly appreciated; any and all remaining errors in the text are my own fault. Many of the reading selections included herein were carefully scanned, edited and typeset by undergraduate students who served as research and editorial assistants on this project: Jaymee Martin-Schnell, Dylan Applin, Samuel Wiepking, Timothy Kriewall, Stephanie Kriewall, Cody Morse, Michaela Otterstatter, and Ethan Jahns deserve special thanks. My home institution, Wisconsin Lutheran College, provided me with considerable time and freedom to develop this book, including a year-long sabbatical leave, for which I am very grateful. During this sabbatical, I received support and encouragement from my trusty colleagues in the Department of Mathematical and Physical Sciences. Also, the Higher Education Initiatives Program of the Wisconsin Space Grant Consortium provided generous funding for this project, as did the Faculty Development Committee of Wisconsin Lutheran College. Greg Schulz has been an invaluable intellectual resource throughout this project. Aaron Jensen conscientiously translated selections of the *Almagest*, included in Volume I of this book, from Heiberg's edition of Ptolemy's Greek manuscript. And Glen Thompson was instrumental in getting this translation project initiated. Starla Siegmann and Jenny

Baker, librarians at the Marvin M. Schwan Library of Wisconsin Lutheran College, were always up to the challenge of speedily procuring obscure resources from remote libraries. I would also like to thank the following individuals who facilitated the complex task of acquiring permissions to reprint the texts included in this book: Jenny Howard at the Liverpool University Press, Elizabeth Sandler, Emilie David and Norma Rosado-Blake at the American Association for the Advancement of Science, Chris Erdmann at the Harvard College Observatory's Wolbach Library, Carmen Pagán at Encyclopædia Britannica, and Michael Fisher, McKenzie Carnahan and and Scarlett Huffman at the Harvard University Press. Cornelia Mutel and Kathryn Hodson very kindly provided digital images for inclusion with the Galileo and Pascal selections from the History of Hydraulics Rare Book Collection at the University of Iowa's IIHR-Hydroscience and Engineering. Also, I would like to thank Jeanine Burke, the acquisition editor at Springer who originally agreed to take on this project with me, and Robert Korec and Tom Spicer who patiently saw it through to publication. Shortly after submitting my book proposal to Springer, I received very encouraging and helpful comments from several anonymous reviewers, for whom I am thankful. I received similar suggestions from the editors of Springer's Undergraduate Lecture Notes in Physics series for which I am likewise grateful. Finally, I would especially like to thank my wife, Cindy, who has provided unwavering encouragement and support for my work from the very start.

Milwaukee, 2015   Kerry Kuehn

Contents

Chapter 1
A New Science of Heat

Profound study of nature is the most fertile source of
mathematical discoveries.

—Joseph Fourier

1.1 Introduction

Jean Baptiste Joseph Fourier (1768–1830) was born in Auxerre, France. His parents died before his ninth year; he was one of the last of their nineteen children. As a young orphan, Fourier was selected to attend the town's military school run by Benedictine monks, and when his application to join the military engineers was turned down he became a novice at the Benedictine abbey of St. Benoît-sur-Loire in 1787. He showed early promise in mathematics, and after relinquishing his novitiate he returned to his old school in Auxerre to teach mathematics, history, philosophy and rhetoric. During this time, Fourier's forceful criticism of government corruption led to official demands for his execution. But after a brief period of imprisonment, and the public execution of Robespierre in July of 1794, Fourier was granted amnesty. After his release, he attended the newly formed École Normal in Paris—a short-lived school where the mathematicians Simon Laplace, Joseph Lagrange and Gaspard Monge were his professors—before transferring (along with many of the Faculty) to the École Polytechnique, an elite military academy. Due to his comparative age and his reputation as a gifted lecturer, Fourier was elevated to an assistant teaching post. He devoted his leisure time to outstanding problems in mathematics, including a proof of Descartes' rule of signs.[1] But this would not last. As an able administrator and outspoken advocate of the French Revolution, Fourier was soon asked to serve as one of Napoleon's scientific advisors on his expedition to Egypt in 1798. He was appointed Secretary of the Institut d'Égypte which was established by Napoleon in Cairo; this role involved both scientific and administrative duties. The discovery of

---

[1] Descartes' rule of signs provides a technique for determining the number of positive real roots of a polynomial. An overview of Fourier's life and work is provided in Grattan-Guinness, I., *Joseph Fourier 1768–1830*, The Massachusettes Institute of Technology, 1972. Grattan-Guinness's text is the source for much of the present introduction.

© Springer International Publishing Switzerland 2016
K. Kuehn, *A Student's Guide Through the Great Physics Texts*,
Undergraduate Lecture Notes in Physics, DOI 10.1007/978-3-319-21828-1_1

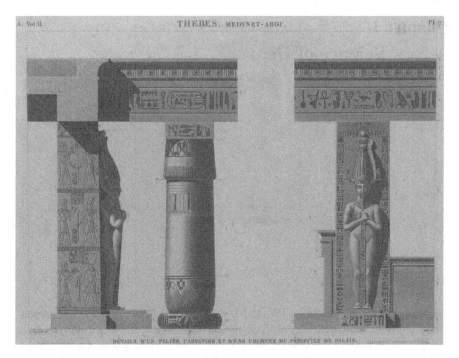

Fig. 1.1 Detail of caryatid pillars; from vol. 2 of the *Description of Egypt*. (image courtesy of the World Digital Library, Washington, DC)

the Rosetta Stone by French troops in 1799 would enable Jean-François Champillon, a skilled linguist and young acquaintance of Fourier's, to finally decipher the Egyptian language.

After Napoleon's premature departure from Egypt and Fourier's subsequent return to France in 1801, he was appointed by Napoleon himself to a leading administrative post as the Prefect of the Department of Isère in Grenoble. Despite his heavy administrative responsibilities in Egypt and Grenoble, Fourier was able to carry out significant scholarly work during this time: his scientific researches on the diffusion of heat were published in 1807, and his researches on Egyptian antiquities were later published as part of the monumental *Description de l'Égypte*—a twenty-one volume scholarly work which Fourier himself had conceived during the French expedition in Egypt (see Fig. 1.1). After Napoleon's abdication and his eventual defeat at Waterloo in 1814, Fourier was made Director of the Bureau of Statistics in Paris, despite the fact that he was a committed Bonapartist. Fourier was elected to the French *Académie des Sciences* in 1817, to the Royal Society in 1823, and to the *Académie Française* and the *Académie de Médicine* in 1827.

The reading selections in the next several chapters of the present volume are from Alexander Freeman's 1878 English translation of Fourier's 1822 *Théorie Analytique de la Chaleur*. This treatise expanded on the famous paper of 1807 in which Fourier employed Newton's law of cooling to advance a mathematically precise theory of heat diffusion through bodies of various shapes, sizes and compositions.[2] Fourier's 1807 paper also introduced a controversial mathematical technique—now known as a *Fourier series*—which allowed him to express the temperature distribution within a solid body as a weighted sum of trigonometric functions (such as sines and cosines) having different spatial frequencies of oscillation. Today, this method of analysis underlies much of electrical engineering, image processing and spectroscopy (to name just three); it also provides a means for understand the Heisenberg uncertainty principal, the basis of modern quantum theory. Fourier begins his 1822 *Analytical Theory of Heat* by situating his scientific work on heat diffusion into the broader framework of natural philosophy.

1.2   Reading: Fourier, *The Analytical Theory of Heat*

Fourier, J., *The Analytical Theory of Heat*, Cambridge University Press, London, 1878.

1.2.1   Preliminary Discourse

Primary causes are unknown to us; but are subject to simple and constant laws, which may be discovered by observation, the study of them being the object of natural philosophy.

Heat, like gravity, penetrates every substance of the universe, its rays occupy all parts of space. The object of our work is to set forth the mathematical laws which this element obeys. The theory of heat will hereafter form one of the most important branches of general physics.

The knowledge of rational mechanics, which the most ancient nations had been able to acquire, has not come down to us, and the history of this science, if we except the first theorems in harmony, is not traced up beyond the discoveries of Archimedes. This great geometer explained the mathematical principles of the equilibrium of solids and fluids. About eighteen centuries elapsed before Galileo, the originator of dynamical theories, discovered the laws of motion of heavy bodies. Within this new science Newton comprised the whole system of the universe. The successors of these philosophers have extended these theories, and given them an admirable perfection: they have taught us that the most diverse phenomena are subject to a small number of fundamental laws which are reproduced in all the acts of nature. It is recognised that the same principles regulate all the movements of the stars, their form, the inequalities of their courses, the equilibrium and the oscillations

---

[2] Newton's law of cooling states that the quantity of heat flowing between two adjacent bodies is simply proportional to their temperature difference.

of the seas, the harmonic vibrations of air and sonorous bodies, the transmission of light, capillary actions, the undulations of fluids, in fine the most complex effects of all the natural forces, and thus has the thought of Newton been confirmed: *quod tam paucis tam multa præstet geometria gloriatur.*[3]

But whatever may be the range of mechanical theories, they do not apply to the effects of heat. These make up a special order of phenomena, which cannot be explained by the principles of motion and equilibrium. We have for a long time been in possession of ingenious instruments adapted to measure many of these effects; valuable observations have been collected; but in this manner partial results only have become known, and not the mathematical demonstration of the laws which include them all.

I have deduced these laws from prolonged study and attentive comparison of the facts known up to this time: all these facts I have observed afresh in the course of several years with the most exact instruments that have hitherto been used.

To found the theory, it was in the first place necessary to distinguish and define with precision the elementary properties which determine the action of heat. I then perceived that all the phenomena which depend on this action resolve themselves into a very small number of general and simple facts; whereby every physical problem of this kind is brought back to an investigation of mathematical analysis. From these general facts I have concluded that to determine numerically the most varied movements of heat, it is sufficient to submit each substance to three fundamental observations. Different bodies in fact do not possess in the same degree the power to *contain* heat, *to receive or transmit it across their surfaces*, nor to *conduct* it through the interior of their masses. These are the three specific qualities which our theory clearly distinguishes and shews how to measure.

It is easy to judge how much these researches concern the physical sciences and civil economy, and what may be their influence on the progress of the arts which require the employment and distribution of heat. They have also a necessary connection with the system of the world, and their relations become known when we consider the grand phenomena which take place near the surface of the terrestrial globe.

In fact the radiation of the sun in which this planet is incessantly plunged, penetrates the air, the earth, and the waters; its elements are divided, change in direction every way, and, penetrating the mass of the globe, would raise its mean temperature more and more, if the heat acquired were not exactly balanced by that which escapes in rays from all points of the surface and expands through the sky.

Different climates, unequally exposed to the action of solar heat, have, after an immense time, acquired the temperatures proper to their situation. This effect is modified by several accessory causes, such as elevation, the form of the ground, the neighbourhood and extent of continents and seas, the state of the surface, the direction of the winds.

---

[3] *Philolophiæ naturalis principia mathematica. Auctoris præratio ad lectorem.* Ao gloriatur geometria quod tam paucis principiis aliunde petitis tam multa præstet. [A. F.]

The succession of day and night, the alternations of the seasons occasion in the solid earth periodic variations, which are repeated every day or every year: but these changes become less and less sensible as the point at which they are measured recedes from the surface. No diurnal variation can be detected at the depth of about 3 m (10 ft.); and the annual variations cease to be appreciable at a depth much less than 60 m . The temperature at great depths is then sensibly fixed at a given place: but it is not the same at all points of the same meridian; in general it rises as the equator is approached.

The heat which the sun has communicated to the terrestrial globe, and which has produced the diversity of climates, is now subject to a movement which has become uniform. It advances within the interior of the mass which it penetrates throughout, and at the same time recedes from the plane of the equator, and proceeds to lose itself across the polar regions.

In the higher regions of the atmosphere the air is very rare and transparent, and retains but a minute part of the heat of the solar rays: this is the cause of the excessive cold of elevated places. The lower layers, denser and more heated by the land and water, expand and rise up: they are cooled by the very fact of expansion. The great movements of the air, such as the trade winds which blow between the tropics, are not determined by the attractive forces of the moon and sun. The action of these celestial bodies produces scarcely perceptible oscillations in a fluid so rare and at so great a distance. It is the changes of temperature which periodically displace every part of the atmosphere.

The waters of the ocean are differently exposed at their surface to the rays of the sun, and the bottom of the basin which contains them is heated very unequally from the poles to the equator. These two causes, ever present, and combined with gravity and the centrifugal force, keep up vast movements in the interior of the seas. They displace and mingle all the parts, and produce those general and regular currents which navigators have noticed.

Radiant heat which escapes from the surface of all bodies, and traverses elastic media, or spaces void of air, has special laws, and occurs with widely varied phenomena. The physical explanation of many of these facts is already known; the mathematical theory which I have formed gives an exact measure of them. It consists, in a manner, in a new catoptrics which has its own theorems, and serves to determine by analysis all the effects of heat direct or reflected.

The enumeration of the chief objects of the theory sufficiently shews the nature of the questions which I have proposed to myself. What are the elementary properties which it is requisite to observe in each substance, and what are the experiments most suitable to determine them exactly? If the distribution of heat in solid matter is regulated by constant laws, what is the mathematical expression of those laws, and by what analysis may we derive from this expression the complete solution of the principal problems? Why do terrestrial temperatures cease to be variable at a depth so small with respect to the radius of the earth? Every inequality in the movement of this planet necessarily occasioning an oscillation of the solar heat beneath the surface, what relation is there between the duration of its period, and the depth at which the temperatures become constant?

What time must have elapsed before the climates could acquire the different temperatures which they now maintain; and what are the different causes which can now vary their mean heat? Why do not the annual changes alone in the distance of the sun from the earth, produce at the surface of the earth very considerable changes in the temperatures?

From what characteristic can we ascertain that the earth has not entirely lost its original heat; and what are the exact laws of the loss?

If, as several observations indicate, this fundamental heat is not wholly dissipated, it must be immense at great depths, and nevertheless it has no sensible influence at the present time on the mean temperature of the climates. The effects which are observed in them are due to the action of the solar rays. But independently of these two sources of heat, the one fundamental and primitive proper to the terrestrial globe, the other due to the presence of the sun, is there not a more universal cause, which determines *the temperature of the heavens*, in that part of space which the solar system now occupies? Since the observed facts necessitate this cause, what are the consequences of an exact theory in this entirely new question; how shall we be able to determine that constant value of *the temperature of space*, and deduce from it the temperature which belongs to each planet?

To these questions must be added others which depend on the properties of radiant heat. The physical cause of the reflection of cold, that is to say the reflection of a lesser degree of heat, is very distinctly known; but what is the mathematical expression of this effect?

On what general principles do the atmospheric temperatures depend, whether the thermometer which measures them receives the solar rays directly, on a surface metallic or unpolished, or whether this instrument remains exposed, during the night, under a sky free from clouds, to contact with the air, to radiation from terrestrial bodies, and to that from the most distant and coldest parts of the atmosphere?

The intensity of the rays which escape from a point on the surface of any heated body varying with their inclination according to a law which experiments have indicated, is there not a necessary mathematical relation between this law and the general fact of the equilibrium of heat; and what is the physical cause of this inequality in intensity?

Lastly, when heat penetrates fluid masses, and determines in them internal movements by continual changes of the temperature and density of each molecule, can we still express, by differential equations, the laws of such a compound effect; and what is the resulting change in the general equations of hydrodynamics?

Such are the chief problems which I have solved, and which have never yet been submitted to calculation. If we consider further the manifold relations of this mathematical theory to civil uses and the technical arts, we shall recognize completely the extent of its applications. It is evident that it includes an entire series of distinct phenomena, and that the study of it cannot be omitted without losing a notable part of the science of nature.

The principles of the theory are derived, as are those of rational mechanics, from a very small number of primary facts, the causes of which are not considered by geometers, but which they admit as the results of common observations confirmed by all experiment.

The differential equations of the propagation of heat express the most general conditions, and reduce the physical questions to problems of pure analysis; and this is the proper object of theory. They are not less rigorously established than the general equations of equilibrium and motion. In order to make this comparison more perceptible, we have always preferred demonstrations analogous to those of the theorems which serve as the foundation of statics and dynamics. These equations still exist, but receive a different form, when they express the distribution of luminous heat in transparent bodies, or the movements which the changes of temperature and density occasion in the interior of fluids. The coefficients which they contain are subject to variations whose exact measure is not yet known, but in all the natural problems which it most concerns us to consider, the limits of temperature differ so little that we may omit the variations of these coefficients.

The equations of the movement of heat, like those which express the vibrations of sonorous bodies, or the ultimate oscillations of liquids, belong to one of the most recently discovered branches of analysis, which it is very important to perfect. After having established these differential equations their integrals must be obtained; this process consists in passing from a common expression to a particular solution subject to all the given conditions. This difficult investigation requires a special analysis founded on new theorems, whose object we could not in this place make known. The method which is derived from them leaves nothing vague and indeterminate in the solutions, it leads them up to the final numerical applications, a necessary condition of every investigation, without which we should only arrive at useless transformations.

The same theorems which have made known to us the equations of the movement of heat, apply directly to certain problems of general analysis and dynamics whose solution has for a long time been desired.

Profound study of nature is the most fertile source of mathematical discoveries. Not only has this study, in offering a determinate object to investigation, the advantage of excluding vague questions and calculations without issue; it is besides a sure method of forming analysis itself, and of discovering the elements which it concerns us to know, and which natural science ought always to preserve: these are the fundamental elements which are reproduced in all natural effects.

We see, for example, that the same expression whose abstract properties geometers had considered, and which in this respect belongs to general analysis, represents as well the motion of light in the atmosphere, as it determines the laws of diffusion of heat in solid matter, and enters into all the chief problems of the theory of probability.

The analytical equations, unknown to the ancient geometers, which Descartes was the first to introduce into the study of curves and surfaces, are not restricted to the properties of figures, and to those properties which are the object of rational mechanics; they extend to all general phenomena. There cannot be a language more universal and more simple, more free from errors and from obscurities, that is to say more worthy to express the invariable relations of natural things.

Considered from this point of view, mathematical analysis is as extensive as nature itself; it defines all perceptible relations, measures times, spaces, forces, temperatures; this difficult science is formed slowly, but it preserves every principle which it has once acquired; it grows and strengthens itself incessantly in the midst of the many variations and errors of the human mind.

Its chief attribute is clearness; it has no marks to express confused notions; It brings together phenomena the most diverse, and discovers the hidden analogies which unite them. If matter escapes us, as that of air and light, by its extreme tenuity, if bodies are placed far from us in the immensity of space, if man wishes to know the aspect of the heavens at successive epochs separated by a great number of centuries, if the actions of gravity and of heat are exerted in the interior of the earth at depths which will be always inaccessible, mathematical analysis can yet lay hold of the laws of these phenomena. It makes them present and measurable, and seems to be a faculty of the human mind destined to supplement the shortness of life and the imperfection of the senses; and what is still more remarkable, it follows the same course in the study of all phenomena; it interprets them by the same language; as if to attest the unity and simplicity of the plan of the universe, and to make still more evident that unchangeable order which presides over all natural causes.

The problems of the theory of heat present so many examples of the simple and constant dispositions which spring from the general laws of nature; and if the order which is established in these phenomena could be grasped by our senses, it would produce in us an impression comparable to the sensation of musical sound.

The forms of bodies are infinitely varied; the distribution of the heat which penetrates them seems to be arbitrary and confused; but all the inequalities are rapidly cancelled and disappear as time passes on. The progress of the phenomenon becomes more regular and simpler, remains finally subject to a definite law which is the same in all cases, and which bears no sensible impress of the initial arrangement.

All observation confirms these consequences. The analysis from which they are derived separates and expresses clearly, 1° the general conditions, that is to say those which spring from the natural properties of heat; 2° the effect, accidental but continued, of the form or state of the surfaces; 3° the effect, not permanent, of the primitive distribution.

In this work we have demonstrated all the principles of the theory of heat, and solved all the fundamental problems. They could have been explained more concisely by omitting the simpler problems, and presenting in the first instance the most general results; but we wished to shew the actual origin of the theory and its gradual progress. When this knowledge has been acquired and the principles thoroughly fixed, it is preferable to employ at once the most extended analytical methods, as we have done in the later investigations. This is also the course which we shall hereafter follow in the memoirs which will be added to this work, and which will form in some manner its complement;[4] and by this means we shall have reconciled, so far as it can depend on ourselves, the necessary development of principles with the precision which becomes the applications of analysis.

The subjects of these memoirs will be, the theory of radiant heat, the problem of the terrestrial temperatures, that of the temperature of dwellings, the comparison of theoretic results with those which we have observed in different experiments, lastly the demonstrations of the differential equations of the movement of heat in fluids.

The work which we now publish has been written a long time since; different circumstances have delayed and often interrupted the printing of it. In this interval, science has been enriched by important observations; the principles of our analysis, which had not at first been grasped, have become better known; the results which we had deduced from them have been discussed and confirmed. We ourselves have applied these principles to new problems, and have changed the form of some of the proofs. The delays of publication will have contributed to make the work clearer and more complete.

The subject of our first analytical investigations on the transfer of heat was its distribution amongst separated masses; these have been preserved in Chap. 4, Sect. 4.2. The problems relative to continuous bodies, which form the theory rightly so called, were solved many years afterwards; this theory was explained for the first time in a manuscript work forwarded to the Institute of France at the end of the year 1807, an extract from which was published in the *Bulletin des Sciences (Société Philomatique*, year 1808, page 112). We added to this memoir, and successively forwarded very extensive notes, concerning the convergence of series, the diffusion of heat in an infinite prism, its emission in spaces void of air, the constructions suitable for exhibiting the chief theorems, and the analysis of the periodic movement at the surface of the earth. Our second memoir, on the propagation of heat, was deposited in the archives of the Institute, on the 28th of September, 1811. It was formed out of the preceding memoir and the notes already sent in; the geometrical constructions and those details of analysis which had no necessary relation to the physical problem were omitted, and to it was added the general equation which expresses the state of the surface. This second work was sent to press in the course of 1821, to be inserted in the collection of the Academy of Sciences. It is printed without any change or addition; the text agrees literally with the deposited manuscript, which forms part of the archives of the Institute.[5]

In this memoir, and in the writings which preceded it, will be found a first explanation of applications which our actual work does not contain; they will be treated in the subsequent memoirs at greater length, and, if it be in our power, with greater

---

[4] These memoirs were never collectively published as a sequel or complement to the *Théorie Analytique de la Chaleur*. But, as will be seen presently, the author had written most of them before the publication of that work in 1822. [A. F.]

[5] It appears as a memoir and supplement in volumes IV. and V. of the *Memoires de l'Académie des Sciences*. For convenience of comparison with, the table of contents of the *Analytical Theory of Heat*, we subjoin the titles and heads of the chapters of the printed memoir: [The titles and heads of chapters have been omitted for brevity.—K.K.]

clearness. The results of our labours concerning the same problems are also indicated in several articles already published. The extract inserted in the *Annales de Chimie et de Physique* shews the aggregate of our researches (Vol. III, page 350, year 1816). We published in the *Annales* two separate notes, concerning radiant heat (Vol. IV, page 128, year 1817, and Vol. VI, page 259, year 1817).

Several other articles of the same collection present the most constant results of theory and observation; the utility and the extent of thermological knowledge could not be better appreciated than by the celebrated editors of the *Annales*.[6]

In the *Bulletin des Sciences* (*Société philomatique* year 1818, page 1, and year 1820, page 60) will be found an extract from a memoir on the constant or variable temperature of dwellings, and an explanation of the chief consequences of our analysis of the terrestrial temperatures.

M. Alexandre de Humboldt, whose researches embrace all the great problems of natural philosophy, has considered the observations of the temperatures proper to the different climates from a novel and very important point of view (Memoir on Isothermal lines, *Société d'Arcueil*, Vol. III, page 462); (Memoir on the inferior limit of perpetual snow, *Annales de Chimie et de Physique*, Vol. V, page 102, year 1817).

As to the differential equations of the movement of heat in fluids[7] mention has been made of them in the annual history of the Academy of Sciences. The extract from our memoir shews clearly its object and principle. (*Analyse des travaux de l'Académie des Sciences*, by M. De Lambre, year 1820.)

The examination of the repulsive forces produced by heat, which determine the statical properties of gases, does not belong to the analytical subject which we have considered. This question connected with the theory of radiant heat has just been discussed by the illustrious author of the *Mécanique céleste*, to whom all the chief branches of mathematical analysis owe important discoveries. (*Connaissance des Temps*, years 1824–1825.)

The new theories explained in our work are united for ever to the mathematical sciences, and rest like them on invariable foundations; all the elements which they at present possess they will preserve, and will continually acquire greater extent. Instruments will be perfected and experiments multiplied. The analysis which we have formed will be deduced from more general, that is to say, more simple and more fertile methods common to many classes of phenomena. For any substances, solid or liquid, for vapours and permanent gases, determinations will be made of all the specific qualities relating to heat, and of the variations of the coefficients which express them.[8] At different stations on the earth observations will be made, of the

---

[6] Gay-Lussac and Arago.

[7] *Mémoires de l'Académie des Sciences, Tome* XII., *Paris*, 1833, contain on pp. 507–514, *Mémoire d'analyse sur le mouvement de la chaleur dans les fluides, par M. Fourier. Lu à l'Académie Royale des Sciences*, 4 *Sep.* 1820. It is followed on pp. 515–530 by *Extrait des notes manuscrites conservées par l'auteur*. The memoir is signed Jh. Fourier, Paris, 1 Sep. 1820, but was published after the death of the author. [A. F.]

[8] *Mémoires de l'Académie des Sciences, Tome* VIII., *Paris* 1829, contain on pp. 581–622, *Mémoire sur la Théorie Analytique de la Chaleur, par M. Fourier*. This was published whilst the author was

temperatures of the ground at different depths, of the intensity of the solar heat and its effects, constant or variable, in the atmosphere, in the ocean and in lakes; and the constant temperature of the heavens proper to the planetary regions will become known.[9] The theory itself will direct all these measures, and assign their precision. No considerable progress can hereafter be made which is not founded on experiments such as these; for mathematical analysis can deduce from general and simple phenomena the expression of the laws of nature; but the special application of these laws to very complex effects demands a long series of exact observations.

1.3   Study Questions

QUES. 1.1. What three scientists laid the foundation of rational mechanics? What phenomena does it encompass? And can the effects of heat be understood using a mechanical theory?

QUES. 1.2. What geophysical problems does Fourier's theory of heat address?

a) Is the applicability of Fourier's theory of heat limited to engineering and civil economy—that is, to practical endeavors?
b) Why doesn't the Earth continuously increase in temperature? Does the concept of equilibrium play a role in Fourier's considerations?
c) What factors influence a region's climate? Are daily, or even seasonal, temperature variations felt beneath the Earth's surface?
d) Why are the higher regions of the atmosphere colder than the lower? What role do heat capacity, gravity and convection play in these phenomena?
e) What is the cause of the trade winds and the ocean currents? What factors determine their direction and strength?

Perpetual Secretary to the Academy. The first only of four parts of the memoir is printed. The contents of all are stated. I. Determines the temperature at any point of a prism whose terminal temperatures are functions of the time, the initial temperature at any point being a function of its distance from one end. II. Examines the chief consequences of the general solution, and applies it to two distinct cases, according as the temperatures of the ends of the heated prism are periodic or not. III. Is historical, enumerates the earlier experimental and analytical researches of other writers relative to the theory of heat; considers the nature of the transcendental equations appearing in the theory; remarks on the employment of arbitrary functions; replies to the objections of M. Poisson; adds some remarks on a problem of the motion of waves. IV. Extends the application of the theory of heat by taking account, in the analysis, of variations in the specific coefficients which measure the capacity of substances for heat, the permeability of solids, and the penetrability of their surfaces. [A. F.]

[9] *Mémoires de l'Académie des Sciences, Tome* VII. *, Paris,* 1827, contain on pp. 569–604, *Mémoire sur les temperatures du globe terrestre et des espaces planétaires, par M. Fourier.* The memoir is entirely descriptive; it was read before the Academy, 20 and 29 Sep. 1824 (*Annales de Chimie et de Physique,* 1824, XXVII. p. 136). [A. F.]

f) What two sources of heat drive all geophysical processes? Are there any other sources of heat in the universe?

QUES. 1.3. What is the "universal language" in which Fourier formulates his theory of heat? And what are the virtues of this language?

a) How are the principles of Fourier's theory derived? Does Fourier consider these principles to be *axioms* or *empirical laws*?
b) What form, then, does Fourier's theory of heat take? What is the benefit of such a formulation?
c) What three measurable properties of a substance are required to achieve a quantitative understanding of the movement of heat? For the temperature ranges typically under consideration, do these properties of bodies vary from place to place?
d) What aspects of a body dictate the particular solution to the differential equation for heat flow through the body?
e) To what seemingly unrelated field of study does Fourier compare the mathematical analysis of the laws of nature? Do you think this is an apt comparison?
f) In what way does Fourier believe his theoretical framework will provide direction for additional scientific studies of heat?

QUES. 1.4. Pedagogically, how does Fourier's book proceed? From the simpler problems to the more general theory, or from the general theory to the simpler? Why does he proceed in this way?

1.4   Exercises

EX. 1.1 (NATURAL PHILOSOPHY ESSAY). Do you agree with Fourier that the objective of natural philosophy is to discover "primary causes"? What does this mean? In particular, how does one discern other causes from primary ones?

EX. 1.2 (FATE OF HEAT). What is the eventual fate of bodies having an initially unequal distribution of heat? Why do you suppose this might be? And are there any exceptions to this rule?

1.5   Vocabulary

1. Radiant heat
2. Catoptrics
3. Hydrodynamics
4. Differential equation
5. Luminous
6. Sonorous

7. Analytical
8. Tenuity
9. Diffusion
10. Emission
11. Prism
12. Propagation
13. Thermological
14. Coefficient
15. Ordinate
16. Homogeneous
17. Heat capacity
18. Interior conductivity
19. Exterior conductivity
20. Diurnal
21. Convergent
22. Definite integral
23. Dynamical theory
24. Isochronism
25. Fluxion
26. Contingent
27. Fortuitous
28. Dilatation

Chapter 2
Mathematics and Temperature

> Mathematical analysis has therefore necessary relations with
> sensible phenomena; its object is not created by human
> intelligence; it is a pre-existent element of the universal order,
> and is not in any way contingent or fortuitous.
>
> —Joseph Fourier

2.1 Introduction

In the previous chapter, we looked at the preliminary discourse of Fourier's *Analytical Theory of Heat*. Herein, he provided an overview of the types of (previously insoluble) problems his theory would address, and how his theory would be formulated: in terms of differential equations. These differential equations govern the flow of heat—and hence the temperature distribution—throughout bodies subjected to sources of heat. In order to arrive at a solution to these differential equations for a given body, one must have knowledge of certain specific qualities of the body. In particular, one must know the body's (i) *specific heat* (its power to contain heat), (ii) *surface conductivity* (its power to receive or transmit heat across its surface), and (iii) *thermal conductivity* (its power to conduct heat through the interior of its mass).Once these qualities are known—along with the thermal conditions existing at the surface of the body—then finding the temperature distribution within the body is reduced to the mathematical process of solving a differential equation given certain boundary conditions. This is not to say it is easy: Fourier would have to develop a new method—the series solution—to solve many such problems. But the technique which Fourier discovered provided a new way of addressing the problem of heat. In the reading selection that follows, Fourier begins to flesh out the mathematical methods outlined in the preliminary discourse. He begins by way of example—describing the distribution of temperature within bodies subjected to various sources of heat.

© Springer International Publishing Switzerland 2016

K. Kuehn, *A Student's Guide Through the Great Physics Texts,*
Undergraduate Lecture Notes in Physics, DOI 10.1007/978-3-319-21828-1_2

2.2   Reading: Fourier, *The Analytical Theory of Heat*

Fourier, J., *The Analytical Theory of Heat*, Cambridge University Press, London, 1878.

2.2.1   Statement of the Object of the Work

1. The effects of heat are subject to constant laws which cannot be discovered without the aid of mathematical analysis. The object of the theory which we are about to explain is to demonstrate these laws; it reduces all physical researches on the propagation of heat, to problems of the integral calculus whose elements are given by experiment. No subject has more extensive relations with the progress of industry and the natural sciences; for the action of heat is always present, it penetrates all bodies and spaces, it influences the processes of the arts, and occurs in all the phenomena of the universe. When heat is unequally distributed among the different parts of a solid mass, it tends to attain equilibrium, and passes slowly from the parts which are more heated to those which are less; and at the same time it is dissipated at the surface, and lost in the medium or in the void. The tendency to uniform distribution and the spontaneous emission which acts at the surface of bodies, change continually the temperature at their different points. The problem of the propagation of heat consists in determining what is the temperature at each point of a body at a given instant, supposing that the initial temperatures are known. The following examples will more clearly make known the nature of these problems.

2. If we expose to the continued and uniform action of a source of heat, the same part of a metallic ring, whose diameter is large, the molecules nearest to the source will be first heated, and, after a certain time, every point of the solid will have acquired very nearly the highest temperature which it can attain. This limit or greatest temperature is not the same at different points; it becomes less and less according as they become more distant from that point at which the source of heat is directly applied.

   When the temperatures have become permanent, the source of heat supplies, at each instant, a quantity of heat which exactly compensates for that which is dissipated at all the points of the external surface of the ring.

   If now the source be suppressed, heat will continue to be propagated in the interior of the solid, but that which is lost in the medium or the void, will no longer be compensated as formerly by the supply from the source, so that all the temperatures will vary and diminish incessantly until they have become equal to the temperatures of the surrounding medium.

3. Whilst the temperatures are permanent and the source remains, if at every point of the mean circumference of the ring an ordinate be raised perpendicular to the plane of the ring, whose length is proportional to the fixed temperature at

that point, the curved line which passes through the ends of these ordinates will represent the permanent state of the temperatures, and it is very easy to determine by analysis the nature of this line. It is to be remarked that the thickness of the ring is supposed to be sufficiently small for the temperature to be sensibly equal at all points of the same section perpendicular to the mean circumference. When the source is removed, the line which bounds the ordinates proportional to the temperatures at the different points will change its form continually. The problem consists in expressing, by one equation, the variable form of this curve, and in thus including in a single formula all the successive states of the solid.

4. Let z be the constant temperature at a point m of the mean circumference, x the distance of this point from the source, that is to say the length of the arc of the mean circumference, included between the point m and the point o which corresponds to the position of the source; z is the highest temperature which the point m can attain by virtue of the constant action of the source, and this permanent temperature z is a function $f(x)$ of the distance x. The first part of the problem consists in determining the function $f(x)$ which represents the permanent state of the solid.

Consider next the variable state which succeeds to the former state as soon as the source has been removed; denote by t the time which has passed since the suppression of the source, and by v the value of the temperature at the point m after the time t. The quantity v will be a certain function $F(x, t)$ of the distance x and the time t; the object of the problem is to discover this function $F(x, t)$, of which we only know as yet that the initial value is $f(x)$, so that we ought to. have the equation $f(x) = F(x, 0)$.

5. If we place a solid homogeneous mass, having the form of a sphere or cube, in a medium maintained at a constant temperature, and if it remains immersed for a very long time, it will acquire at all its points a temperature differing very little from that of the fluid. Suppose the mass to be withdrawn in order to transfer it to a cooler medium, heat will begin to be dissipated at its surface; the temperatures at different points of the mass will not be sensibly the same, and if we suppose it divided into an infinity of layers by surfaces parallel to its external surface, each of those layers will transmit, at each instant, a certain quantity of heat to the layer which surrounds it. If it be imagined that each molecule carries a separate thermometer, which indicates its temperature at every instant, the state of the solid will from time to time be represented by the variable system of all these thermometric heights. It is required to express the successive states by analytical formulæ, so that we may know at any given instant the temperatures indicated by each thermometer, and compare the quantities of heat which flow during the same instant, between two adjacent layers, or into the surrounding medium.

6. If the mass is spherical, and we denote by x the distance of a point of this mass from the centre of the sphere, by t the time which has elapsed since the commencement of the cooling, and by v the variable temperature of the point m, it is easy to see that all points situated at the same distance x from the centre

of the sphere have the same temperature v. This quantity v is a certain function $F(x, t)$ of the radius x and of the time t; it must be such that it becomes constant whatever be the value of x, when we suppose t to be nothing; for by hypothesis, the temperature at all points is the same at the moment of emersion. The problem consists in determining that function of x and t which expresses the value of v.

7. In the next place it is to be remarked, that during the cooling, a certain quantity of heat escapes, at each instant, through the external surface, and passes into the medium. The value of this quantity is not constant; it is greatest at the beginning of the cooling. If however we consider the variable state of the internal spherical surface whose radius is x, we easily see that there must be at each instant a certain quantity of heat which traverses that surface, and passes through that part of the mass which is more distant from the centre. This continuous flow of heat is variable like that through the external surface, and both are quantities comparable with each other; their ratios are numbers whose varying values are functions of the distance x, and of the time t which has elapsed. It is required to determine these functions.

8. If the mass, which has been heated by a long immersion in a medium, and whose rate of cooling we wish to calculate, is of cubical form, and if we determine the position of each point m by three rectangular co-ordinates x, y, z, taking for origin the centre of the cube, and for axes lines perpendicular to the faces, we see that the temperature v of the point m after the time t, is a function of the four variables x, y, z, and t. The quantities of heat which flow out at each instant through the whole external surface of the solid, are variable and comparable with each other; their ratios are analytical functions depending on the time t, the expression of which must be assigned.

9. Let us examine also the case in which a rectangular prism of sufficiently great thickness and of infinite length, being submitted at its extremity to a constant temperature, whilst the air which surrounds it is maintained at a less temperature, has at last arrived at a fixed state which it is required to determine. All the points of the extreme section at the base of the prism have, by hypothesis, a common and permanent temperature. It is not the same with a section distant from the source of heat; each of the points of this rectangular surface parallel to the base has acquired a fixed temperature, but this is not the same at different points of the same section, and must be less at points nearer to the surface exposed to the air. We see also that, at each instant, there flows across a given section a certain quantity of heat, which always remains the same, since the state of the solid has become constant. The problem consists in determining the permanent temperature at any given point of the solid, and the whole quantity of heat which, in a definite time, flows across a section whose position is given.

10. Take as origin of co-ordinates x, y, z, the centre of the base of the prism, and as rectangular axes, the axis of the prism itself, and the two perpendiculars on the sides: the permanent temperature v of the point m, whose co-ordinates are x, y, z, is a function of three variables $F(x, y, z)$: it has by hypothesis a constant value, when we suppose x nothing, whatever be the values of y and

z. Suppose we take for the unit of heat that quantity which in the unit of time would emerge from an area equal to a unit of surface, if the heated mass which that area bounds, and which is formed of the same substance as the prism, were continually maintained at the temperature of boiling water, and immersed in atmospheric air maintained at the temperature of melting ice.

We see that the quantity of heat which, in the permanent state of the rectangular prism, flows, during a unit of time, across a certain section perpendicular to the axis, has a determinate ratio to the quantity of heat taken as unit. This ratio is not the same for all sections: it is a function $\phi(x)$ of the distance x, at which the section is situated. It is required to find an analytical expression of the function $\phi(x)$.

11. The foregoing examples suffice to give an exact idea of the different problems which we have discussed.

The solution of these problems has made us understand that the effects of the propagation of heat depend in the case of every solid substance, on three elementary qualities, which are, its capacity for heat, its own conductibility, and the exterior conductibility.

It has been observed that if two bodies of the same volume and of different nature have equal temperatures, and if the same quantity of heat be added to them, the increments of temperature are not the same; the ratio of these increments is the inverse ratio of their capacities for heat. In this manner, the first of the three specific elements which regulate the action of heat is exactly defined, and physicists have for a long time known several methods of determining its value. It is not the same with the two others; their effects have often been observed, but there is but one exact theory which can fairly distinguish, define, and measure them with precision.

The proper or interior conductibility of a body expresses the facility with which heat is propagated in passing from one internal molecule to another. The external or relative conductibility of a solid body depends on the facility with which heat penetrates the surface, and passes from this body into a given medium, or passes from the medium into the solid. The last property is modified by the more or less polished state of the surface; it varies also according to the medium in which the body is immersed; but the interior conductibility can change only with the nature of the solid.

These three elementary qualities are represented in our formulæ by constant numbers, and the theory itself indicates experiments suitable for measuring their values. As soon as they are determined, all the problems relating to the propagation of heat depend only on numerical analysis. The knowledge of these specific properties may be directly useful in several applications of the physical sciences; it is besides an element in the study and description of different substances. It is a very imperfect knowledge of bodies which ignores the relations which they have with one of the chief agents of nature. In general, there is no mathematical theory which has a closer relation than this with public economy, since it serves to give clearness and perfection to the practice of the numerous arts which are founded on the employment of heat.

12. The problem of the terrestrial temperatures presents one of the most beautiful applications of the theory of heat; the general idea to be formed of it is this. Different parts of the surface of the globe are unequally exposed to the influence of the solar rays; the intensity of their action depends on the latitude of the place; it changes also in the course of the day and in the course of the year, and is subject to other less perceptible inequalities. It is evident that, between the variable state of the surface and that of the internal temperatures a necessary relation exists, which may be derived from theory. We know that, at a certain depth below the surface of the earth, the temperature at a given place experiences no annual variation: this permanent underground temperature becomes less and less according as the place is more and more distant from the equator. We may then leave out of consideration the exterior envelope, the thickness of which is incomparably small with respect to the earth's radius, and regard our planet as a nearly spherical mass, whose surface is subject to a temperature which remains constant at all points on a given parallel, but is not the same on another parallel. It follows from this that every internal molecule has also a fixed temperature determined by its position. The mathematical problem consists in discovering the fixed temperature at any given point, and the law which the solar heat follows whilst penetrating the interior of the earth.

This diversity of temperature interests us still more, if we consider the changes which succeed each other in the envelope itself on the surface of which we dwell. Those alternations of heat and cold which are reproduced every day and in the course of every year, have been up to the present time the object of repeated observations. These we can now submit to calculation, and from a common theory derive all the particular facts which experience has taught us. The problem is reducible to the hypothesis that every point of a vast sphere is affected by periodic temperatures; analysis then tells us according to what law the intensity of these variations decreases according as the depth increases, what is the amount of the annual or diurnal changes at a given depth, the epoch of the changes, and how the fixed value of the underground temperature is deduced from the variable temperatures observed at the surface.

13. The general equations of the propagation of heat are partial differential equations, and though their form is very simple the known methods do not furnish any general mode of integrating them; we could not therefore deduce from them the values of the temperatures after a definite time. The numerical interpretation of the results of analysis is however necessary, and it is a degree of perfection which it would be very important to give to every application of analysis to the natural sciences. So long as it is not obtained, the solutions may be said to remain incomplete and useless, and the truth which it is proposed to discover is no less hidden in the formulæ of analysis than it was in the physical problem itself. We have applied ourselves with much care to this purpose, and we have been able to overcome the difficulty in all the problems of which we have treated, and which contain the chief elements of the theory of heat. There is not one of the problems whose solution does not provide convenient and exact means for discovering the numerical values of the temperatures acquired, or

those of the quantities of heat which have flowed through, when the values of the time and of the variable coordinates are known. Thus will be given not only the differential equations which the functions that express the values of the temperatures must satisfy; but the functions themselves will be given under a form which facilitates the numerical applications.

14. In order that these solutions might be general, and have an extent equal to that of the problem, it was requisite that they should accord with the initial state of the temperatures, which is arbitrary. The examination of this condition shews that we may develop in convergent series, or express by definite integrals, functions which are not subject to a constant law, and which represent the ordinates or irregular or discontinuous lines. This property throws a new light on the theory of partial differential equations, and extends the employment of arbitrary functions by submitting them to the ordinary processes of analysis.

15. It still remained to compare the facts with theory. With this view, varied and exact experiments were undertaken, whose results were in conformity with those of analysis, and gave them an authority which one would have been disposed to refuse to them in a new matter which seemed subject to so much uncertainty. These experiments confirm the principle from which we started, and which is adopted by all physicists in spite of the diversity of their hypotheses on the nature of heat.

16. Equilibrium of temperature is effected not only by way of contact, it is established also between bodies separated from each other, which are situated for a long time in the same region. This effect is independent of contact with a medium; we have observed it in spaces wholly void of air. To complete our theory it was necessary to examine the laws which radiant heat follows, on leaving the surface of a body. It results from the observations of many physicists and from our own experiments, that the intensities of the different rays, which escape in all directions from any point in the surface of a heated body, depend on the angles which their directions make with the surface at the same point. We have proved that the intensity of a ray diminishes as the ray makes a smaller angle with the element of surface, and that it is proportional to the sine of that angle.[1] This general law of emission of heat which different observations had already indicated, is a necessary consequence of the principle of the equilibrium of temperature and of the laws of propagation of heat in solid bodies.

Such are the chief problems which have been discussed in this work; they are all directed to one object only, that is to establish clearly the mathematical principles of the theory of heat, and to keep up in this way with the progress of the useful arts, and of the study of nature.

17. From what precedes it is evident that a very extensive class of phenomena exists, not produced by mechanical forces, but resulting simply from the presence and accumulation of heat. This part of natural philosophy cannot be

---

[1] 1 *Mem. Acad. d. Sc.* Tome V. Paris, 1826, pp. 179–213. [A. F.].

connected with dynamical theories, it has principles peculiar to itself, and is founded on a method similar to that of other exact sciences. The solar heat, for example, which penetrates the interior of the globe, distributes itself therein according to a regular law which does not depend on the laws of motion, and cannot be determined by the principles of mechanics. The dilatations which the repulsive force of heat produces, observation of which serves to measure temperatures, are in truth dynamical effects; but it is not these dilatations which we calculate, when we investigate the laws of the propagation of heat.

18. There are other more complex natural effects, which depend at the same time on the influence of heat, and of attractive forces: thus, the variations of temperatures which the movements of the sun occasion in the atmosphere and in the ocean, change continually the density of the different parts of the air and the waters. The effect of the forces which these masses obey is modified at every instant by a new distribution of heat, and it cannot be doubted that this cause produces the regular winds, and the chief currents of the sea; the solar and lunar attractions occasioning in the atmosphere effects but slightly sensible, and not general displacements. It was therefore necessary, in order to submit these grand phenomena to calculation, to discover the mathematical laws of the propagation of heat in the interior of masses.

19. It will be perceived, on reading this work, that heat attains in bodies a regular disposition independent of the original distribution, which may be regarded as arbitrary.

In whatever manner the heat was at first distributed, the system of temperatures altering more and more, tends to coincide sensibly with a definite state which depends only on the form of the solid. In the ultimate state the temperatures of all the points are lowered in the same time, but preserve amongst each other the same ratios: in order to express this property the analytical formulæ contain terms composed of exponentials and of quantities analogous to trigonometric functions.

Several problems of mechanics present analogous results, such as the isochronism of oscillations, the multiple resonance of sonorous bodies. Common experiments had made these results remarked, and analysis afterwards demonstrated their true cause. As to those results which depend on changes of temperature, they could not have been recognised except by very exact experiments; but mathematical analysis has outrun observation, it has supplemented our senses, and has made us in a manner witnesses of regular and harmonic vibrations in the interior of bodies.

20. These considerations present a singular example of the relations which exist between the abstract science of numbers and natural causes.

When a metal bar is exposed at one end to the constant action of a source of heat, and every point of it has attained its highest temperature, the system of fixed temperatures corresponds exactly to a table of logarithms; the numbers are the elevations of thermometers placed at the different points, and the logarithms are the distances of these points from the source. In general, heat distributes itself in the interior of solids according to a simple law expressed by a partial differential equation common to physical problems of different order. The irradiation of heat has an evident relation to the tables of sines, for the rays which depart from the same point of a heated surface, differ very much from each other, and their intensity is rigorously proportional to the sine of the angle which the direction of each ray makes with the element of surface.

If we could observe the changes of temperature for every instant at every point of a solid homogeneous mass, we should discover in these series of observations the properties of recurring series, as of sines and logarithms; they would be noticed for example in the diurnal or annual variations of temperature of different points of the earth near its surface.

We should recognise again the same results and all the chief elements of general analysis in the vibrations of elastic media, in the properties of lines or of curved surfaces, in the movements of the stars, and those of light or of fluids. Thus the functions obtained by successive differentiations, which are employed in the development of infinite series and in the solution of numerical equations, correspond also to physical properties. The first of these functions, or the fluxion properly so called, expresses in geometry the inclination of the tangent of a curved line, and in dynamics the velocity of a moving body when the motion varies; in the theory of heat it measures the quantity of heat which flows at each point of a body across a given surface. Mathematical analysis has therefore necessary relations with sensible phenomena; its object is not created by human intelligence; it is a pre-existent element of the universal order, and is not in any way contingent or fortuitous; it is imprinted throughout all nature.

21. Observations more exact and more varied will presently ascertain whether the effects of heat are modified by causes which have not yet been perceived, and the theory will acquire fresh perfection by the continued comparison of its results with the results of experiment; it will explain some important phenomena which we have not yet been able to submit to calculation; it will shew how to determine all the thermometric effects of the solar rays, the fixed or variable temperature which would be observed at different distances from the equator, whether in the interior of the earth or beyond the limits of the atmosphere, whether in the ocean or in different regions of the air. From it will be derived the mathematical knowledge of the great movements which result from the influence of heat combined with that of gravity. The same principles will serve to measure the conductivities, proper or relative, of different bodies, and their specific capacities, to distinguish all the causes which modify the emission of heat at the surface of solids, and to perfect thermometric instruments.

The theory of heat will always attract the attention of mathematicians, by the rigorous exactness of its elements and the analytical difficulties peculiar to it, and above all by the extent and usefulness of its applications; for all its consequences concern at the same time general physics, the operations of the arts, domestic uses and civil economy.

2.3   Study Questions

QUES. 2.1. How does Fourier's theory address the problem of heat?

QUES. 2.2. What is the temperature distribution within a metallic ring held above a source of heat, such as a candle?

a) Does the ring reach a uniform temperature? Why or why not? How may the temperature distribution be expressed mathematically? What assumption does Fourier make?
b) What happens to the temperature of the ring when the heat source is suddenly extinguished? What, then, is the ultimate goal of Fourier's theory?

QUES. 2.3. What is the temperature of a solid sphere alternately immersed and removed from a warm fluid?

a) Does the sphere attain a uniform temperature immediately upon immersion? Does it ever attain a uniform temperature? If so, what is its value?
b) When the sphere is removed from the warm fluid, is its rate of cooling uniform? Is its temperature uniform during the cooling process? Do any two points on its surface differ in temperature?
c) More generally, how may the sphere's temperature distribution be expressed mathematically? Does the temperature distribution have any symmetries?
d) Is the behavior of a cube different than that of a sphere? Does the temperature distribution have any symmetries?

QUES. 2.4. What is the temperature of a long rectangular prism one end of which is held at a high temperature?

a) Does the prism attain a uniform temperature immediately upon immersion? Does it ever attain a uniform temperature?
b) Is the temperature constant along the length of the prism? What about within any cross-section of the prism?
c) Does the temperature distribution have any symmetries? Is the temperature distribution time-dependent?
d) What does Fourier select as the unit of heat? Why might he have chosen this unit?
e) In its final state, is the heat flowing through any cross-section of the prism time-dependent, constant, or perhaps even zero? Is it the same for every cross-sectional area along the length of the prism?

QUES. 2.5. Upon what elementary quantities doe the propagation of heat through a body depend? How are each of these quantities defined? What can a knowledge of these quantities provide?

QUES. 2.6. What are some significant problems that Fourier's theory of heat allows one to solve? What does a solution to such problems entail? Are all such problems solvable analytically?

QUES. 2.7. Must bodies be in physical contact in order to achieve equilibrium? Must there be anything at all between them? What law governs the intensity of radiant heat from a surface? In particular, how does its intensity depend on its emission angle?

QUES. 2.8. Is Fourier's theory of heat a dynamical theory of heat? That is: does his theory of heat follow from Newton's three laws of motion? Is there any relationship at all between Newton's three laws of motion and, for instance, atmospheric or oceanic currents?

QUES. 2.9. Does the final state of a body depend on its initial distribution of temperatures? Upon what does the final state depend? Are any other phenomena in nature independent of initial conditions?

QUES. 2.10. In what way does mathematical analysis, and derivatives in particular, relate to sensible phenomena?

a) In geometry, what quantity does the first derivative of a line express?
b) In dynamics, what quantity does the first derivative of the position of a moving body express?
c) In the theory of heat, what quantity does the first derivative of the temperature distribution inside a body express?

2.4 Exercises

EX. 2.1 (NATURE AND MATHEMATICS ESSAY). Consider the Fourier quote at the outset of this chapter. What does he mean when he asserts that mathematics "is a pre-existent element of the universal order" and that it "is not created by human intelligence"? Do you agree with these assertions, or do you have a different view of the origin of mathematics? Can you defend your view?

EX. 2.2 (TEMPERATURE AND SYMMETRY). Make sketches which clearly indicate the temperature distribution within objects under the following conditions: (i) a thin ring of metal heated for a long time at a single point along its circumference, (ii) a solid homogeneous sphere immersed for a long time in fluid at 50°C, (iii) the same homogeneous sphere shortly after it has been removed from the fluid and placed in the air at 20°C, and (iv) a long, thin rectangular prism in 20°C air, one end of which is heated to 50°C for a long time.

EX. 2.3 (HEAT CAPACITY AND THE METHOD OF MIXTURES). According to the *caloric theory*, heat is an invisible and imponderable (weightless) fluid which, upon entering a substance raises its temperature, and upon leaving a substance lowers its temperature. Different substances have different abilities to store heat—different *heat capacities*. The mathematical relationship between the amount of heat added to a substance, ΔQ, the heat capacity of the substance, C, and the change in temperature of the substance, ΔT, is given by

$$\Delta Q = C \Delta T \qquad (2.1)$$

The *method of mixtures* is an experimental technique used to measure the relative heat capacities of two substances, one of which is a fluid. Suppose, for instance, that a 1 pound sheet of lead foil is loosely rolled up and then suspended in the steam rising from a tea kettle until it finally reaches $212°F$. It is then plunged into a 1-pound water bath initially at $57°F$. After the lead foil and water achieve thermal equilibrium, the water is measured to be $62°F$. If 1 unit of heat is now defined as the amount of heat required to raise 1 pound of water by $1°F$, then what is the heat capacity of 1 pound of lead? (ANSWER: $\frac{1}{30}$)

Ex. 2.4 (LATENT HEAT OF ICE MELTING). According to the caloric theory, when heat enters a substance it may cause it to undergo a phase transformation *instead of raising its temperature*. After such a phase transformation, the added heat or *caloric* is hidden within the substance in the form of *latent heat*. For example, when a certain quantity of heat is added to ice at $0°C$, it melts and becomes water at $0°C$. This latent heat becomes manifest—it reappears and must be removed—when water is refrozen into ice. In order to determine the latent heat associated with the melting of ice, consider the following two laboratory experiments.

Experiment 1: 100 grains of ice at $0°C$ are added to a 5000-grain bath of water at $55°C$. After equilibration, the final temperature is $52.3°C$.

Experiment 2: 1000 grains of water at $5°C$ are poured into a 5000-grain bath of water at $56°C$. After equilibration, the final temperature is $47.5°C$.

According to convention, a grain is a unit of weight approximately equivalent to 65 mg,[2] and a unit of heat is the amount required to raise one grain of water by $1°C$.

a) In experiment 1, what was the change in temperature of the ice? of the (initially) warm water bath?

b) In experiment 2, what was the change in temperature of the (initially) cool water? of the (initially) warm water bath? How many units of heat did the (initially) cool water absorb? And how many units of heat left the (initially) warm water bath? (ANSWER: 42,500 units)

c) In experiment 1, how many units of heat left the (initially) hot water bath? How much of this heat went into raising the temperature of the melted ice? And how much went into melting the ice? What, then, is the latent heat of one grain of ice? (ANSWER: 82.3)

---

[2] The grain is part of the traditional English weight system; it is still used today by apothecaries, dentists, and ammunition manufacturers.

2.5   Vocabulary

1. Propagation
2. Equilibrium
3. Dissipate
4. Incessant
5. Ordinate
6. Commencement
7. Emersion
8. Prism
9. Conductibility
10. Diurnal
11. Facilitate
12. Requisite
13. Shew
14. Conform
15. Dynamical
16. Exponential
17. Logarithm
18. Differentiation
19. Tangent
20. Contingent
21. Fortuitous
22. Thermometric

Chapter 3
Steam Engines and Heat Flow

Wherever there exists a difference of temperature, motive-power can be produced.

—Sadi Carnot

3.1 Introduction

Nicolas Léonard Sadi Carnot (1796–1832) was born in the Palais du Petit-Luxembourg in Paris. His father, a famous mathematician and minister of war under Napoleon, named his son after the eminent medieval Persian poet Sadi of Shiraz.[1] Carnot attended the Charlemagne Lycée before enrolling at the age of 16 in the École Polytechnique, the elite military academy which attracted many famous scientists and mathematicians during the age of the French Revolution: Biot, Arago, Laplace, Fourier, Gay-Lussac, Ampère, and Poisson. After his graduation in 1814, Carnot served as a cadet sub-lieutenant in the engineer corps at Metz, a position which he found increasingly frustrating and confining after Napoleon's defeat at Waterloo and his father's consequent retirement and exile to Magdeburg, Germany. So in 1819 Carnot took an examination and was appointed a lieutenant in the French general staff. After attaining a furlough he increasingly turned his attention to the study of scientific matters, particularly the writings of Pascal and the design of steam engines. This latter interest would eventually lead to his only publication in 1824. Carnot's *Réflexions sur la puissance mortice du feu* went largely unnoticed until after his death from cholera at the age of 36. Several years later, Clausius and Kelvin developed Carnot's seminal ideas on heat engines into one of the foundations of modern science: the second law of thermodynamics. The reading selections that follow are from an 1897 English translation by Robert Henry Thurston (Fig. 3.1).

---

[1] Much of the present introduction was derived from a biographical sketch of Sadi Carnot written by his younger brother, Hippolyte, which can be found in Chap. II of Carnot, S., *Reflections on the Motive Power of Heat*, second ed., John Wiley & Sons and Chapman & Hall, New York and London, 1897.

© Springer International Publishing Switzerland 2016
K. Kuehn, *A Student's Guide Through the Great Physics Texts*,
Undergraduate Lecture Notes in Physics, DOI 10.1007/978-3-319-21828-1_3

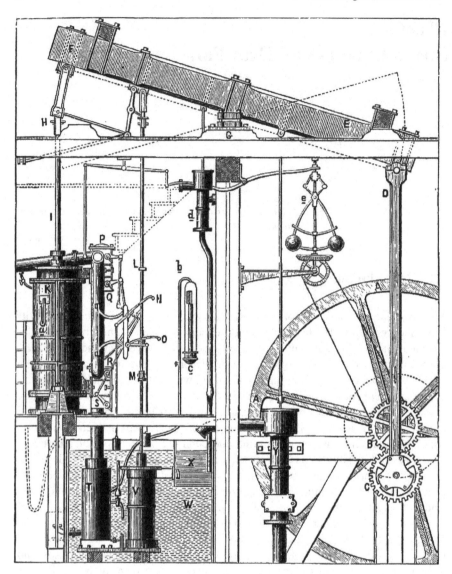

Fig. 3.1 Watt's steam engine, 1781; Fig. 27 from R.H. Thurston, *A History of the Growth of the Steam Engine*, 1878

3.2   Reading: Carnot, *Reflections on the Motive Power of Heat, and on Machines Fitted to Develop that Power*

Carnot, S., *Reflections on the Motive Power of Heat*, second ed., John Wiley & Sons and Chapman & Hall, New York and London, 1897.[2]

Every one knows that heat can produce motion. That it possesses vast motive-power no one can doubt, in these days when the steam-engine is everywhere so well known.

To heat also are due the vast movements which take place on the earth. It causes the agitations of the atmosphere, the ascension of clouds, the fall of rain and of meteors, the currents of water which channel the surface of the globe, and of which man has thus far employed but a small portion. Even earthquakes and volcanic eruptions are the result of heat.

From this immense reservoir we may draw the moving force necessary for our purposes. Nature, in providing us with combustibles on all sides, has given us the power to produce, at all times and in all places, heat and the impelling power which is the result of it. To develop this power, to appropriate it to our uses, is the object of heat engines.

The study of these engines is of the greatest interest, their importance is enormous, their use is continually increasing, and they seem destined to produce a great revolution in the civilized world.

Already the steam-engine works our mines, impels our ships, excavates our ports and our rivers, forges iron, fashions wood, grinds grains, spins and weaves our cloths, transports the heaviest burdens, *etc.* It appears that it must some day serve as a universal motor, and be substituted for animal power, waterfalls, and air currents.

Over the first of these motors it has the advantage of economy, over the two others the inestimable advantage that it can be used at all times and places without interruption.

If, some day, the steam-engine shall be so perfected that it can be set up and supplied with fuel at small cost, it will combine all desirable qualities, and will afford to the industrial arts a range the extent of which can scarcely be predicted. It is not merely that a powerful and convenient motor that can be procured and carried anywhere is substituted for the motors already in use, but that it causes rapid extension in the arts in which it is applied, and can even create entirely new arts.

The most signal service that the steam-engine has rendered to England is undoubtedly the revival of the working of the coal-mines, which had declined, and threatened to cease entirely, in consequence of the continually increasing difficulty

---

[2] Sadi Carnot's *Réflexions sur la puissance motrice du feu* (Paris, Bachelier 1824) was long ago completely exhausted. As but a small number of copies were printed, this remarkable work remained long unknown to the earlier writers on Thermodynamics. It was therefore for the benefit of savants unable to study a work out of print, as well as to render honor to the memory of Sadi Carnot, that the new publishers of the *Annales Scientifique de l'École Normale supérieure* (ii. series,t. 1, 1872) published a new edition, from which this translation is reproduced.

of drainage, and of raising the coal.[3] We should rank second the benefit to iron manufacture, both by the abundant supply of coal substituted for wood just when the latter had begun to grow scarce, and by the powerful machines of all kinds, the use of which the introduction of the steam-engine has permitted or facilitated.

Iron and heat are, as we know, the supporters, the bases, of the mechanic arts. It is doubtful if there be in England a single industrial establishment of which the existence does not depend on the use of these agents, and which does not freely employ them. To take away to-day from England her steam-engines would be to take away at the same time her coal and iron. It would be to dry up all her sources of wealth, to ruin all on which her prosperity depends, in short, to annihilate that colossal power. The destruction of her navy, which she considers her strongest defence, would perhaps be less fatal.

The safe and rapid navigation by steamships may be regarded as an entirely new art due to the steam-engine. Already this art has permitted the establishment of prompt and regular communications across the arms of the sea, and on the great rivers of the old and new continents. It has made it possible to traverse savage regions where before we could scarcely penetrate. It has enabled us to carry the fruits of civilization over portions of the globe where they would else have been wanting for years. Steam navigation brings nearer together the most distant nations. It tends to unite the nations of the earth as inhabitants of one country. In fact, to lessen the time, the fatigues, the uncertainties, and the dangers of travel—is not this the same as greatly to shorten distances?[4]

The discovery of the steam-engine owed its birth, like most human inventions, to rude attempts which have been attributed to different persons, while the real author is not certainly known. It is, however, less in the first attempts that the principal discovery consists, than in the successive improvements which have brought steam-engines to the condition in which we find them to-day. There is almost as great a distance between the first apparatus in which the expansive force of steam was displayed and the existing machine, as between the first raft that man ever made and the modern vessel.

If the honor of a discovery belongs to the nation in which it has acquired its growth and all its developments, this honor cannot be here refused to England. Savery, Newcomen, Smeaton, the famous Watt, Woolf, Trevithick, and some other English engineers, are the veritable creators of the steam-engine. It has acquired at

---

[3] It may be said that coal-mining has increased tenfold in England since the invention of the steam-engine. It is almost equally true in regard to the mining of copper, tin, and iron. The results produced in a half-century by the steam-engine in the mines of England are to-day paralleled in the gold and silver mines of the New World—mines of which the working declined from day to day, principally on account of the insufficiency of the motors employed in the draining and the extraction of the minerals.

[4] We say, to lessen the dangers of journeys. In fact, although the use of the steam-engine on ships is attended by some danger which has been greatly exaggerated, this is more than compensated by the power of following always an appointed and well-known route, of resisting the force of the winds which would drive the ship towards the shore, the shoals, or the rocks.

their hands all its successive degrees of improvement. Finally, it is natural that an invention should have its birth and especially be developed, be perfected, in that place where its want is most strongly felt.

Notwithstanding the work of all kinds done by steam-engines, notwithstanding the satisfactory condition to which they have been brought to-day, their theory is very little understood, and the attempts to improve them are still directed almost by chance.

The question has often been raised whether the motive power of heat[5] is unbounded, whether the possible improvements in steam-engines have an assignable limit,—a limit which the nature of things will not allow to be passed by any means whatever; or whether, on the contrary, these improvements may be carried on indefinitely. We have long sought, and are seeking to-day, to ascertain whether there are in existence agents preferable to the vapor of water for developing the motive power of heat; whether atmospheric air, for example, would not present in this respect great advantages. We propose now to submit these questions to a deliberate examination.

The phenomenon of the production of motion by heat has not been considered from a sufficiently general point of view. We have considered it only in machines the nature and mode of action of which have not allowed us to take in the whole extent of application of which it is susceptible. In such machines the phenomenon is, in a way, incomplete. It becomes difficult to recognize its principles and study its laws.

In order to consider in the most general way the principle of the production of motion by heat, it must be considered independently of any mechanism or any particular agent. It is necessary to establish principles applicable not only to steam-engines[6] but to all imaginable heat-engines, whatever the working substance and whatever the method by which it is operated.

Machines which do not receive their motion from heat, those which have for a motor the force of men or of animals, a waterfall, an air-current, *etc.*, can be studied even to their smallest details by the mechanical theory. All cases are foreseen, all imaginable movements are referred to these general principles, firmly established, and applicable under all circumstances. This is the character of a complete theory. A similar theory is evidently needed for heat-engines. We shall have it only when the laws of Physics shall be extended enough, generalized enough, to make known beforehand all the effects of heat acting in a determined manner on anybody.

We will suppose in what follows at least a superficial knowledge of the different parts which compose an ordinary steam-engine; and we consider it unnecessary to explain what are the furnace, boiler, steam-cylinder, piston, condenser, *etc.*

---

[5] We use here the expression motive power to express the useful effect that a motor is capable of producing. This effect can always be likened to the elevation of a weight to a certain height. It has, as we know, as a measure, the product of the weight multiplied by the height to which it is raised.

[6] We distinguish here the steam-engine from the heat-engine in general. The latter may make use of any agent whatever, of the vapor of water or of any other, to develop the motive power of heat.

The production of motion in steam-engines is always accompanied by a circumstance on which we should fix our attention. This circumstance is the re-establishing of equilibrium in the caloric; that is, its passage from a body in which the temperature is more or less elevated, to another in which it is lower. What happens in fact in a steam-engine actually in motion? The caloric developed in the furnace by the effect of the combustion traverses the walls of the boiler, produces steam, and in some way incorporates itself with it. The latter carrying it away, takes it first into the cylinder, where it performs some function, and from thence into the condenser, where it is liquefied by contact with the cold water which it encounters there. Then, as a final result, the cold water of the condenser takes possession of the caloric developed by the combustion. It is heated by the intervention of the steam as if it had been placed directly over the furnace. The steam is here only a means of transporting the caloric. It fills the same office as in the heating of baths by steam, except that in this case its motion is rendered useful.

We easily recognize in the operations that we have just described the re-establishment of equilibrium in the caloric, its passage from a more or less heated body to a cooler one. The first of these bodies, in this case, is the heated air of the furnace; the second is the condensing water. The re-establishment of equilibrium of the caloric takes place between them, if not completely, at least partially, for on the one hand the heated air, after having performed its function, having passed round the boiler, goes out through the chimney with a temperature much below that which it had acquired as the effect of combustion; and on the other hand, the water of the condenser, after having liquefied the steam, leaves the machine with a temperature higher than that with which it entered.

The production of motive power is then due in steam-engines not to an actual consumption of caloric, but *to its transportation from a warm body to a cold body*, that is, to its re-establishment of equilibrium—an equilibrium considered as destroyed by any cause whatever, by chemical action such as combustion, or by any other. We shall see shortly that this principle is applicable to any machine set in motion by heat.

According to this principle, the production of heat alone is not sufficient to give birth to the impelling power: it is necessary that there should also be cold; without it, the heat would be useless. And in fact, if we should find about us only bodies as hot as our furnaces, how can we condense steam? What should we do with it if once produced? We should not presume that we might discharge it into the atmosphere, as is done in some engines;[7] the atmosphere would not receive it. It does receive it under the actual condition of things, only because it fulfills the office of a vast

---

[7] Certain engines at high pressure throw the steam out into the atmosphere instead of the condenser. They are used specially in places where it would be difficult to procure a stream of cold water sufficient to produce condensation.

condenser, because it is at a lower temperature; otherwise it would soon become fully charged, or rather would be already saturated.[8]

Wherever there exists a difference of temperature, wherever it has been possible for the equilibrium of the caloric to be re-established, it is possible to have also the production of impelling power. Steam is a means of realizing this power, but it is not the only one. All substances in nature can be employed for this purpose, all are susceptible of changes of volume, of successive contractions and dilatations, through the alternation of heat and cold. All are capable of overcoming in their changes of volume certain resistances, and of thus developing the impelling power. A solid body—a metallic bar for example—alternately heated and cooled increases and diminishes in length, and can move bodies fastened to its ends. A liquid alternately heated and cooled increases and diminishes in volume, and can overcome obstacles of greater or less size, opposed to its dilatation. An aeriform fluid is susceptible of considerable change of volume by variations of temperature. If it is enclosed in an expansible space, such as a cylinder provided with a piston, it will produce movements of great extent. Vapors of all substances capable of passing into a gaseous condition, as of alcohol, of mercury, of sulphur, *etc.*, may fulfill the same office as vapor of water. The latter, alternately heated and cooled, would produce motive power in the shape of permanent gases, that is, without ever returning to a liquid state. Most of these substances have been proposed, many even have been tried, although up to this time perhaps without remarkable success.

We have shown that in steam-engines the motive power is due to a re-establishment of equilibrium in the caloric; this takes place not only for steam-engines, but also for every heat-engine—that is, for every machine of which caloric is the motor. Heat can evidently be a cause of motion only by virtue of the changes of volume or of form which it produces in bodies.

These changes are not caused by uniform temperature, but rather by alternations of heat and cold. Now to heat any substance whatever requires a body warmer than the one to be heated; to cool it requires a cooler body. We supply caloric to the first of these bodies that we may transmit it to the second by means of the intermediary substance. This is to re-establish, or at least to endeavor to re-establish, the equilibrium of the caloric.

---

[8] The existence of water in the liquid state here necessarily assumed, since without it the steam-engine could not be fed, supposes the existence of a pressure capable of preventing this water from vaporizing, consequently of a pressure equal or superior to the tension of vapor at that temperature. If such a pressure were not exerted by the atmospheric air, there would be instantly produced a quantity of steam sufficient to give rise to that tension, and it would be necessary always to overcome this pressure in order to throw out the steam from the engines into the new atmosphere. Now this is evidently equivalent to overcoming the tension which the steam retains after its condensation, as effected by ordinary means.

If a very high temperature existed at the surface of our globe, as it seems certain that it exists in its interior, all the waters of the ocean would be in a state of vapor in the atmosphere, and no portion of it would be found in a liquid state.

It is natural to ask here this curious and important question: Is the motive power of heat invariable in quantity, or does it vary with the agent employed to realize it as the intermediary substance, selected as the subject of action of the heat?

It is clear that this question can be asked only in regard to a given quantity of caloric,[9] the difference of the temperatures also being given. We take, for example, one body A kept at a temperature of 100° and another body B kept at a temperature of 0°, and ask what quantity of motive power can be produced by the passage of a given portion of caloric (for example, as much as is necessary to melt a kilogram of ice) from the first of these bodies to the second. We inquire whether this quantity of motive power is necessarily limited, whether it varies with the substance employed to realize it, whether the vapor of water offers in this respect more or less advantage than the vapor of alcohol, of mercury, a permanent gas, or any other substance. We will try to answer these questions, availing ourselves of ideas already established.

We have already remarked upon this self-evident fact, or fact which at least appears evident as soon as we reflect on the changes of volume occasioned by heat: *wherever there exists a difference of temperature, motive-power can be produced.* Reciprocally, wherever we can consume this power, it is possible to produce a difference of temperature, it is possible to occasion destruction of equilibrium in the caloric. Are not percussion and the friction of bodies actually means of raising their temperature, of making it reach spontaneously a higher degree than that of the surrounding bodies, and consequently of producing a destruction of equilibrium in the caloric, where equilibrium previously existed? It is a fact proved by experience, that the temperature of gaseous fluids is raised by compression and lowered by rarefaction. This is a sure method of changing the temperature of bodies, and destroying the equilibrium of the caloric as many times as may be desired with the same substance. The vapor of water employed in an inverse manner to that in which it is used in steam-engines can also be regarded as a means of destroying the equilibrium of the caloric. To be convinced of this we need but to observe closely the manner in which motive power is developed by the action of heat on vapor of water. Imagine two bodies A and B, kept each at a constant temperature, that of A being higher than that of B. These two bodies, to which we can give or from which we can remove the heat without causing their temperatures to vary, exercise the functions of two unlimited reservoirs of caloric. We will call the first the furnace and the second the refrigerator.

If we wish to produce motive power by carrying a certain quantity of heat from the body A to the body B we shall proceed as follows:

(1) To borrow caloric from the body A to make steam with it—that is, to make this body fulfil the function of a furnace, or rather of the metal composing the boiler in ordinary engines—we here assume that the steam is produced at the same temperature as the body A.

---

[9] It is considered unnecessary to explain here what is quantity of caloric or quantity of heat (for we employ these two expressions indifferently), or to describe how we measure these quantities by the calorimeter. Nor will we explain what is meant by latent heat, degree of temperature, specific heat, *etc.* The reader should be familiarized with these terms through the study of the elementary treatises of physics or of chemistry.

(2) The steam having been received in a space capable of expansion, such as a cylinder furnished with a piston, to increase the volume of this space, and consequently also that of the steam. Thus rarefied, the temperature will fall spontaneously, as occurs with all elastic fluids; admit that the rarefaction may be continued to the point where the temperature becomes precisely that of the body B.

(3) To condense the steam by putting it in contact with the body B, and at the same time exerting on it a constant pressure until it is entirely liquefied. The body B fills here the place of the injection-water in ordinary engines, with this difference, that it condenses the vapor without mingling with it, and without changing its own temperature.[10]

The operations which we have just described might have been performed in an inverse direction and order. There is nothing to prevent forming vapor with the caloric of the body B, and at the temperature of that body, compressing it in such a way as to make it acquire the temperature of the body A, finally condensing it by contact with this latter body, and continuing the compression to complete liquefaction.

By our first operations there would have been at the same time production of motive power and transfer of caloric from the body A to the body B. By the inverse operations there is at the same time expenditure of motive power and return of caloric from the body B to the body A. But if we have acted in each case on the same quantity of vapor, if there is produced no loss either of motive power or caloric, the quantity of motive power produced in the first place will be equal to that which would have been expended in the second, and the quantity of caloric passed in the first case from the body A to the body B would be equal to the quantity which passes back again in the second from the body B to the body A; so that an indefinite number of alternative operations of this sort could be carried on without in the end having either produced motive power or transferred caloric from one body to the other.

---

[10] We may perhaps wonder here that the body B being at the same temperature as the steam is able to condense it. Doubtless this is not strictly possible, but the slightest difference of temperature will determine the condensation, which suffices to establish the justice of our reasoning. It is thus that, in the differential calculus, it is sufficient that we can conceive the neglected quantities indefinitely reducible in proportion to the quantities retained in the equations, to make certain of the exact result.

The body B condenses the steam without changing its own temperature—this results from our supposition. We have admitted that this body may be maintained at a constant temperature. We take away the caloric as the steam furnishes it. This is the condition in which the metal of the condenser is found when the liquefaction of the steam is accomplished by applying cold water externally, as was formerly done in several engines. Similarly, the water of a reservoir can be maintained at a constant level if the liquid flows out at one side as it flows in at the other.

One could even conceive the bodies A and B maintaining the same temperature, although they might lose or gain certain quantities of heat. If, for example, the body A were a mass of steam ready to become liquid, and the body B a mass of ice ready to melt, these bodies might, as we know, furnish or receive caloric without thermometric change.

Now if there existed any means of using heat preferable to those which we have employed, that is, if it were possible by any method whatever to make the caloric produce a quantity of motive power greater than we have made it produce by our first series of operations, it would suffice to divert a portion of this power in order by the method just indicated to make the caloric of the body *B* return to the body *A* from the refrigerator to the furnace, to restore the initial conditions, and thus to be ready to commence again an operation precisely similar to the former, and so on: this would be not only perpetual motion, but an unlimited creation of motive power without consumption either of caloric or of any other agent whatever. Such a creation is entirely contrary to ideas now accepted, to the laws of mechanics and of sound physics. It is inadmissible. We should then conclude that *the maximum of motive power resulting from the employment of steam is also the maximum of motive power realizable by any means whatever.* We will soon give a second more rigorous demonstration of this theory. This should be considered only as an approximation. (See page 39.)

We have a right to ask, in regard to the proposition just enunciated, the following questions: What is the sense of the word *maximum* here? By what sign can it be known that this maximum is attained? By what sign can it be known whether the steam is employed to greatest possible advantage in the production of motive power?

Since every re-establishment of equilibrium in the caloric may be the cause of the production of motive power, every re-establishment of equilibrium which shall be accomplished without production of this power should be considered as an actual loss. Now, very little reflection would show that all change of temperature which is not due to a change of volume of the bodies can be only a useless reestablishment of equilibrium in the caloric.[11] The necessary condition of the maximum is, then, *that in the bodies employed to realize the motive power of heat there should not occur any change of temperature which may not be due to a change of volume.* Reciprocally, every time that this condition is fulfilled the maximum will be attained. This principle should never be lost sight of in the construction of heat-engines; it is its fundamental basis. If it cannot be strictly observed, it should at least be departed from as little as possible.

Every change of temperature which is not due to a change of volume or to chemical action (an action that we provisionally suppose not to occur here) is necessarily due to the direct passage of the caloric from a more or less heated body to a colder body. This passage occurs mainly by the contact of bodies of different temperatures; hence such contact should be avoided as much as possible. It cannot probably be avoided entirely, but it should at least be so managed that the bodies brought in contact with each other differ as little as possible in temperature. When we just now

---

[11] We assume here no chemical action between the bodies employed to realize the motive power of heat. The chemical action which takes place in the furnace is, in some sort, a preliminary action,— an operation destined not to produce immediately motive power, but to destroy the equilibrium of the caloric, to produce a difference of temperature which may finally give rise to motion.

supposed, in our demonstration, the caloric of the body A employed to form steam, this steam was considered as generated at the temperature of the body A; thus the contact took place only between bodies of equal temperatures; the change of temperature occurring afterwards in the steam was due to dilatation, consequently to a change of volume. Finally, condensation took place also without contact of bodies of different temperatures. It occurred while exerting a constant pressure on the steam brought in contact with the body B of the same temperature as itself. The conditions for a maximum are thus found to be fulfilled. In reality the operation cannot proceed exactly as we have assumed. To determine the passage of caloric from one body to another, it is necessary that there should be an excess of temperature in the first, but this excess may be supposed as slight as we please. We can regard it as insensible in theory, without thereby destroying the exactness of the arguments.

A more substantial objection may be made to our demonstration, thus: When we borrow caloric from the body A to produce steam, and when this steam is afterwards condensed by its contact with the body B, the water used to form it, and which we considered at first as being of the temperature of the body A, is found at the close of the operation at the temperature of the body B. It has become cool. If we wish to begin again an operation similar to the first, if we wish to develop a new quantity of motive power with the same instrument, with the same steam, it is necessary first to re-establish the original condition—to restore the water to the original temperature. This can undoubtedly be done by at once putting it again in contact with the body A; but there is then contact between bodies of different temperatures, and loss of motive power.[12] It would be impossible to execute the inverse operation, that is, to return to the body A the caloric employed to raise the temperature of the liquid.

This difficulty may be removed by supposing the difference of temperature between the body A and the body B indefinitely small. The quantity of heat necessary to raise the liquid to its former temperature will be also indefinitely small and unimportant relatively to that which is necessary to produce steam—a quantity always limited.

The proposition found elsewhere demonstrated for the case in which the difference between the temperatures of the two bodies is indefinitely small, may be easily extended to the general case. In fact, if it operated to produce motive power by the passage of caloric from the body A to the body Z, the temperature of this latter body being very different from that of the former, we should imagine a series of bodies B, C, D... of temperatures intermediate between those of the bodies A, Z, and selected

---

[12] This kind of loss is found in all steam-engines. In fact, the water destined to feed the boiler is always cooler than the water which it already contains. There occurs between them a useless re-establishment of equilibrium of caloric. We are easily convinced, *à posteriori*, that this reestablishment of equilibrium causes a loss of motive power if we reflect that it would have been possible to previously heat the feed-water by using it as condensing-water in a small accessory engine, when the steam drawn from the large boiler might have been used, and where the condensation might be produced at a temperature intermediate between that of the boiler and that of the principal condenser. The power produced by the small engine would have cost no loss of heat, since all that which had been used would have returned into the boiler with the water of condensation.

so that the differences from A to B, from B to C, *etc.*, may all be indefinitely small. The caloric coming from A would not arrive at Z till after it had passed through the bodies B, C, D, *etc.*, and after having developed in each of these stages maximum motive power. The inverse operations would here be entirely possible, and the reasoning of page 36 would be strictly applicable.

According to established principles at the present time, we can compare with sufficient accuracy the motive power of heat to that of a waterfall. Each has a maximum that we cannot exceed, whatever may be, on the one hand, the machine which is acted upon by the water, and whatever, on the other hand, the substance acted upon by the heat. The motive power of a waterfall depends on its height and on the quantity of the liquid; the motive power of heat depends also on the quantity of caloric used, and on what may be termed, on what in fact we will call, the *height of its fall*,[13] that is to say, the difference of temperature of the bodies between which the exchange of caloric is made. In the waterfall the motive power is exactly proportional to the difference of level between the higher and lower reservoirs. In the fall of caloric the motive power undoubtedly increases with the difference of temperature between the warm and the cold bodies; but we do not know whether it is proportional to this difference. We do not know, for example, whether the fall of caloric from 100 to 50° furnishes more or less motive power than the fall of this same caloric from 50 to 0. It is a question which we propose to examine hereafter.

3.3   Study Questions

QUES. 3.1. What are some of the most important uses of the steam-engine? Which nation-state, at the time of Carnot, had most fully developed its potential? And what is the goal of Carnot's work?

QUES. 3.2. What are the basic components—and the principle of operation—of a steam-engine?

a) What are the purposes of the boiler, steam-cylinder, piston and condenser?
b) What is caloric? Where is it developed? In which direction does it flow? And where does it end up? What role does the steam itself play?
c) What is the cause of the motive power of steam-engines? Is heat alone (from a hot body) necessary to produce a steam-engine's motive power? Do all steam-engines require a condenser?

QUES. 3.3. Does the motive power of a heat-engine depend on the particular intermediary substance employed?

a) Is steam the only feasible intermediate-substance for a heat-engine? If not, what other substances have been employed instead of steam?

---

[13] The matter here dealt with being entirely new, we are obliged to employ expressions not in use as yet, and which perhaps are less clear than is desirable.

b) In principle, what physical property must any body have in order to serve as the intermediate-substance in a heat-engine?

c) What purpose, then, do the hot- and cold- bodies (*e.g.* the boiler and condenser) serve?

d) How does Carnot define a unit of heat? And does the motive power of a steam engine depend on the temperatures of the hot and cold bodies?

QUES. 3.4. If nature always acts in such a way as to *restore* the equilibrium of the caloric, then what types of processes might *destroy* such equilibrium? What is required? Can equilibrium be destroyed by a heat-engine? If so, how?

QUES. 3.5. Is it possible for a heat-engine to be more efficient—to produce more motive power when operating between a pair of reservoirs—than a reversible heat-engine?

a) Carefully describe the sequence of operations involved in Carnot's steam engine cycle. Which steps involve temperature changes? Which involve the flow of caloric?

b) Is the amount of work done *by* the steam equal to the amount of work done *on* the steam during an entire cycle? Does the engine accomplish any useful work?

c) If the cycle of operations is reversed, in which way does caloric flow? Does the reversed heat engine accomplish useful work?

d) What is meant by a *reversible* heat-engine? In particular, is *any* heat-engine which is run in reverse a *reversible* heat-engine? If not, then what *is* the mark of a reversible heat-engine?

e) Suppose that a forward-running reversible heat-engine drives a second reversible heat-engine backwards. Is there any overall transport of caloric between the hot and cold reservoirs between which they (both) operate?

f) Would your answer to the previous question be the same if the backwards-running engine were somehow *more* efficient than a reversible heat-engine? What does this imply about the possibility of such a more-efficient heat-engine?

QUES. 3.6. Ideally, what physical process should accompany any and every flow of caloric within a heat-engine so as to maximize the heat-engine's efficiency in generating motive power? What processes or situations should one avoid, and why? Are these optimization conditions met in Carnot's proposed engine cycle?

QUES. 3.7. In what sense is a heat-engine analogous to a water-fall? In particular, does the power generated by a water-fall depend only upon the height of the fall? Does the efficiency of a heat engine depend only upon the temperature difference between its hot and cold reservoirs?

3.4 Exercises

EX. 3.1 (IDEAL GAS LAWS). An ideal gas is one whose constituent molecules are so widely separated as to be effectively non-interacting. *Charles' law* states that the temperature of an ideal gas which is maintained at a constant pressure is directly proportional to its volume:

$$\frac{V}{T} = \text{Constant} \qquad\qquad \text{(Charles' law)} \qquad (3.1)$$

Boyle's law states that the pressure of an ideal gas which is maintained at a constant temperature is inversely proportional to its volume:

$$PV = \text{Constant} \qquad\qquad \text{(Boyle's law)} \qquad (3.2)$$

As an exercise, suppose that you are scuba diving in 55° F water. When you are 50 ft below the water's surface, you take a gulp of air from your tank. The regulator on your tank simplifies this task by ensuring that the air which you take in is at the same pressure as the ambient water pressing against your body. (a) By what factor does the volume of this inhaled air increase as it heats up to your body temperature, 98.6° F? (HINT: be sure to convert to Kelvin.) (b) Suppose that you now make the mistake of holding your breath as you ascend to the water's surface. Assuming the temperature does not change as you ascend, by what factor does the volume of air in your lungs increase by the time your reach the surface?

EX. 3.2 (WORK, PRESSURE AND THE VAN DER WAALS EQUATION OF STATE). A gas exerts a force on the walls of its confining chamber. As a result, the gas expands until the walls of the chamber—stretched like a spring—produce a force large enough to check its expansion. For pliable containers (such as latex balloons) the relative expansion can be quite large, while for very rigid containers (such as steel gas cylinders) the expansion is miniscule. In any case, when the chamber expands, the gas does *work* on the chamber walls. Recall that the work done by a force is defined as the product of the force and the distance through which it travels. Since the pressure of a gas is defined as the force it exerts on a unit area of the confining chamber, the work may be expressed in terms of the gas pressure and its change in volume. In particular, if the expansion is slow,[14] then the work done by the gas on the chamber wall(s) during its expansion can be written as a definite integral of the pressure over the change in volume from V_1 to V_2:

$$W = \int_{V_1}^{V_2} p\,dV. \qquad (3.3)$$

---

[14] Equation 3.3 for the work done by an expanding gas can only be used when the pressure exerted by the gas is nearly the same as the external pressure attempting (in vain) to restrain its expansion. During such an expansion, the gas pressure is uniform throughout. Equation. 3.3 cannot be used when the gas expands rapidly against a much weaker restraining force, for example, when a balloon is popped and an initially high pressure gas rushes into a surrounding low-pressure gas. There is no work done by the gas in such a "free expansion." For a more sophisticated treatment of this topic, and its relationship to entropy, see Rudolph Clausius' discussion of reversible and irreversible processes in Chap. 6 of the present volume.

Fig. 3.2 The pressure of a
gas drops when it expands
isothermally

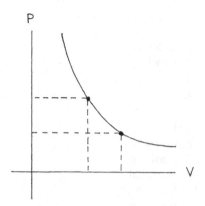

For many gasses, the pressure, temperature and molar volume[15] are related by the
Van der Waals equation of state:

$$\left(p + \frac{a}{v^2}\right)(v - b) = RT \tag{3.4}$$

Here, R is the *universal gas constant*, and a and b are empirically determined gas-
dependent parameters. They account for (a) the slight diminution of pressure due
to attraction between pairs of molecules, and (b) the slightly increased volume due
to the size of the gas molecules themselves. For an ideal (*i.e.* noninteracting) gas, a
and b are taken to be zero.

As an exercise, consider a particular gas for which $a = 1 \times 10^{-4}$ and $b =
3 \times 10^{-5}$ (using SI units). 0.3 moles of this gas are confined in a cylindrical chamber
fitted with a movable piston at one end. The initial pressure of the gas is 144 psi.
It undergoes a slow expansion (by pushing out the piston) from 1 to 1.5 L while
being maintained at a constant temperature. This process is represented graphically
in Fig. 3.2.

a) What is the temperature of this gas during this isothermal expansion? How much
   work does this gas perform on the piston as it expands? (ANSWER: 402 J)
b) In order to maintain the gas at a constant temperature, heat must be added to the
   gas during its expansion. For a real gas, some of the added heat energy goes into
   work done on the piston and some of it goes into changing the arrangement of
   the (mutually attracting) gas molecules.[16] For an ideal gas, on the other hand,
   all of the added heat energy goes into the work performed during an isothermal
   expansion. In this case, how much heat must be added during the aforementioned
   expansion?

---

[15] Molar volume is the volume per mole of gas.

[16] The division of heat into internal energy and external work will be discussed in more detail by
Rudolph Clausius; see Chap. 6.

3.5 Vocabulary

1. Ascension
2. Impel
3. Appropriate
4. Procure
5. Ascertain
6. Impel
7. Condenser
8. Aeriform
9. Perpetual
10. Inadmissible
11. Enunciate
12. Dilatation

Chapter 4
Carnot's Cycle

The motive power of heat is independent of the agents employed to realize it; its quantity is fixed solely by the temperatures of the bodies between which is effected, finally, the transfer of the caloric.

—Sadi Carnot

4.1 Introduction

In the previous reading selection, from his *Reflections on the Motive Power of Heat*, Carnot claimed that the fundamental reason why all heat-engines are able to generate motive power is because caloric, or heat, flows from hot to cold bodies. This tendency of nature to "restore the equilibrium in the caloric" is a *necessary*, but not a *sufficient* condition to accomplish work. For example, no useful work is accomplished when a red-hot iron is simply plunged into a bath of cold water, despite the fact that heat flows from the hot iron into the cold water. On the other hand, by heating up steam in a boiler chamber, allowing the high-pressure steam to drive a piston upwards, and then allowing the steam to cool back down in a condenser chamber, one may accomplish useful work. In other words, by carefully managing the flow of caloric from hot to cold bodies, one can accomplish useful work. A steam-engine is a well-known example of this, but *any* process that converts heat-flow into useful work is called a heat-engine.

Now, how much motive power can a heat-engine generate? The *efficiency* of a heat engine, η, is defined as the ratio of the work that it accomplishes during a complete engine cycle, W, to the heat flowing into it from the hot reservoir during the same cycle, Q_h:

$$\eta = \frac{W}{Q_h}. \tag{4.1}$$

How might a heat engine attain the highest possible efficiency? Does a heat-engine's efficiency depend on its construction? On the type of intermediate substance employed? On the temperature of the boiler? Of the condenser? Carnot argued that all these factors may, in fact, affect the efficiency of a heat-engine. But in principle, there exists a special class of heat-engines which possess the highest possible efficiency. These special heat-engines are the *reversible* heat-engines—those in which

© Springer International Publishing Switzerland 2016

K. Kuehn, *A Student's Guide Through the Great Physics Texts*,
Undergraduate Lecture Notes in Physics, DOI 10.1007/978-3-319-21828-1_4

there is (essentially) no useless heat flow between various internal components, and thus in which all of the heat flow from the hot to the cold reservoir provides useful work.

Why must a reversible heat-engine have the highest possible efficiency? Carnot employs a proof by contradiction to demonstrate this point: he supposes, for the sake of argument, that a heat engine exists which is *more* efficient than a reversible heat engine. He then proves that such a heat-engine could be driven backwards (by a reversible heat engine running forwards) in such a way as to drive heat from a cold to a hot body without expending any work. But this would violate a fundamental principle of nature: its tendency to restore equilibrium in the caloric. Thus, such a more-efficient heat-engine could not be possible. In the following reading selection, Carnot provides a more exact formulation of this argument. He does so by introducing what is now known as the Carnot-cycle—a cycle of operations which are performed on a heat engine in such a way as to accomplish the maximum amount of work.

4.2   Reading: Carnot, *Reflections on the Motive Power of Heat, and on Machines Fitted to Develop that Power*

Carnot, S., *Reflections on the Motive Power of Heat*, second ed., John Wiley & Sons and Chapman & Hall, New York and London, 1897.

We shall give here a second demonstration of the fundamental proposition enunciated on page 38, and present this proposition under a more general form than the one already given.

When a gaseous fluid is rapidly compressed its temperature rises. It falls, on the contrary, when it is rapidly dilated. This is one of the facts best demonstrated by experiment. We will take it for the basis of our demonstration.

If, when the temperature of a gas has been raised by compression, we wish to reduce it to its former temperature without subjecting its volume to new changes, some of its caloric must be removed. This caloric might have been removed in proportion as pressure was applied, so that the temperature of the gas would remain constant. Similarly, if the gas is rarefied we can avoid lowering the temperature by supplying it with a certain quantity of caloric. Let us call the caloric employed at such times, when no change of temperature occurs, *caloric due to change of volume*. This denomination does not indicate that the caloric appertains to the volume: it does not appertain to it any more than to pressure, and might as well be called *caloric due to the change of pressure*. We do not know what laws it follows relative to the variations of volume: it is possible that its quantity changes either with the nature of the gas, its density, or its temperature. Experiment has taught us nothing on this subject. It has only shown us that this caloric is developed in greater or less quantity by the compression of the elastic fluids.

Fig. 4.1 An air-filled vessel
fitted with a movable piston
is alternately isolated from,
and brought into contact with,
hot and cold reservoirs when
executing the four steps of a
Carnot cycle.—[*K.K.*]

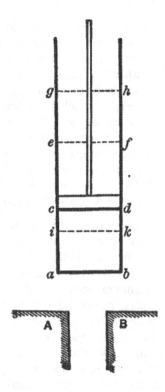

This preliminary idea being established, let us imagine an elastic fluid, atmospheric air for example, shut up in a cylindrical vessel, *abcd* (Fig. 4.1), provided
with a movable diaphragm or piston, *cd*. Let there be also two bodies, *A* and *B*,
kept each at a constant temperature, that of *A* being higher than that of *B*. Let us
picture to ourselves now the series of operations which are to be described:

(1) Contact of the body *A* with the air enclosed in the space *abcd* or with the wall
    of this space—a wall that we will suppose to transmit the caloric readily. The
    air becomes by such contact of the same temperature as the body *A*; *cd* is the
    actual position of the piston.
(2) The piston gradually rises and takes the position *ef*. The body *A* is all the time
    in contact with the air, which is thus kept at a constant temperature during the
    rarefaction. The body *A* furnishes the caloric necessary to keep the temperature
    constant.
(3) The body *A* is removed, and the air is then no longer in contact with any body
    capable of furnishing it with caloric. The piston meanwhile continues to move,
    and passes from the position *ef* to the position *gh*. The air is rarefied without
    receiving caloric, and its temperature falls. Let us imagine that it falls thus till it
    becomes equal to that of the body *B*; at this instant the piston stops, remaining
    at the position *gh*.

(4) The air is placed in contact with the body B; it is compressed by the return of the piston as it is moved from the position gh to the position cd. This air remains, however, at a constant temperature because of its contact with the body B, to which it yields its caloric.

(5) The body B is removed, and the compression of the air is continued, which being then isolated, its temperature rises. The compression is continued till the air acquires the temperature of the body A. The piston passes during this time from the position cd to the position ik.

(6) The air is again placed in contact with the body A. The piston returns from the position ik to the position ef; the temperature remains unchanged.

(7) The step described under number 3 is renewed, then successively the steps 4, 5, 6, 3, 4, 5, 6, 3, 4, 5; and so on.

In these various operations the piston is subject to an effort of greater or less magnitude, exerted by the air enclosed in the cylinder; the elastic force of this air varies as much by reason of the changes in volume as of changes of temperature. But it should be remarked that with equal volumes, that is, for the similar positions of the piston, the temperature is higher during the movements of dilatation than during the movements of compression. During the former the elastic force of the air is found to be greater, and consequently the quantity of motive power produced by the movements of dilatation is more considerable than that consumed to produce the movements of compression. Thus we should obtain an excess of motive power—an excess which we could employ for any purpose whatever. The air, then, has served as a heat-engine; we have, in fact, employed it in the most advantageous manner possible, for no useless re-establishment of equilibrium has been effected in the caloric.

All the above-described operations may be executed in an inverse sense and order. Let us imagine that, after the sixth period, that is to say the piston having arrived at the position ef, we cause it to return to the position ik, and that at the same time we keep the air in contact with the body A. The caloric furnished by this body during the sixth period would return to its source, that is, to the body A, and the conditions would then become precisely the same as they were at the end of the fifth period. If now we take away the body A, and if we cause the piston to move from ef to cd, the temperature of the air will diminish as many degrees as it increased during the fifth period, and will become that of the body B. We may evidently continue a series of operations the inverse of those already described. It is only necessary under the same circumstances to execute for each period a movement of dilatation instead of a movement of compression, and reciprocally.

The result of these first operations has been the production of a certain quantity of motive power and the removal of caloric from the body A to the body B. The result of the inverse operations is the consumption of the motive power produced and the return of the caloric from the body B to the body A; so that these two series of operations annul each other, after a fashion, one neutralizing the other.

The impossibility of making the caloric produce a greater quantity of motive power than that which we obtained from it by our first series of operations, is now easily proved. It is demonstrated by reasoning very similar to that employed at page 38; the reasoning will here be even more exact. The air which we have used to develop the motive power is restored at the end of each cycle of operations exactly to the state in which it was at first found, while, as we have already remarked, this would not be precisely the case with the vapor of water.[1]

We have chosen atmospheric air as the instrument which should develop the motive power of heat, but it is evident that the reasoning would have been the same for all other gaseous substances, and even for all other bodies susceptible of change of temperature through successive contractions and dilatations, which comprehends all natural substances, or at least all those which are adapted to realize the motive power of heat. Thus we are led to establish this general proposition:

The motive power of heat is independent of the agents employed to realize it; its quantity is fixed solely by the temperatures of the bodies between which is effected, finally, the transfer of the caloric.

We must understand here that each of the methods of developing motive power attains the perfection of which it is susceptible. This condition is found to be fulfilled if, as we remarked above, there is produced in the body no other change of temperature than that due to change of volume, or, what is the same thing in other words, if there is no contact between bodies of sensibly different temperatures.

Different methods of realizing motive power may be taken, as in the employment of different substances, or in the use of the same substance in two different states—for example, of a gas at two different densities.

This leads us naturally to those interesting researches on the aeriform fluids—researches which lead us also to new results in regard to the motive power of heat, and give us the means of verifying, in some particular cases, the fundamental proposition above stated.[2]

We readily see that our demonstration would have been simplified by supposing the temperatures of the bodies A and B to differ very little. Then the movements of the piston being slight during the periods 3 and 5, these periods might have been suppressed without influencing sensibly the production of motive power. A very

---

[1] We tacitly assume in our demonstration, that when a body has experienced any changes, and when after a certain number of transformations it returns to precisely its original state, that is, to that state considered in respect to density, to temperature, to mode of aggregation—let us suppose, I say, that this body is found to contain the same quantity of heat that it contained at first, or else that the quantities of heat absorbed or set free in these different transformations are exactly compensated. This fact has never been called in question. It was first admitted without reflection, and verified afterwards in many cases by experiments with the calorimeter. To deny it would be to overthrow the whole theory of heat to which it serves as a basis. For the rest, we may say in passing, the main principles on which the theory of heat rests require the most careful examination. Many experimental facts appear almost inexplicable in the present state of this theory.

[2] We will suppose, in what follows, the reader to be *au courant* with the later progress of modern Physics in regard to gaseous substances and heat.

little change of volume should suffice in fact to produce a very slight change of temperature, and this slight change of volume may be neglected in presence of that of the periods 4 and 6, of which the extent is unlimited.

If we suppress periods 3 and 5, in the series of operations above described, it is reduced to the following:

(1) Contact of the gas confined in *abcd* (Fig. 4.1) with the body *A*, passage of the piston from *cd* to *ef*.
(2) Removal of the body *A*, contact of the gas confined in *abef* with the body *B*, return of the piston from *ef* to *cd*.
(3) Removal of the body *B*, contact of the gas with the body *A*, passage of the piston from *cd* to *ef*, that is, repetition of the first period, and so on.

The motive power resulting from the *ensemble* of operations 1 and 2 will evidently be the difference between that which is produced by the expansion of the gas while it is at the temperature of the body *A*, and that which is consumed to compress this gas while it is at the temperature of the body *B*.

Let us suppose that operations 1 and 2 be performed on two gases of different chemical natures but under the same pressure—under atmospheric pressure, for example. These two gases will behave exactly alike under the same circumstances, that is, their expansive forces, originally equal, will remain always equal, whatever may be the variations of volume and of temperature, provided these variations are the same in both. This results obviously from the laws of Mariotte and MM. Gay-Lussac and Dalton—laws common to all elastic fluids, and in virtue of which the same relations exist for all these fluids between the volume, the expansive force, and the temperature.

Since two different gases at the same temperature and under the same pressure should behave alike under the same circumstances, if we subjected them both to the operations above described, they should give rise to equal quantities of motive power.

Now this implies, according to the fundamental proposition that we have established, the employment of two equal quantities of caloric; that is, it implies that the quantity of caloric transferred from the body *A* to the body *B* is the same, whichever gas is used. The quantity of caloric transferred from the body *A* to the body *B* is evidently that which is absorbed by the gas in its expansion of volume, or that which this gas relinquishes during compression. We are led, then, to establish the following proposition:

> When a gas passes without change of temperature from one definite volume and pressure to another volume and another pressure equally definite, the quantity of caloric absorbed or relinquished is always the same, whatever may be the nature of the gas chosen as the subject of the experiment.

Take, for example, 1 l of air at the temperature of 100° and under the pressure of one atmosphere. If we double the volume of this air and wish to maintain it at the temperature of 100°, a certain quantity of heat must be supplied to it. Now this quantity will be precisely the same if, instead of operating on the air, we operate upon carbonic-acid gas, upon nitrogen, upon hydrogen, upon vapor of water or of alcohol, that is, if we double the volume of 1 liter of these gases taken at the temperature of 100° and under atmospheric pressure.

It will be the same thing in the inverse sense if, instead of doubling the volume of gas, we reduce it one half by compression. The quantity of heat that the elastic fluids set free or absorb in their changes of volume has never been measured by any direct experiment, and doubtless such an experiment would be very difficult, but there exists a datum which is very nearly its equivalent. This has been furnished by the theory of sound. It deserves much confidence because of the exactness of the conditions which have led to its establishment. It consists in this:

Atmospheric air should rise 1 °C when by sudden compression it experiences a reduction of volume of $\frac{1}{116}$.[3]

Experiments on the velocity of sound having been made in air under the pressure of 760 mm of mercury and at the temperature of 6°, it is only to these two circumstances that our datum has reference. We will, however, for greater facility, refer it to the temperature 0°, which is nearly the same.

Air compressed $\frac{1}{116}$, and thus heated 1°, differs from air heated directly 1° only in its density. The primitive volume being supposed to be V, the compression of $\frac{1}{116}$ reduces it to $V - \frac{1}{116}V$.

Direct heating under constant pressure should, according to the rule of M. Gay-Lussac, increase the volume of air $\frac{1}{267}$ above what it would be at 0°: so the air is, on the one hand, reduced to the volume $V - \frac{1}{116}V$; on the other, it is increased to $V + \frac{1}{267}V$.

The difference between the quantities of heat which the air possesses in both cases is evidently the quantity employed to raise it directly 1°; so then the quantity of heat that the air would absorb in passing from the volume $V - \frac{1}{116}V$ to the volume $V + \frac{1}{267}V$ is equal to that which is required to raise it 1°.

Let us suppose now that, instead of heating 1° the air subjected to a constant pressure and able to dilate freely, we inclose it within an invariable space, and that in this condition we cause it to rise 1° in temperature. The air thus heated 1° will differ from the air compressed $\frac{1}{116}$ only by its $\frac{1}{116}$ greater volume. So then the quantity of heat that the air would set free by a reduction of volume of $\frac{1}{116}$ is equal to that which would be required to raise it 1 °C under constant volume. As the differences between the volumes $V - \frac{1}{116}V$, V, and $V + \frac{1}{267}V$ are small relatively to the volumes themselves, we may regard the quantities of heat absorbed by the air in passing from the first of these volumes to the second, and from the first to the third, as sensibly proportional to the changes of volume. We are then led to the establishment of the following relation:

---

[3] M. Poisson, to whom this figure is due, has shown that it accords very well with the result of an experiment of MM. Clement and Desormes on the return of air into a vacuum, or rather, into air slightly rarefied. It also accords very nearly with results found by MM. Gay-Lussac and Welter.

The quantity of heat necessary to raise $1°$ air under constant pressure is to the quantity of heat necessary to raise $1°$ the same air under constant volume, in the ratio of the numbers

$$\frac{1}{116} + \frac{1}{267} \text{ to } \frac{1}{116};$$ (4.2)

or, multiplying both by 116×267, in the ratio of the numbers $267 + 116$ to 267.

This, then, is the ratio which exists between the capacity of air for heat under constant pressure and its capacity under constant volume. If the first of these two capacities is expressed by unity, the other will be expressed by the number $\frac{267}{267+116}$ or very nearly 0.700; their difference, $1 - 0.700$ or 0.300, will evidently express the quantity of heat which will produce the increase of volume in the air when it is heated $1°$ under constant pressure.

According to the law of MM. Gay-Lussac and Dalton, this increase of volume would be the same for all other gases; according to the theory demonstrated on page 87,[4] the heat absorbed by these equal increases of volume is the same for all the elastic fluids, which leads to the establishment of the following proposition:

> The difference between specific heat under constant pressure and specific heat under constant volume is the same for all gases.

It should be remarked here that all the gases are considered as taken under the same pressure, atmospheric pressure for example, and that the specific heats are also measured with reference to the volumes.

4.3 Study Questions

QUES. 4.1. What happens when a gaseous fluid is rapidly compressed? Can a compressed gas be returned to its former temperature without it being dilated? If so, how? Similarly, can a dilated gas be returned to its former temperature without it being compressed?

QUES. 4.2. In what sense is Carnot's cycle done "in the most advantageous manner possible"?

a) Carefully describe the series of operations in Carnot's heat-engine cycle. In particular, in which stage(s) does the temperature remain constant? In which stage(s) is caloric supplied or removed? Where does the caloric come from (or go to)?
b) Is the temperature of the gas greater when the gas is expanding or when it is contracting? Is the force exerted by the piston greater when the piston is moving outwards or inwards?

---

[4] Carnot refers here to a page number in his own Treatise—[K.K.]

c) Is the work done *by* the gas during an expansion equal to the work done *on* the gas during a compression? What does this imply?

d) After a complete Carnot cycle, is the gas in the same state as it was initially? Are the hot and cold reservoirs in the same state(s)? Has any caloric flowed? If so, in which direction?

e) What specific condition(s) must be met so as to render the Carnot cycle reversible? How is this related to its efficiency?

QUES. 4.3. What happens if Carnot's cycle is performed in reverse order? In particular, is more work done *on* the gas or *by* the gas during a complete cycle? And in which direction does caloric flow?

QUES. 4.4. Does Carnot's cycle require air (or any other particular substance, for that matter) as the working substance? If not, then what (alone) determines the efficiency of Carnot's reversible engine cycle? Has Carnot proved this proposition? What assertion does Carnot make regarding the nature of the gases employed in a Carnot cycle?

QUES. 4.5. What is the relationship between the specific heat capacities of a gas when measured at a *constant pressure* and when measured at a *constant volume*? How does Carnot calculate this number empirically? Does it matter what kind of gas is employed?

4.4 Exercises

EX. 4.1 (ADIABATIC COMPRESSION OF AN IDEAL GAS). When a gaseous substance is compressed, its temperature rises. If the compression is sufficiently rapid, then it does not have time to exchange heat (or caloric) with its surroundings—the compression is said to be done *adiabatically*. For example, when a volume of air is adiabatically compressed by a factor of 1/116, its temperature is found to rise by 1 °C. Of course since the temperature of this rapidly compressed gas is elevated above that of its surroundings, heat will eventually flow out of the gas unless it is perfectly insulated from its surroundings. But to the extent that a substance is perfectly thermally isolated form its surroundings, all processes which it undergoes are adiabatic processes. During an adiabatic expansion (or compression) of an ideal gas, the pressure and volume are related by

$$PV^\gamma = \text{constant}, \tag{4.3}$$

where $\gamma = c_p/c_v$ is the ratio of the heat capacities of the ideal gas when measured at constant pressure and at constant volume, respectively. Similarly, during an adiabatic expansion (or compression) of an ideal gas, the volume and temperature are related by

$$TV^{\gamma-1} = \text{constant}, \tag{4.4}$$

According to Carnot, the value of γ is measured to be $(267 + 116)/267$. According to more recent theoretical considerations, $\gamma = 7/5$ (or $5/3$) for any diatomic (or monatomic) ideal gas.

As an exercise, consider air confined in a 1-liter chamber at 397 K and having a pressure of 144 psi. (a) Compare the slopes, dP/dV, of the adiabatic (Eq. 4.3) and isothermal (Eq. 3.2) curves under these conditions. Which is greater? Is this always the case? (b) If the air is allowed to expand isothermally from 1.0 to 1.5 l, what will be its final pressure and temperature? What if this process is carried out adiabatically?

Ex. 4.2 (CARNOT CYCLE). Suppose that air is shut up in a 1-liter cylindrical vessel provided with a movable diaphragm piston (see Fig. 4.1). Its initial temperature and pressure are 397 K and 144 psi, respectively. The air is made to undergo a Carnot cycle, which consists of four steps. First, it undergoes an isothermal expansion at 397 K (the temperature of reservoir A) until it reaches a volume of 1.5 l. Second, it is removed from reservoir A and is allowed to expand adiabatically until it reaches a temperature of 300 K (the temperature of reservoir B). Third, it is brought into contact with reservoir B and is isothermally compressed. Finally, it is removed from reservoir B and is adiabatically compressed until it returns to its initial state.

a) What is the pressure, volume and temperature of the air (taken as a diatomic ideal gas) at the beginning (and end) of each step comprising this carnot cycle? (HINT: Use Eqs. 4.3 and 4.4.) How many moles of gas are in this cylinder?
b) How much work is done by (or on) the gas during each of the isothermal steps? Where (according to Carnot) does the heat drawn in from the hot reservoir end up?
c) If this engine were run *backwards*, then how much work would be done *on* the gas during each of the isothermal steps? Where (according to Carnot) would the heat drawn in from the cold reservoir end up?
d) While maintaining the hot and cold reservoirs at the same temperatures, can any other working substance be employed so as to construct a higher-efficiency-engine than the one just constructed? Justify your answer.

4.5   Vocabulary

1. Enunciate
2. Dilate
3. Appertain
4. Diaphragm
5. Dilation
6. Aeriform
7. *Au courant*
8. Datum

Chapter 5
Engines as Thermometers

> *This may justly be termed an absolute scale, since its characteristic is quite independent of the physical properties of any specific substance.*
>
> —William Thomson (Lord Kelvin)

5.1 Introduction

William Thomson (1824–1907)—later known as Lord Kelvin—was born in Belfast, Northern Ireland.[1] His father, James Thomson, educated his sons at home in both classical languages and modern subjects. At the age of 10, Kelvin began to attend classes at the University of Glasgow, where his father was a Professor of Mathematics. Alongside his studies of Roman antiquities, astronomy, logic, and moral and natural philosophy, Kelvin acquired an abiding interest in the analytical techniques of French mathematicians such as Laplace, Lagrange and Fourier. After reading Fourier's *La Théorie Analytique de la Chaleur* on a family trip to Germany in 1840, Kelvin wrote his first original paper entitled "On Fourier's Expansion of Functions in Trigonometric Series."[2] In fact, many of Kelvin's initial publications were inspired by Fourier's *Analytical Theory of Heat*.[3] In 1841, at the age of 17, Kelvin left Glasgow to attend St. Peter's College, Cambridge. As an undergraduate, he published 16 mathematical and physical papers. Among these was one in which he proposed the now-famous *method of electric images* for determining the electric field in the vicinity of a charged body situated near a conducting surface. He graduated in 1845, and in 1846 the faculty of his former University of Glasgow unanimously elected him to the post of Professor of Natural Philosophy. The essay

---

[1] Much of the biographical information on Lord Kelvin was obtained from Gray, A., *Lord Kelvin: An Account of His Scientific Life and Work*, J.M. Dent & Co. and E.P. Dutton & Co., London and New York, 1908.

[2] *Cambridge Mathematical Journal*, vol. ii, May 1841.

[3] Selections from Fourier's *Analytical Theory of Heat* are included in Chaps. 1–2 of the present volume.

© Springer International Publishing Switzerland 2016
K. Kuehn, *A Student's Guide Through the Great Physics Texts*,
Undergraduate Lecture Notes in Physics, DOI 10.1007/978-3-319-21828-1_5

which he was assigned to submit (as a formal requirement for induction to this post) proposed a limit to the age of the earth based on its present temperature. Kelvin quickly set to work establishing a physical laboratory at Glasgow—the first of its kind in Britain—and he remained in his post at the University for over half a century. In 1892, Kelvin received the title Baron Kelvin of Largs, after the river Kelvin which flowed near the University. He was buried just south of the grave of Sir Isaac Newton in Westminster Abbey.

Like many natural philosophers of his day, Kelvin's scientific interests were broad, encompassing mathematics, astronomy, thermodynamics, hydrodynamics, electrodynamics and terrestrial dynamics. He served as a key scientific advisor in the successful deployment of an Atlantic telegraph cable between Ireland and Newfoundland, for which he was knighted by Queen Victoria in 1866.[4] In 1872, he designed and built an ocean-tide predicting machine which was essentially an analog computer consisting of carefully arranged pulleys, cranks and rotating cylinders.[5] Kelvin's vortex theory of the atom, which treated the atom as a tiny smoke-ring-like structure arising within an all-pervasive æthereal medium, once enjoyed popular support among leading scientists.[6] Kelvin's vortex theory of the atom is now seen as a mere curiosity, albeit one whose ideas bear a curious resemblance to recent theories which treat fundamental particles as extended loop-like structures.

Today, Kelvin is best known for his work on the development of an absolute temperature scale whose units bear his name. What is the difference between an absolute and a practical temperature scale? Practical temperature scales, such as the one developed in 1742 by Swedish astronomer Anders Celsius, are defined in terms of the physical properties of a particular substance, such as the melting and vaporization points of water. Is it possible to conceive of a temperature scale which in no way references the particular properties of any substance? Kelvin said yes. Inspired by the work of Sadi Carnot,[7] Kelvin recognized how the principles of operation of heat engines—in particular the efficiency of the idealized *Carnot cycle*—might be used to define a temperature scale which is independent of the properties of any particular substance. His ideas were originally published in 1848 in the June 5 *Cambridge Philosophical Society Proceedings* and also in the October *Philosophical Magazine*. The following reading selection is from an edited version of this paper which was republished by Kelvin in 1881 in his collection of *Mathematical and Physical Papers*.

---

[4] See, for example, Bart, D., and J. Bart, Sir William Thomson, on the 150th Anniversary of the Atlantic Cable, *Antique Wireless Association Review*, *21*, 2008.

[5] See "The Tides", Kelvin's Evening Lecture to the British Association at the Southampton Meeting, Friday, August 25th, 1882, contained in Eliot, C. W. (Ed.), *Scientific Papers: Physics, Chemistry, Astronomy, Geology*, vol. 30, P.F. Collier & Son, New York, 1910.

[6] Albert Michelson, America's first nobel-prize winning scientist, enthusiastically endorsed Kelvin's vortex theory of the atom when explaining his famous speed-of-light measurements in his book *Light Waves and their Uses*; see Chap. 34 of volume III of the present work.

[7] Dover Publications has reprinted Carnot, S., *Reflections on the Motive Power of Heat*, second ed., John Wiley & Sons and Chapman & Hall, New York and London, 1897.

5.2 Reading: Kelvin, *On an Absolute Thermometric Scale Founded on Carnot's Theory of the Motive Power of Heat, and Calculated from Regnault's Observations*

Thomson, S. W., *Mathematical and Physical Papers*, Cambridge University Press, 1882. Art. XXXIX, pages 100–106.[8],[9]

The determination of temperature has long been recognized as a problem of the greatest importance in physical science. It has accordingly been made a subject of most careful attention, and, especially in late years, of very elaborate and refined experimental researches;[10] and we are thus at present in possession of as complete a practical solution of the problem as can be desired, even for the most accurate investigations. The theory of thermometry is however as yet far from being in so satisfactory a state. The principle to be followed in constructing a thermometric scale might, at first sight seem to be obvious, as it might appear that a perfect thermometer would indicate equal additions of heat, as corresponding to equal elevations of temperature, estimated by the numbered divisions of its scale. It is however now recognized (from the variations in the specific heats of bodies) as an experimentally demonstrated fact that thermometry under this condition is impossible, and we are left without any principle on which to found an absolute thermometric scale.

Next in importance to the primary establishment of an absolute scale, independently of the properties of any particular kind of matter, is the fixing upon an arbitrary system of thermometry, according to which results of observations made by different experimenters, in various positions and circumstances, may be exactly compared. This object is very fully attained by means of thermometers constructed and graduated according to the clearly defined methods adopted by the best instrument-makers of the present day, when the rigorous experimental processes which have been indicated, especially by Regnault, for interpreting their indications in a comparable way, are followed. The particular kind of thermometer which is

---

[8] Published in 1824 in a work entitled *Réflections sur la Puissance Motrice du Feu*, by M.S. Carnot. Having never met with the original work, it is only through a paper by M. Clapeyron, on the same subject, published in the *Journal de l'École Polytechnique*, Vol. XIV, 1834, and translated in the first volume of Taylor's *Scientific Memoirs*, that the Author has become acquainted with Carnot's Theory.—W.T. [Note of Nov. 5th, 1881. A few months later through the kindness of my late colleague Prof. Lewis Gordon, I received a copy of Carnot's original work and was thus enabled to give to the Royal Society of Edinburgh my "Account of Carnot's theory" which is reprinted as Art. XLI. below. The original work has since been republished, with a biographical notice, Paris, 1878.].

[9] An account of the first part of a series of researches undertaken by M. Regnault by order of the French Government, for ascertaining the various physical data of importance in the Theory of the Steam Engine, is just published in the *Mémpires de l'Institut*, of which it constitutes the twenty-first volume (1847). The second part of the researches has not yet been published. [Note of Nov. 5, 1881. The continuation of these researches has now been published: thus we have for the whole series, Vol. I. in 1847; Vol II. in 1862; and Vol. III. in 1870.].

[10] A very important section of Regnault's work is devoted to this object.

least liable to uncertain variations of any kind is that founded on the expansion of air, and this is therefore generally adopted as the standard for the comparison of thermometers of all constructions. Hence the scale which is at present employed for estimating temperature is that of the air-thermometer; and in accurate researches care is always taken to reduce to this scale the indications of the instrument actually used, whatever may be its specific construction and graduation.

The principle according to which the scale of the air-thermometer is graduated is simply that equal absolute expansions of the mass of air or gas in the instrument, under a constant pressure, shall indicate equal differences of the numbers on the scale; the length of a "degree" being determined by allowing a given number for the interval between the freezing- and the boiling-points. Now it is found by Regnault that various thermometers, constructed with air under different pressures, or with different gases, give indications which coincide so closely, that, unless when certain gases, such as sulphurous acid, which approach the physical condition of vapours at saturation, are made use of, the variations are inappreciable.[11] This remarkable circumstance enhances very much the practical value of the air-thermometer; but still a rigorous standard can only be defined by fixing upon a certain gas at a determinate pressure, as the thermometric substance. Although we have thus a strict principle for constructing a *definite* system for the estimation of temperature, yet as reference is essentially made to a specific body as the standard thermometric substance, we cannot consider that we have arrived at an *absolute* scale, and we can only regard, in strictness, the scale actually adopted as *an arbitrary series of numbered points of reference sufficiently close for the requirements of practical thermometry.*

In the present state of physical science, therefore, a question of extreme interest arises: *Is there any principle on which an absolute thermometric scale can be founded?* It appears to me that Carnot's theory of the motive power of heat enables us to give an affirmative answer.

The relation between motive power and heat, as established by Carnot, is such that *quantities of heat*, and *intervals of temperature*, are involved as the sole elements in the expression for the amount of mechanical effect to be obtained through the agency of heat; and since we have, independently, a definite system for the measurement of quantities of heat, we are thus furnished with a measure for intervals according to which absolute differences of temperature may be estimated. To make this intelligible, a few words in explanation of Carnot's theory must be given; but for a full account of this most valuable contribution to physical science, the reader is referred to either of the works mentioned above (the original treatise by Carnot, and Clapeyron's paper on the same subject.)

---

[11] Regnault, *Relation des Expériences,* &c., Fourth Memoir, First Part. The differences, it is remarked by Regnault, would be much more sensible if the graduation were effected on the supposition that the coefficients of expansion of the different gases are equal, instead of being founded on the principle laid down in the text, according to which the freezing-and boiling-points are experimentally determined for each thermometer.

In the present state of science no operation is known by which heat can be absorbed, without either elevating the temperature of matter, or becoming latent and producing some alteration in the physical condition of the body into which it is absorbed; and the conversion of heat (or *caloric*) into mechanical effect is probably impossible,[12] certainly undiscovered. In actual engines for obtaining mechanical effect through the agency of heat, we must consequently look for the source of power, not in any absorption and conversion, but merely in a transmission of heat. Now Carnot, starting from universally acknowledged physical principles, demonstrates that it is by the *letting down* of heat from a hot body to a cold body, through the medium of an engine (a steam-engine, or an air-engine for instance), that mechanical effect is to be obtained; and conversely, he proves that the same amount of heat may, by the expenditure of an equal amount of labouring force, be *raised* from the cold to the hot body (the engine being in this case *worked backwards*); just as mechanical effect may be obtained by the descent of water let down by a water-wheel, and by spending labouring force in turning the wheel backwards, or in working a pump, water may be elevated to a higher level. The amount of mechanical effect to be obtained by the transmission of a given quantity of heat, through the medium of any kind of engine in which the economy is perfect, will depend, as Carnot demonstrates, not on the specific nature of the substance employed as the medium of transmission of heat in the engine, but solely on the interval between the temperature of the two bodies between which the heat is transferred.

Carnot examines in detail the ideal construction of an air-engine and of a steam-engine, in which, besides the condition of perfect economy being satisfied, the machine is so arranged, that at the close of a complete operation the substance (air in one case and water in the other) employed is restored to precisely the same physical condition as at the commencement. He thus shews on what elements, capable of experimental determination, either with reference to air, or with reference to a liquid and its vapour, the absolute amount of mechanical effect due to the transmission of a unit of heat from a hot body to a cold body, through any given interval of the thermometric scale, may be ascertained. In M. Clapeyron's paper various experimental data, confessedly very imperfect, are brought forward, and the amounts of mechanical effect due to a unit of heat descending a degree of the air-thermometer, in various parts of the scale, are calculated from them, according to Carnot's expressions. The results so obtained indicate very decidedly, that what we may with much propriety call *the value of a degree* (estimated by the mechanical effect to be obtained from

---

[12] This opinion seems to be nearly universally held by those who have written on the subject. A contrary opinion however has been advocated by Mr Joule of Manchester; some very remarkable discoveries which he has made with reference to the *generation* of heat by the friction of fluids in motion, and some known experiments with magneto-electric machines, seeming to indicate an actual conversion of mechanical effect into caloric. No experiment however is adduced in which the converse operation is exhibited; but it must be confessed that as yet much is involved in mystery with reference to these fundamental questions of natural philosophy.

the descent of a unit of heat through it) of the air-thermometer depends on the part of the scale in which it is taken, being less for high than for low temperatures.[13]

The characteristic property of the scale which I now propose is, that all degrees have the same value; that is, that a unit of heat descending from a body A at the temperature $T°$ of this scale, to a body B at the temperature $(T-1)°$, would give out the same mechanical effect, whatever be the number T. This may justly be termed an absolute scale, since its characteristic is quite independent of the physical properties of any specific substance.

To compare this scale with that of the air-thermometer, the *values* (according to the principle of estimation stated above) of degrees of the air-thermometer must be known. Now an expression, obtained by Carnot from the consideration of his ideal steam-engine, enables us to calculate these values, when the latent heat of a given volume and the pressure of saturated vapour at any temperature are experimentally determined. The determination of these elements is the principal object of Regnault's great work, already referred to, but at present his researches are not complete. In the first part, which alone has been as yet published, the latent heats of a given weight, and the pressures of saturated vapour, at all temperatures between 0 and 230° (Cent. of the air-thermometer), have been ascertained; but it would be necessary in addition to know the densities of saturated vapour at different temperatures, to enable us to determine the latent heat of a given volume at any temperature. M. Regnault announces his intention of instituting researches for this object; but till the results are made known, we have no way of completing the data necessary for the present problem, except by estimating the density of saturated vapour at any temperature (the corresponding pressure being known by Regnault's researches already published) according to the approximate laws of compressibility and expansion (the laws of Mariotte and Gay-Lussac, or Boyle and Dalton). Within the limits of natural temperature in ordinary climates, the density of saturated vapour is actually found by Regnault (*Études Hygrométriques* in the *Annales de Chimie*) to verify very closely these laws; and we have reason to believe from experiments which have been made by Gay-Lussac and others, that as high as the temperature 100° there can be no considerable deviation; but our estimate of the density of saturated vapour, founded on these laws, may be very erroneous at such high temperatures as 230°. Hence a completely satisfactory calculation of the proposed scale cannot be made till after the additional experimental data shall have been obtained; but with the data which we actually possess, we may make an approximate comparison of the new scale with that of the air-thermometer, which at least between 0 and 100° will be tolerably satisfactory.

---

[13] This is what we might anticipate, when we reflect that infinite cold must correspond to a finite number of degrees of the air-thermometer below zero; since, if we push the strict principle of graduation, stated above, sufficiently far, we should arrive at a point corresponding to the volume of air being reduced to nothing, which would be marked as −273° of the scale (−100/0.366, if 0.366 be the coefficient of expansion); and therefore −273° of the air-thermometer is a point which cannot be reached at any finite temperature, however low.

The labour of performing the necessary calculations for effecting a comparison of the proposed scale with that of the air-thermometer, between the limits 0 and 230° of the latter, has been kindly undertaken by Mr William Steele, lately of Glasgow College, now of St Peter's College, Cambridge. His results in tabulated forms were laid before the Society, with a diagram, in which the comparison between the two scales is represented graphically. In the first table,[14] the amounts of mechanical effect due to the descent of a unit of heat through the successive degrees of the air-thermometer are exhibited. The unit of heat adopted is the quantity necessary to elevate the temperature of a kilogramme of water from 0 to 1° of the air-thermometer; and the unit of mechanical effect is a metre-kilogramme; that is, a kilogramme raised a metre high.

In the second table, the temperatures according to the proposed scale, which correspond to the different degrees of the air-thermometer from 0 to 230°, are exhibited. [The arbitrary points which coincide on the two scales are 0 and 100°].

Note.—If we add together the first hundred numbers given in the first table, we find 135.7 for the amount of work due to a unit of heat descending from a body *A* at 100° to *B* at 0°. Now 79 such units of heat would, according to Dr Black (his result being very slightly corrected by Regnault), melt a kilogramme of ice. Hence if the heat necessary to melt a pound of ice be now taken as unity, and if a *metre-pound* be taken as the unit of mechanical effect, the amount of work to be obtained by the descent of a unit of heat from 100 to 0° is 79 × 135.7, or 10,700 nearly. This is the same as 35,100 ft lb, which is a little more than the work of a one-horse-power engine (33,000 ft lb) in a minute; and consequently, if we had a steam-engine working with perfect economy at one-horse-power, the boiler being at the temperature 100°, and the condenser kept at 0° by a constant supply of ice, rather less than a pound of ice would be melted in a minute.

[*Note of Nov. 4, 1881.* This paper was wholly founded on Carnot's uncorrected theory, according to which the quantity of heat taken in in the hot pan of the engine, (the boiler of the steam engine for instance), was supposed to be equal to that abstracted from the cold part (the condenser of the steam engine), in a complete period of the regular action of the engine, when every varying temperature, in every part of the apparatus, has become strictly periodic. The reconciliation of Carnot's theory with what is now known to be the true nature of heat is fully described in Article XLVIII. below; and in §§24–41 of that article, are shewn in detail the consequently required corrections of the thermodynamic estimates of the present article.[15] These corrections however do not in any way affect the absolute scale for thermometry which forms the subject of the present article. Its relation to the practically more convenient scale (agreeing with air thermometers nearly enough for most purposes,

---

[14] [Note of Nov. 4, 1881. This table (reduced from metres to feet) was repeated in my "Account of Carnot's Theory of the Motive power of Heat," republished as Article XLI. below, in §38 of which it will be found.].

[15] check this KKK.

throughout the range from the lowest temperatures hitherto measured, to the highest
that can exist so far as we know) which I gave subsequently, Dynamical Theory of
Heat (Art. XLVIII, below), Part VI., §§99, 100; *Trans. R. S. E.,* May, 1854: and Article 'Heat,' §§35–38, 47–67, *Encyclopædis Britannica,* is shewn in the following
formula:

$$\theta = 100 \, \frac{\log t - \log 273}{\log 373 - \log 273},$$

where θ and t are the reckonings of one and the same temperature, according to my
first and according to my second thermodynamic absolute scale.]

5.3 Study Questions

QUES. 5.1. What is the difference between a *practical* and an *absolute* temperature
scale?

a) What problem arises from defining temperature using the heat capacity of a
   particular substance?
b) At the time of Kelvin, what type of thermometer served as a standard against
   which all other thermometers were compared? Why was this? On what physical
   property of air was this thermometer based? Do different gases yield different
   results?
c) In what sense was this thermometer a *practical*, as opposed to an *absolute*, temperature scale? What interesting questions does this temperature scale raise? And
   wherein does Kelvin seek the answer to this question?
d) According to Carnot's theory, what two elements completely determine the
   amount of mechanical work done by a (reversible) heat engine? What does this
   imply?

QUES. 5.2. How can an absolute temperature scale be created?

a) What are the two known effects of adding heat to a substance? What was the
   common opinion regarding caloric (the substance of heat)? In particular, is it a
   conserved quantity?
b) How, then do engines accomplish mechanical work? What analogy does Kelvin
   (and Carnot) employ to illustrate this process? Can mechanical work, in turn,
   pump heat from a cold to a hot reservoir?
c) For an engine whose economy (or efficiency) is perfect, what determines the
   relationship between heat flow and mechanical work? Does it depend upon the
   particular substance of the engine? What does this imply?
d) Are degrees of the air-thermometer all of the same size? By what standard is the
   size of a degree to be measured? What if one uses the amount of heat absorbed
   by a particular substance in order to change by 1°? What if one uses the amount
   of work done by a Carnot engine when heat falls, so to speak, through 1°?
e) How does Kelvin define his own temperature scale? In what sense is his an
   absolute scale?

5.4   Exercises

EX. 5.1 (CONSTANT PRESSURE AIR THERMOMETRY). Galileo is said to have invented the first air thermometer consisting of a glass bulb with a long neck open at the bottom. The air was heated and the bottom of the neck was dipped into an open pool of colored liquid. When the heat source was removed and the air subsequently cooled, some of the liquid from the pool was drawn up into the neck. The level of the liquid in the neck thereafter provided a measure of the temperature of the air residing in the bulb.[16] As an exercise, suppose that the neck of such an air thermometer (whose inner diameter is 1 mm) is dipped into a pool of colored liquid. The bulb itself holds 0.2 cubic centimeters of air and is initially in thermal equilibrium with boiling water. Assume that air obeys *Charles' law* (see Eq. 3.1), and that a unit of volume of air expands to 1.3665 units when raised from 0 to 100 °C—as discovered by Regnault.

(a) By how much does the liquid rise when the bulb is brought into thermal equi-librium with ice water? After such a calibration procedure, what is the spacing of 1° marks on the neck of the air thermometer?

(b) Theoretically, by how many degrees would the bulb need to be cooled in order to reduce its volume to zero? Does the air thermometer then provide a *practical* or an *absolute* temperature scale? In particular, does the scale thus constructed depend on the particular physical properties of the gas employed?

EX. 5.2 (CONSTANT VOLUME GAS THERMOMETRY LABORATORY). By measuring the pressure of a fixed volume of gas at several temperatures on a practical temperature scale, one can estimate the lowest attainable temperature on that scale. To do so, record the pressure of a gas-filled bulb[17] submerged in (i) boiling water, (ii) ice water, (iii) a methanol-dry ice slurry, and (iv) liquid nitrogen. To avoid an explosion, be sure not to over-pressurize your bulb; use a valve if necessary to equalize the gas pressure with atmospheric pressure when inserting the bulb into boiling water. Then seal the valve and proceed to acquire low-temperature data. Next, make a plot of the gas pressure as a function of temperature (in Celsius). Does the gas obey Boyle's law? Fit a curve to your data and extrapolate to zero pressure; at what value would zero gas pressure occur on your temperature scale? Does gas thermometry thereby provide an absolute temperature scale?

EX. 5.3 (WATER-WHEELS, HEAT-ENGINES AND ABSOLUTE TEMPERATURE SCALES). Carnot compared the work done by a heat-engine to the work done by falling water. In this exercise we will explore this analogy and how Kelvin used it to conceive

---

[16] Unfortunately, such an air thermometer is of limited utility, since the level of the liquid depends not only on the temperature of the air in the bulb, but also on the local atmospheric pressure which presses down on the surface of the liquid pool. Thus, it must be used in conjunction with a barometer so to compensate for varying atmospheric pressures.

[17] For these laboratory experiments, you might consider the Absolute Zero Demonstration apparatus (Model #WLS1828-89), available form Sargent Welch, Chicago, IL.

of an absolute temperature scale. When a water-wheel is placed beneath a waterfall it is able to accomplish work. For instance, it can raise a stone attached to a rope wound around the axle of the water-wheel. More specifically, when a mass (m) of water descends, the water-wheel can accomplish an amount of work (W) given by

$$W = mg(h_2 - h_1). \tag{5.1}$$

Here g is the local gravitational acceleration and the quantity in parentheses represents the distance the water descends. The work accomplished by one waterwheel could, in turn, be expended so as to drive a second water-wheel backwards, raising a mass of water back up. The most efficient water wheel would be one having no internal friction in its gears. Such a reversible water-wheel, when driven by a falling mass of water, would be able to drive a second reversible water wheel backwards in such a way as to raise an identical mass of water back to the same height. If a water-wheel could be constructed which was (somehow) more efficient than a reversible water-wheel, then it could be driven in such a way as to lift a larger quantity of water than the quantity which drove it in the first place, and this would violate the principe of conservation of energy.[18]

Similarly, Carnot supposed that a heat-engine is able to accomplish work when a portion of caloric falls, so to speak, through a temperature difference.[19] Generally speaking, the work accomplished depends on the construction of the particular heat-engine. But for a *reversible* heat-engine—one in which there is no useless flow of heat—the work accomplished depends only on the quantity of heat, Q, which drives the engine and the temperature difference through which it descends:

$$W = Q(T_h - T_c). \tag{5.2}$$

This law follows from the assumption that the flow of caloric—a universal substance identified with heat—drives all heat-engines. The particular working substance of the heat-engine (*e.g.* steam or alcohol) merely serves to convey the caloric from the high temperature to the low temperature reservoir. Just as water is neither created nor destroyed by the waterfall, so also caloric is neither created nor destroyed by the engine. According to this *caloric theory of heat*, one unit of absolute temperature can now be defined in terms of the efficiency of a reversible heat-engine:

$$\eta_{rev} = T_h - T_c. \tag{5.3}$$

Equation 5.3 follows from Eq. 5.2 and the definition of a heat-engine's efficiency (Eq. 4.1). Notice that *the efficiency of a reversible heat-engine is the same over*

---

[18] Herman von Helmholtz discusses water-wheels, work and the conservation of energy in his famous lecture entitled *On the Conservation of Force*; this lecture can be found in Chaps. 9–11 of Volume III.

[19] This is discussed in Carnot, S., *Reflections on the Motive Power of Heat*, second ed., John Wiley & Sons and Chapman & Hall, New York and London, 1897; see Chap. 3 of the present volume.

the entire temperature scale defined in this way, provided it is operated between two reservoirs whose temperatures differ by the same amount, for example 1° of absolute temperature. This was Kelvin's original method of establishing an absolute temperature scale.

But what if, as suggested by Joule (and later conceded by Kelvin), heat is not an indestructible substance? What if it is *consumed* by a heat-engine in the act of performing work? According to this *mechanical theory of heat*, the amount of heat drawn in from the hot reservoir in one cycle of the engine, Q_h, and the amount of heat ejected to the cold reservoir during the same cycle, Q_c, differ by exactly the amount of work done by the engine during a cycle, W.

a) Assuming the mechanical theory of heat to be true, show that the efficiency of a heat-engine may now be expressed as

$$\eta = 1 - \frac{Q_c}{Q_h}. \tag{5.4}$$

   Notice that Eq. 5.4 is the efficiency of *all* heat-engines, not just reversible heat-engines.
b) For a *reversible* heat-engine, Eq. 5.4 must be capable of being expressed as a function of the temperatures of the hot and cold reservoirs alone (as opposed to the physical properties of any particular working substance, such as steam). In other words

$$\frac{Q_c}{Q_h} = f(T_h, T_c) \tag{5.5}$$

   To determine the functional form of $f(T_h, T_c)$, consider two identical reversible heat-engines arranged between three reservoirs maintained at temperatures T_h, T_m, and T_c, as depicted schematically in Fig. 5.1. Heat Q_h from the hot reservoir enters the first heat-engine, which does work W_1 and ejects heat Q_m into the middle reservoir. The second engine is adjusted so that it draws heat Q_m from the middle reservoir, does work W_2 and ejects heat Q_c into the cold reservoir. Now show that the function $f(T_h, T_c)$ must satisfy the condition

$$f(T_h, T_c) = f(T_h, T_m) \cdot f(T_m, T_c). \tag{5.6}$$

Next, show that Eq. 5.6 is true only if $f(T_h, T_c)$ takes the form

$$f(T_h, T_c) = \left(\frac{T_c}{T_h}\right)^n \tag{5.7}$$

where n is some number. Finally, show that the efficiency of a reversible heat-engine can thus be expressed as

$$\eta_{rev} = 1 - \left(\frac{T_c}{T_h}\right)^n \tag{5.8}$$

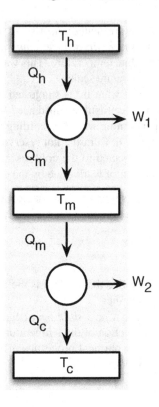

Fig. 5.1 Two reversible heat engines arranged so that the cold reservoir of the first serves as the hot reservoir of the second

Like our previous Eq. 5.3, Eq. 5.8 may be used to establish an absolute temperature scale. Suppose that a reversible heat-engine draws heat from a reservoir maintained at the temperature of boiling water and ejects heat into a reservoir maintained at the temperature of ice water. Suppose that, experimentally, this reversible heat-engine is found to absorb one Joule of heat from the hot reservoir and to accomplish 0.268 Joules of work during a single cycle.

c) How much heat flows into the cold reservoir during one cycle of this heat-engine? And what is the efficiency of this engine operating between these two temperatures? (ANSWER: 26.8 %)

d) Suppose we wish to establish a temperature scale having just 10° between the boiling and freezing points of ice. (Arbitrarily) choose $n = 1$ and use Eq. 5.8 to determine the absolute temperatures of the hot and cold reservoirs on this scale. What are they? What if we instead chose $n = 2$?

e) Are either of these two absolute temperature scales equivalent to the Kelvin temperature scale in use today? If not, how would you modify either of these to produce Kelvin's scale?

f) According to Eq. 5.8, is the efficiency of a reversible heat-engine the same over the entire absolute temperature scale, provided it is operating between two reservoirs whose temperatures differ by the same amount (*e.g.* between 150° and 100°, or between 100° and 50°)?

g) Finally, how might you design a reversible heat-engine so as to maximize its efficiency? Might this explain why diesel fuel engines are generally more efficient than gasoline engines?

5.5 Vocabulary

1. Saturated
2. Conversely
3. Condenser
4. Periodic

Chapter 6
The Second Law of Thermodynamics

The algebraic sum of all the transformations occurring in a cyclical process can only be positive, or, as an extreme case, equal to nothing.

—Rudolph Clausius

6.1 Introduction

Rudolph Clausius (1822–1888) was one of seventeen children born to a Pastor in Köslin, Pomerania, which was then a province in the Kingdom of Prussia. As a boy he attended the Gymnasium founded by his father at Stettin, from which he graduated in 1840. He enrolled at the Frederic Wilhelm University of Berlin, where he was attracted to history, physics and mathematics. After graduating in 1844, he taught courses in mathematics and physics at the Frederic-Werder Gymnasium for a year. In 1848, he earned his doctorate from the University of Halle for his work on the refraction of sunlight by Earth's atmosphere. He went on to serve as a Professor at the Swiss Federal Institute of Technology, at the University of Züric, at the University of Würzburg, and finally at the University of Bonn. During the time of the Franco-Prussian war, Clausius organized an ambulance corps comprised of university students; he was wounded in battle and was awarded the Iron Cross.

Clausius was one of the principle architects of the science of thermodynamics. In 1850 he published a groundbreaking paper entitled *On the Moving Force of Heat and the Laws of Heat which May Be Deduced Therefrom*. In this paper he criticized the caloric theory of heat—upon which Carnot's work had been founded—on the grounds that it violated the first law of thermodynamics. Heat, Clausius argued, was not an indestructible substance, but rather a form of energy which could be transformed into work, and *vice versa*. The conversion of mechanical work into heat (*via* friction) had been explored as early as 1798 by Count Rumford while boring canons, and more recently by Julius Mayer and James Joule in the 1840s. Indeed, when proposing his general principle of the conservation of energy in 1847,

© Springer International Publishing Switzerland 2016
K. Kuehn, *A Student's Guide Through the Great Physics Texts*,
Undergraduate Lecture Notes in Physics, DOI 10.1007/978-3-319-21828-1_6

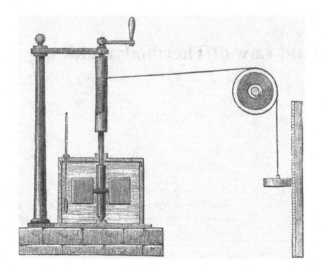

Fig. 6.1 An engraving of James Joule's method for measuring the mechanical equivalent of heat, from *Harper's New Monthly Magazine*, No. 231, August, 1869. A thermometer measures the heating of water when stirred by rotating paddles which are driven by a falling weight

Hermann Helmholtz recounts the careful experiments of Joule, who had measured the heating of a bath of water stirred by a descending weight (see Fig. 6.1).[1]

By 1854, Clausius had expanded these ideas into what is today known as the second law of thermodynamics. In so doing, he provided the first mathematical definition of *entropy*, a term which he himself coined. In addition, Clausius' study of phase transitions led to what is now known as the Clausius-Clapeyron equation.[2] And his work on the kinetic theory of gasses led him to develop a mathematical formula for the *mean free path*—the average distance which a particle travels within a gas before striking a neighboring particle. The reading selection included in this chapter is from an 1867 collection of Clausius' memoirs translated into English by T. Archer Hirst. It was originally published in Poggendorff's *Annalen* in May 1862, as well as in the *Philosophical Magazine* and the *Journal des Mathématiques* of Paris.

---

[1] Helmholtz's famous essay entitled *On the Conservation of Force* can be found in Chaps. 9–11 of volume III.

[2] The Clausius-Clapeyron equation relates the latent heat developed during a first-order phase transition to the slope of the coexistence curve separating the phases (for example, the liquid and solid phases of water).

6.2   Reading: Clausius, *the Mechanical Theory of Heat*

Clausius, R., *Mechanical Theory of Heat, with its Applications to the Steam-Engine and to the Physical Properties of Bodies*, John Van Voorst, London, 1867. Sixth Memoir.

In a memoir published in the year 1854,[3] wherein I sought to simplify to some extent the form of the developments I had previously published, I deduced, from my fundamental proposition *that heat cannot, by itself, pass from a colder into a warmer body,* a theorem which is closely allied to, but does not entirely coincide with, the one first deduced by S. Carnot from considerations of a different kind, based upon the older views of the nature of heat. It has reference to the circumstances under which work can be transformed into heat, and conversely, heat converted into work; and I have called it the *Theorem of the Equivalence of Transformations*. I did not, however, there communicate the entire theorem in the general form in which I had deduced it, but confined myself on that occasion to the publication of a part which can be treated separately from the rest, and is capable of more strict proof.

In general, when a body changes its state, work is performed *externally* and *internally* at the same time,—the exterior work having reference to the forces which extraneous bodies exert upon the body under consideration, and the interior work to the forces exerted by the constituent molecules of the body in question upon each other. The interior work is for the most part so little known, and connected with another equally unknown quantity[4] in such a way, that in treating of it we are obliged in some measure to trust to probabilities; whereas the exterior work is immediately accessible to observation and measurement, and thus admits of more strict treatment. Accordingly, since, in my former paper, I wished to avoid everything that was hypothetical, I entirely excluded the interior work, which I was able to do by confining myself to the consideration of *cyclical processes*—that is to say, operations in which the modifications which the body undergoes are so arranged that the body finally returns to its original condition. In such operations the interior work which is performed during the several modifications, partly in a positive sense and partly in a negative sense, neutralizes itself, so that nothing but exterior work remains, for which the theorem in question can then be demonstrated with mathematical strictness, starting from the above-mentioned fundamental proposition.

I have delayed till now the publication of the remainder of my theorem, because it leads to a consequence which is considerably at variance with the ideas hitherto generally entertained of the heat contained in bodies, and I therefore thought it desirable to make still further trial of it. But as I have become more and more convinced in the

---

[3] Clausius' 1854 publication, entitled "On a modified form of the second Fundamental Theorem in the Mechanical Theory of heat," appears as the Fourth Memoir of Clausius' 1867 *Mechanical Theory of Heat.*—[*K.K.*]

[4] [In fact with the increase of the heat actually present in the body.—1864.]

course of years that we must not attach too great weight to such ideas, which in part are founded more upon usage than upon a scientific basis, I feel that I ought to hesitate no longer, but to submit to the scientific public the theorem of the equivalence of transformations in its complete form, with the theorems which attach themselves to it. I venture to hope that the importance which these theorems, supposing them to be true, possess in connexion with the theory of heat will be thought to justify their publication in their present hypothetical form.

I will, however, at once distinctly observe that, whatever hesitation may be felt in admitting the truth of the following theorems, the conclusions arrived at in my former paper, in reference to cyclical processes, are not at all impaired.

1. I will begin by briefly stating the theorem of the equivalence of transformations, as I have already developed it, in order to be able to connect with it the following considerations.

> When a body goes through a cyclical process, a certain amount of exterior work may be produced, in which case a certain quantity of heat must be simultaneously expended; or, conversely, work may be expended and a corresponding quantity of heat may be gained. This may be expressed by saying:—*Heat can be transformed into work, or work into heat, by a cyclical process.*

There may also be another effect of a cyclical process: heat may be transferred from one body to another, by the body which is undergoing modification absorbing heat from the one body and giving it out again to the other. In this case the bodies between which the transfer of heat takes place are to be viewed merely as heat reservoirs, of which we are not concerned to know anything except the temperatures. If the temperatures of the two bodies differ, heat passes, either from a warmer to a colder body, or from a colder to a warmer body, according to the direction in which the transference of heat takes place. Such a transfer of heat may also be designated, for the sake of uniformity, *a transformation,* inasmuch as it may be said that *heat of one temperature is transformed into heat of another temperature.*

The two kinds of transformations that have been mentioned are related in such a way that one presupposes the other, and that they can mutually replace each other. If we call transformations which can replace each other equivalent, and seek the mathematical expressions which determine the amount of the transformations in such a manner that equivalent transformations become equal in magnitude, we arrive at the following expression:—

> *If the quantity of heat Q of the temperature t is produced from work, the equivalence-value of this transformation is*

$$\frac{Q}{T};\tag{6.1}$$

> *and if the quantity of heat Q passes from a body whose temperature is t_1 into another whose temperature is t_2, the equivalence-value of this transformation is*

$$Q\left(\frac{1}{T_2} - \frac{1}{T_1}\right);\tag{6.2}$$

where T is a function of the temperature which is independent of the kind of process by means of which the transformation is effected, and T_1 and T_2 denote the values of this function which correspond to the temperatures t_1 and t_2. I have shown by separate considerations that T is in all probability nothing more than the *absolute temperature*.

These two expressions further enable us to recognize the positive or negative sense of the transformations. In the first, Q is taken as positive when work is transformed into heat, and as negative when heat is transformed into work. In the second, we may always take Q as positive, since the opposite senses of the transformations are indicated by the possibility of the difference $\frac{1}{T_2} - \frac{1}{T_1}$ being either positive or negative. It will thus be seen that the passage of heat from a higher to a lower temperature is to be looked upon as a positive transformation, and its passage from a lower to a higher temperature as a negative transformation. If we represent the transformations which occur in a cyclical process by these expressions, the relation existing between them can be stated in a simple and definite manner. If the cyclical process is *reversible*, the transformations which occur therein must be partly positive and partly negative, and the equivalence-values of the positive transformations must be together equal to those of the negative transformations, so that the algebraic sum of all the equivalence-values becomes $= 0$. If the cyclical process is not reversible, the equivalence-values of the positive and negative transformations are not necessarily equal, but they can only differ in such a way that the positive transformations predominate. The theorem respecting the equivalence-values of the transformations may accordingly be stated thus:—

> The algebraic sum of all the transformations occurring in a cyclical process can only be positive, or, as an extreme case, equal to nothing.

The mathematical expression for this theorem is as follows. Let dQ be an element of the heat given up by the body to any reservoir of heat during its own changes (heat which it may absorb from a reservoir being here reckoned as negative), and T the absolute temperature of the body at the moment of giving up this heat, then the equation

$$\int \frac{dQ}{T} = 0 \qquad\qquad (6.3)$$

must be true for every reversible cyclical process, and the relation

$$\int \frac{dQ}{T} \geq 0 \qquad\qquad (6.4)$$

must hold good for every cyclical process which is in any way possible.

2. Although the necessity of this theorem admits of strict mathematical proof if we start from the fundamental proposition above quoted, it thereby nevertheless retains an abstract form, in which it is with difficulty embraced by the mind, and we feel compelled to seek for the precise physical cause, of which this theorem is a consequence. Moreover, since there is no essential difference between interior

and exterior work, we may assume almost with certainty that a theorem which is so generally applicable to exterior work cannot be restricted to this alone, but that, where exterior work is combined with interior work, it must be capable of application to the latter also.

Considerations of this nature led me, in my first investigations on the mechanical theory of heat, to assume a general law respecting the dependence of the active force of heat on temperature, among the immediate consequences of which is the theorem of the equivalence of transformations in its more complete form, and which at the same time leads to other important conclusions. This law I will at once quote, and will endeavour to make its meaning clear by the addition of a few comments. As for the reasons for supposing it to be true, such as do not at once appear from its internal probability will gradually become apparent in the course of this paper. It is as follows:—

> *In all cases in which the heat contained in a body does mechanical work by overcoming resistances, the magnitude of the resistances which it is capable of overcoming is proportional to the absolute temperature.*

In order to understand the significance of this law, we require to consider more closely the processes by which heat can perform mechanical work. These processes always admit of being reduced to the alteration in some way or another of the arrangement of the constituent parts of a body. For instance, bodies are expanded by heat, their molecules being thus separated from each other: in this case the mutual attractions of the molecules on the one hand, and external opposing forces on the other, in so far as any such are in operation, have to be overcome. Again, the state of aggregation of bodies is altered by heat, solid bodies being rendered liquid, and both solid and liquid bodies being rendered aëriform: here likewise internal forces, and in general external forces also, have to be overcome. Another case which I will also mention, because it differs so widely from the foregoing, and therefore shows how various are the modes of action which have here to be considered, is the transfer of electricity from one body to the other, constituting the thermo-electric current, which takes place by the action of heat on two heterogeneous bodies in contact.

In the cases first mentioned, the arrangement of the molecules is altered. Since, even while a body remains in the same state of aggregation, its molecules do not retain fixed unvarying positions, but are constantly in a state of more or less extended motion, we may, when speaking of the *arrangement of the molecules* at any particular time, understand either the arrangement which would result from the molecules being fixed in the actual positions they occupy at the instant in question, or we may suppose such an arrangement that each molecule occupies its mean position. Now the effect of heat always tends to loosen the connexion between the molecules, and so to increase their mean distances from one another. In order to be able to represent this mathematically, we will express the degree in which the molecules of a body are separated from each other, by introducing a new magnitude, which we will call the *disgregation* of the body, and by help of which we can define the effect of heat as simply *tending to increase the disgregation*. The way in which a definite measure of this magnitude can be arrived at will appear from the sequel.

In the case last mentioned, an alteration in the arrangement of the electricity takes place, an alteration which can be represented and taken into calculation in a way corresponding to the alteration of the position of the molecules, and which, when it occurs, we will consider as always included in the general expression *change of arrangement*, or *change of disgregation*.

It is evident that each of the changes that have been named may also take place in the reverse sense, if the effect of the opposing forces is greater than that of the heat. We will assume as likewise self-evident that, for the production of work, a corresponding quantity of heat must always be expended, and conversely, that, by the expenditure of work, an equivalent quantity of heat must be produced.

3. If we now consider more closely the various cases which occur in relation to the forces which are operative in each of them, the case of the expansion of a permanent gas presents itself as particularly simple. We may conclude from certain properties of the gases that the mutual attraction of their molecules at their mean distances is very small, and therefore that only a very slight resistance is offered to the expansion of a gas, so that the resistance of the sides of the containing vessel must maintain equilibrium with almost the whole effect of the heat. Accordingly the externally sensible pressure of a gas forms an approximate measure of the separative force of the heat contained in the gas; and hence, according to the foregoing law, this pressure must be nearly proportional to the absolute temperature. The internal probability of the truth of this result is indeed so great, that many physicists since Gay-Lussac and Dalton have without hesitation presupposed this proportionality, and have employed it for calculating the absolute temperature.

In the above-mentioned case of thermo-electric action, the force which exerts an action contrary to that of the heat is likewise simple and easily determined. For at the point of contact of two heterogeneous substances, such a quantity of electricity is driven from the one to the other by the action of the heat, that the opposing force resulting from the electric tension suffices to hold the force exerted by the heat in equilibrium. Now in a former memoir "On the application of the Mechanical Theory of Heat to the Phenomena of Thermal Electricity",[5] I have shown that, in so far as changes in the arrangement of the molecules are not produced at the same time by the changes of temperature, the difference of tension produced by heat must be proportional to the absolute temperature, as is required by the foregoing law.

In the other cases that are quoted, as well as in most others, the relations are less simple, because in them an essential part is played by the forces exerted by the molecules upon one another, forces which, as yet, are quite unknown. It results, however, from the mere consideration of the external resistances which

---

[5] Poggendorff's *Annalen*, vol. xc. p. 513.

heat is capable of overcoming, that in general its force increases with the temperature. If we wish, for instance, to prevent the expansion of a body by means of external pressure, we are obliged to employ a greater pressure the more the body is heated; hence we may conclude, without having a knowledge of the interior forces, that the total amount of the resistances which can be overcome in expansion, increases with the temperature. We cannot, however, directly ascertain whether it increases exactly in the proportion required by the foregoing law, without knowing the interior forces. On the other hand, if this law be regarded as proved on other grounds, we may reverse the process, and employ it for the determination of the interior forces exerted by the molecules.

The forces exerted upon one another by the molecules are not of so simple a kind that each molecule can be replaced by a mere point; for many cases occur in which it can be easily seen that we have not merely to consider the distances of the molecules, but also their relative positions. If we take, for example, the melting of ice, there is no doubt that interior forces, exerted by the molecules upon each other, are overcome, and accordingly increase of disgregation takes place; nevertheless the centres of gravity of the molecules are on the average not so far removed from each other in the liquid water as they were in the ice, for the water is the more dense of the two. Again, the peculiar behaviour of water in contracting when heated above 0° C., and only beginning to expand when its temperature exceeds 4°, shows that likewise in liquid water, in the neighbourhood of its melting-point, increase of disgregation is not accompanied by increase of the mean distances of its molecules.

In the case of the interior forces, it would accordingly be difficult—even if we did not want to measure them, but only to represent them mathematically—to find a fitting expression for them which would admit of a simple determination of magnitude. This difficulty, however, disappears if we take into calculation, not the forces themselves, but the *mechanical work* which, in any change of arrangement, is required to overcome them. The expressions for the quantities of work are simpler than those for the corresponding forces; for the quantities of work can be all expressed, without further secondary statements, by numbers which, having reference to the same unit, can be added together, or subtracted from one another, however various the forces may be to which they refer.

It is therefore convenient to alter the form of the above law by introducing, instead of the forces themselves, the work done in overcoming them. In this form it reads as follows:—

> *The mechanical work which can be done by heat during any change of the arrangement of a body is proportional to the absolute temperature at which this change occurs.*

4. The law does not speak of the work which the heat *does*, but of the work which it *can do*; and similarly, in the first form of the law, it is not of the resistances which the heat overcomes, but of those which it *can overcome* that mention is made. This distinction is necessary for the following reasons:—
Since the exterior forces which act upon a body while it is undergoing a change of arrangement may vary very greatly, it may happen that the heat, while causing

a change of arrangement, has not to overcome the whole resistance which it would be possible for it to overcome. A well-known and often-quoted example of this is afforded by a gas which expands under such conditions that it has not to overcome an opposing pressure equal to its own expansive force, as, for instance, when the space filled by the gas is made to communicate with another which is empty, or contains a gas of lower pressure. In order in such cases to determine the force of the heat, we must evidently not consider the resistance which actually is overcome, but that which can be overcome.

Also in changes of arrangement of the opposite kind, that is, where the action of heat is overcome by the opposing forces, a similar distinction may require to be made, but in this case only as far as this—that the total amount of the forces by which the action of the heat is overcome may be greater than the active force of the heat, but not smaller.

Cases in which these differences occur may be thus characterized. When a change of arrangement takes place so that the force and counterforce are equal, the change can likewise take place in the reverse direction under the influence of the same forces. But if it occurs so that the overcoming force is greater than that which is overcome, the change cannot take place in the opposite direction under the influence of the same forces. We may say that the change has occurred in the first case in a *reversible* manner, and in the second case in an *irreversible* manner.

Strictly speaking, the overcoming force must always be more powerful than the force which it overcomes; but as the excess of force does not require to have any assignable value, we may think of it as becoming continually smaller and smaller, so that its value may approach to nought as nearly as we please. Hence it may be seen that the case in which the changes take place reversibly is a limit which in reality is never quite reached, but to which we can approach as nearly as we please. We may therefore, in theoretical discussions, still speak of this case as one which really exists; indeed, as a limiting case it possesses special theoretical importance.

I will take this opportunity of mentioning another process in which this distinction is likewise to be observed. In order for one body to impart heat to another by conduction or radiation (in the case of radiation, wherein mutual communication of heat takes place, it is to be understood that we speak here of a body which gives out more heat than it receives), the body which parts with heat must be warmer than the body which takes up heat; and hence the passage of heat between two bodies of different temperature can take place in one direction only, and not in the contrary direction. The only case in which the passage of heat can occur equally in both directions is when it takes place between bodies of equal temperature. Strictly speaking, however, the communication of heat from one body to another of the same temperature is not possible; but since the difference of temperature may be as small as we please, the case in which it is equal to nothing, and the passage of heat accordingly reversible, is a limiting case which may be regarded as theoretically possible.

6.3   Study Questions

QUES. 6.1. Are there any limitations on the types of cyclical processes that can occur in nature?

a) What is the difference between the external and internal work done on a body? Which is more accessible to observation and measurement?
b) What is meant by a *cyclical process*? And what is the virtue of limiting one's attention to cyclical processes alone?
c) What does Clausius mean by a *transformation*? What is being transformed? What are the two classes of transformations that Clausius considers? And how does Clausius assign an *equivalence value* to each of these types of transformations?
d) When a quantity work is transformed into a quantity of heat, Q (for instance when a blacksmith's hammer strikes an anvil) is the sense of this transformation positive, negative or zero?
e) When a quantity of heat passes from a high-temperature to a low-temperature body, is the sense of this transformation positive, negative, or zero? What about when passing from a low to a high-temperature body? Does this latter case ever, in fact, happen?
f) What general rule governs the equivalence values of transformations? In particular, must the equivalence value of each particular transformation be positive? What about the sum of the equivalence values of all transformations making up a cyclical process?
g) How is Clausius' *Theorem of the Equivalence of Transformations* presented mathematically? And how can it be understood physically?

QUES. 6.2. Does the pressure exerted by a gas on the walls of a chamber depend on the mutual attraction or repulsion of its molecules?

a) What does Clausius mean by the term *disgregation*? What effect does the addition of heat to a substance have on the disgregation of its molecules? Is this always the case?
b) What limits the expansion of a gas (as opposed to a solid or liquid)? Do unconfined gases resist expansion at all? How is the pressure of a confined gas typically determined?
c) Why does the pressure of a gas depend linearly on its temperature? And why do other substances exhibit a more complicated dependence on temperature?
d) How might one infer the strength of the interior forces acting between the molecules of substance? Why might one wish to do so? Must one completely understand the strength of these forces when analyzing cyclical processes?

QUES. 6.3. What is the relationship between the various *forces* acting on a substance and the *reversibility* of a transformation caused by these forces?

a) What is the difference between a reversible and an irreversible change of volume of a gas? If an expanding gas does not accomplish any work, then is this expansion a reversible process? Give an example of such an expansion.
b) What is the difference between a reversible and an irreversible transfer of heat from one body to another? If the flow of heat does not accomplish work, then is this heat flow a reversible process? Give an example of such a heat flow.
c) What happens when two bodies at different temperatures are brought into contact? What happens to a moveable piston which separates two chambers containing gases at different pressures?
d) In what sense is a temperature analogous to a force? How is the reversibility of a transformation governed by the relationship between forces and counter-forces?

6.4 Exercises

EX. 6.1 (REVERSIBLE AND IRREVERSIBLE COMPRESSION). Suppose a cylindrical chamber contains a compressible substance of mass M. The chamber is fitted with a movable piston with area A at one end. Force F is applied to the piston, compressing the substance a distance Δx in time Δt.

a) If this process is done slowly, then is this process reversible?
b) How much work is done by F per unit time? Can this work be expressed as the product of the (internal) pressure of the substance and its change of volume?
c) Is there a change in disgregation of the molecules of the substance during this process? If so, is it positive, negative or zero?
d) Is the equivalence-value of this transformation positive, negative, or zero?
e) Finally, would your answers to each of the previous questions change if the process was done rapidly instead of slowly?

EX. 6.2 (COPPER BLOCK AND BATH). A 1-kg block of copper at 373 K is gently placed into an enormous bath of water at 293 K.

a) What is the equivalence-value of the transformation (in Joules per Kelvin) that the copper block undergoes as it comes to equilibrium with the bath at 293 K? (Hint: You may assume that the specific heat capacity of copper is constant as the block cools.)
b) How much heat flows from the block into the bath during the equilibration process? And what is the equivalence-value of the transformation that the water bath undergoes? (ANSWER: $+105$ J/K)
c) What is the sum of the equivalence values of these transformations? Is this equilibration process reversible?

d) Is energy conserved in this process? Does the principle of conservation of energy *necessitate* this equilibration process? If not, then why *does* the block equilibrate with the water?

e) Would your answer to each of the previous questions change if the copper block was initially at the same temperature as the water, but was *dropped* into the water bath from a height of 100 m? If so, how?

6.5   Vocabulary

1. Deduce
2. Hitherto
3. Hypothetical
4. Endeavour
5. Aggregation
6. Aëriform
7. Heterogeneous
8. Disgregation
9. Presuppose
10. Simultaneous
11. Thermo-electric
12. Irreversible

Chapter 7
Work, Heat, and Irreversibility

The excess of force may then give rise to motions of considerable velocity in the parts of the body under consideration, and these motions may subsequently be changed into the molecular motions which we call heat.

—Rudolph Clausius

7.1 Introduction

In Arts. 1–4 of his sixth memoir on *The Mechanical Theory of Heat*, Clausius introduced the so-called *equivalence values* of several transformations that a substance might undergo. For example, when a small quantity of heat, dQ, flows into (or out of) a substance maintained at a constant temperature T, an equivalence value dQ/T is assigned to this transformation. Most importantly, Clausius asserted that, while the equivalence values of individual transformations may be either positive or negative, when a substance undergoes a *cyclic process comprised of reversible transformations*, these equivalence values must sum to zero:

$$\int \frac{dQ}{T} = 0. \tag{7.1}$$

Now what is meant by "a cyclic process comprised of reversible transformations"? A reversible transformation is one which is carried out in such a way that the substance is always nearly in equilibrium—for example, when a gas expands against a movable piston which is itself backed by nearly the same pressure, or when heat flows from a warm body into another one which is at nearly the same temperature. A cyclic process is a process comprised of a set of transformations which eventually return the substance back to its initial state (*i.e.* the same temperature, the same pressure, and most importantly, the same internal energy). Consequently, during a cyclic processes any mechanical work done *on* the substance must be either (i) expended as work done *by* the substance, or (ii) expelled from the substance in the form of heat. This follows form the first law of thermodynamics. For a non-cyclic process, on the other hand, the internal energy of a substance may change. This is the topic to

© Springer International Publishing Switzerland 2016
K. Kuehn, *A Student's Guide Through the Great Physics Texts,*
Undergraduate Lecture Notes in Physics, DOI 10.1007/978-3-319-21828-1_7

which Clausius now turns in Art. 5 of his sixth memoir. Then, in Art. 11, he broadens the discussion to examine *irreversible* transformations, such as the conduction of heat between bodies at significantly different temperatures.

7.2   Reading: Clausius, *The Mechanical Theory of Heat*

Clausius, R., *Mechanical Theory of Heat, with its Applications to the Steam-Engine and to the Physical Properties of Bodies*, John Van Voorst, London, 1867. Sixth Memoir, continued.

(5) We will now deduce the mathematical expression for the above law, treating in the first place the case in which the change of condition undergone by the body under consideration takes place *reversibly*. The result at which we shall arrive for this case will easily admit of subsequent generalization, so as to include also the cases in which a change occurs irreversibly.

Let the body be supposed to undergo an infinitely small change of condition, whereby the quantity of heat contained in it, and also the arrangement of its constituent particles, may be altered. Let the quantity of heat contained in it be expressed by H, and the change of this quantity by dH. Further, let the work, both interior and exterior together, performed by the heat in the change of arrangement be denoted by dL, a magnitude which may be either positive or negative according as the active force of the heat overcomes the forces acting in the contrary direction, or is overcome by them. We obtain the heat expended to produce this quantity of work by multiplying the work by the thermal-equivalent of a unit of work which we may call A; hence it is $A\,dL$.

The sum $dH + A\,dL$ is the quantity of heat which the body must receive from without, and must accordingly withdraw from another body during the change of condition. We have, however, already represented by dQ the infinitely small quantity of heat imparted to another body by the one which is undergoing modification, hence we must represent in a corresponding manner, by $-dQ$, the heat which it withdraws from another body. We thus obtain the equation

$$-dQ = dH + A\,dL,$$

or[1]

$$dQ + dH + A\,dL = 0 \tag{7.6}$$

---

[1] In my previous memoirs I have separated from one another the *interior* the *exterior* work performed by the heat during the change of condition of the body. If the former be denoted by dI, and the latter by dW, the above equation becomes

$$dQ + dI + A\,dI + A\,dW = 0. \tag{7.2}$$

In order now to be able to introduce the disgregation also into the formulæ, we must first settle how we are to determine it as a mathematical quantity.

By disgregation is represented, as stated in Art. 2, the degree of dispersion of the body. Thus, for example, the disgregation of a body is greater in the liquid state than in the solid, and greater in the aëriform than in the liquid state. Further, if part of a given quantity of matter is solid and the rest liquid, the disgregation is greater the greater the proportion of the whole mass that is liquid; and similarly, if one part is liquid and the remainder aëriform, the disgregation is greater the larger the aëriform portion. The disgregation of a body is fully determined when the arrangement of its constituent particles is given; but, on the other hand, we cannot say conversely that the arrangement of the constituent particles is determined when the magnitude of the disgregation is known. It might, for example, happen that the disgregation of a given quantity of matter should be the same when one part was solid and one part aëriform, as when the whole mass was liquid.

We will now suppose that, with the aid of heat, the body changes its condition, and we will provisionally confine ourselves to such changes of condition as can occur in a continuous and reversible manner, and we will also assume

---

Since, however, the increase in the quantity of heat actually contained in a body, and the heat consumed by interior work during a change of condition, are magnitudes of which we commonly do not know the individual values, but only the sum of those values, and which resemble each other in being fully determined as soon as we know the initial and final conditions of the body, without our requiring to know how it has passed from the one to the other, I have thought it advisable to introduce a function which shall represent the sum of these two magnitudes, and which I have denoted by U. Accordingly

$$dU = dH + A\,dI = 0. \tag{7.3}$$

and hence the foregoing equation becomes

$$dQ + dU + A\,dW = 0; \tag{7.4}$$

and if we suppose the last equation integrated for any finite alteration of condition, we have

$$Q + U + AW = 0. \tag{7.5}$$

These are the equations which I have used in my memoirs published in 1850 and in 1854, partly in the particular form which they assume for the permanent gases, and partly in the general form in which they are here given, with no other difference than that I there took the positive and negative quantities of heat in the opposite sense to what I have done here, in order to attain greater correspondence with the Eq. (6.3I) given in Art. 1. The function U which I introduced is capable of manifold application in the theory of heat, and, since its introduction, has been the subject of very interesting mathematical developments by W. Thomson and by Kirchhoff (see *Philosophical Magazine*, S. 4. vol. ix. p. 528, and Poggendorff's *Annalen*, vol. ciii. p. 177). Thomson has called it "mechanical energy of a body in a given state," and Kirchhoff "Wirkungsfunction." Although I consider my original definition of it as representing the *sum of the heat added to the quantity already present and of that expended in interior work*, starting from any given initial state (pp. 29 and 118),[2] as perfectly exact, I can still have no objection to make against an abbreviated mode of expression. See the Appendix A. On Terminology at the end of this memoir (Included in Chap. 8 of the present volume.—[K.K.])

that the body has a uniform temperature throughout. Since the increase of disgregation is the action by means of which heat performs work, it follows that the quantity of work must bear a definite ratio to the quantity by which the disgregation is increased; we will therefore fix the still arbitrary determination of the magnitude of disgregation so that, at any given temperature, the increase of disgregation shall be proportional to the work which the heat can thereby perform. The influence of the temperature is determined by the foregoing law. For if the same change of disgregation takes place at different temperatures, the corresponding work must be proportional to the absolute temperature. Accordingly, let Z be the disgregation of the body, and dZ an infinitely small change of it, and let dL be the corresponding infinitely small quantity of work, we can then put

$$dL = KT\,dZ$$

or

$$dZ = \frac{dL}{KT}$$

where K is a constant dependent on the unit, hitherto left undetermined, according to which Z is to be measured. We will choose this unit of measure so that $K = \frac{1}{A}$, and the equation becomes

$$dZ = \frac{A\,dL}{T} \tag{7.7}$$

If we suppose this expression integrated, from any initial condition in which Z has the value Z_0, we get

$$Z = Z_0 + A \int \frac{dL}{T} \tag{7.8}$$

The magnitude Z is thus determined, with the exception of a constant dependent upon the initial condition that is chosen.

If the temperature of the body is not everywhere the same, we can regard it as divided into any number we choose of separate parts, refer the elements dZ and dL in Eq. (7.7) to anyone of them, and at once substitute for T the value of the absolute temperature of that part. If we then unite by summation the infinitely small changes of disgregation of the separate parts, or by integration, if there is an infinite number of them, we obtain the similarly infinitely small change of disgregation of the entire body, and from this we can obtain, likewise by integration, any desired finite change of disgregation.

We will now return to Eq. (7.6), and by help of Eq. (7.7) we will eliminate from it the element of work dL. Thus we get

$$dQ + dH + T\,dZ = 0 \tag{7.9}$$

or, dividing by T,

$$\frac{dQ + dH}{T} + dZ = 0 \tag{7.10}$$

If we suppose this equation integrated for a finite change of condition, we have

$$\int \frac{dQ + dH}{T} + \int dZ = 0 \qquad (7.11)$$

Supposing the body not to be of uniform temperature throughout, we may imagine it broken up again into separate parts, make the elements dQ, dH, and dZ in Eq. (7.10) refer, in the first instance, to one part only, and for T put the absolute temperature of this part. The symbols of integration in (7.11) are then to be understood as embracing the changes of all the parts. We must here remark that cases in which one continuous body is of different temperatures at different parts, so that a passage of heat immediately takes place by conduction from the warmer to the colder parts, must be for the present disregarded, because such a passage of heat is not reversible, and we have provisionally confined ourselves to the consideration of reversible changes.

Equation (7.11) is the required mathematical expression of the above law, *for all reversible changes of condition of a body*; and it is clearly evident that it also remains applicable, if a series of successive changes of condition be considered instead of a single one.

[Articles 6–10 have been omitted for the sake of brevity.]

(11) We must now examine the manner in which the foregoing theorem is modified when we give up the condition that all changes of condition are to take place reversibly.

From what has been said in Art. 4 concerning non-reversible changes of condition, it is easy to perceive that the following must be a general property of all three kinds of transformations. A negative transformation can never occur without a simultaneous positive transformation whose equivalence-value is at least as great; on the other hand, positive transformations are not necessarily accompanied by negative transformations of equal value, but may take place in conjunction with smaller negative transformations, or even without any at all.

If heat is to be transformed into work, which is a negative transformation, a positive change of disgregation must take place at the same time, which cannot be smaller in amount than that determinate magnitude which we regard as equivalent. In the positive transformation of work into heat, on the other hand, the state of things is different. If the force of heat is overcome by opposing forces, so that a negative change of disgregation is brought about, we know that in this case the overcoming forces may be greater than is required to produce the particular result. The excess of force may then give rise to motions of considerable velocity in the parts of the body under consideration, and these motions may subsequently be changed into the molecular motions which we call heat, so that in the end more work comes to be transformed into heat than corresponds to the negative change of disgregation brought about. In many operations, especially in friction, the transformation of work into heat may take place even quite independently of any simultaneous negative transformation.

The relation in which the third kind of transformation, namely change of disgregation, stands to considerations of this nature, is implied in what has been already said. The positive change of disgregation may indeed be greater, but cannot be smaller, than the accompanying transformation of heat into work; and the negative change of disgregation may be smaller, but cannot be greater, than the transformation of work into heat.

Finally, in so far as regards the second kind of transformation, or the passage of heat between bodies of different temperatures, I have thought myself justified in assuming as a fundamental proposition what, according to all that we know of heat, must be regarded as well-established, namely, that the passage from a lower to a higher temperature, which counts as a negative transformation, cannot take place of itself—that is, without a simultaneous positive transformation. On the other hand, the passage of heat in the contrary direction, from a higher to a lower temperature, may very well take place without a simultaneous negative transformation.

Taking these circumstances into consideration, we will now return once more to the consideration of the development by means of which we arrived at Eq. (7.11) in Art. 5. Equation (7.7), which occurs in the same Article, expresses the relation in which an infinitely small change of disgregation must stand to the work simultaneously performed by the heat, under the condition that the change takes place in a reversible manner. In case this last condition need not be fulfilled, the change of disgregation may be greater, provided it is positive, than the value calculated from the work; and if negative, it may be, when taken absolutely, smaller than that value, but in this case also it would algebraically have to be stated as greater. Instead of Eq. (7.7), we must therefore write

$$dZ \geq \frac{A\,dL}{T} \qquad (7.12)$$

Applying this to Eq. (7.6), we obtain, instead of Eq. (7.10),

$$\frac{dQ + dH}{T} + dZ \geq 0 \qquad (7.13)$$

The further question now arises, what influence would it have on the formulæ, if a direct passage of heat took place between parts of different temperature within the body in question.

In case the body is not uniform temperature throughout, the differential expression occurring in Eq. (7.13) must not be referred to the entire body, but only to a portion whose temperature may be considered as the same throughout; so that if the temperature of the body varies continuously, the number of parts must be assumed as infinite. In integrating, the expressions which apply to the separate parts may be united again to a single expression for the whole body, by extending the integral, not only to the changes of one part, but to the changes of all the parts. In forming this integral, we must now have regard to the passage of heat taking place between the different parts.

It must here be remarked that dQ is an element of the heat which the body under consideration gives up to, or absorbs from, an external body which serves only as a reservoir of heat, and that this element does not come into question now that we are discussing the passage of heat between the different parts of the body itself. This transfer of heat is mathematically expressed by a decrease in the quantity of heat H in one part, and an equivalent increase in another part; and accordingly we require to direct our attention only to the term $\frac{dH}{T}$ in the differential expression (7.13). If we now suppose that the infinitely small quantity of heat dH leaves one part of the body whose temperature is T_1 and passes into another part whose temperature is T_2, there result the two following infinitely small terms,

$$-\frac{dH}{T_1} \quad \text{and} \quad +\frac{dH}{T_2}$$

which must be contained in the integral; and since T_1 must be greater than T_2, it follows that the positive term must in any case be greater than the negative term, and that consequently the algebraic sum of both is positive. The same thing applies equally to every other element of heat transferred from one part to another; and the change which the integral of the whole differential expression occurring in (7.13) undergoes, on account of this transfer of heat, can therefore only consist in the addition of a positive quantity to the value which would else have been obtained. But since, as results from Eq. (7.13), the last value which would be obtained, without taking this direct transfer of heat into consideration, cannot be less than nothing; this can still less be the case when it has been increased by another positive quantity.

We may therefore write as a general expression, including all the circumstances which occur in non-reversible changes, the following, instead of Eq. (7.11):—

$$\int \frac{dQ + dH}{T} + \int dZ \geq 0 \tag{7.14}$$

The theorem which in Art. 1 was enunciated in reference to cyclical processes only, and was represented by the expression (6.4), has thus assumed a more general form, and may be enunciated thus:—

The algebraic sum of all the transformations occurring during any change of condition whatever can only be positive, or, as an extreme case, equal to nothing.

In my previous paper I have spoken of two transformations with opposite signs, which neutralize each other in the algebraic sum, as *compensating* transformations. The foregoing theorem may therefore be enunciated still more briefly as follows:—

Uncompensated transformations can only be positive.

(12) In conclusion, we will submit the integral

$$\int \frac{dH}{T},$$

which has been frequently used above, to a somewhat closer consideration. We will call this integral, when it is taken from any given initial condition to the condition actually existing, *the transformation-value of the heat actually present in the body when calculated from the given initial condition*. That is, when in any way whatever work is transformed into heat, or heat into work, and the quantity of heat present in the body is thereby altered, the increment or decrement of this integral gives the equivalence-value of the transformations which have taken place. Further, if transfers of heat take place between parts of different temperature within the body itself, or within a system of bodies, the equivalence-value of these transfers of heat is likewise expressed by the increment or decrement of this integral, if it is extended to the whole system of bodies under consideration.

In order to be able actually to perform the integration which has been indicated, we must know the relation between the quantity of heat H and the temperature T. If we call the mass of the body m, and its real (capacity for heat) c, we have, for a change of temperature throughout amounting to dT, the equation

$$dH = mc\,dT \tag{7.15}$$

According to what has been said above, the real (capacity for heat) of a body is independent of the arrangement of its particles; and since an arrangement is known, namely, that in perfect gases, for which we must regard it as established, partly by existing experimental data, and partly as the result of theoretical considerations, that the real (capacity for heat) is independent of temperature, we may assume the same thing for the other states of aggregation, and may regard the real (capacity for heat) as always *constant*. Thence it follows that the amount of heat present in a body is simply proportional to its absolute temperature, inasmuch as we can write

$$H = mcT \tag{7.16}$$

Even when the body is not homogeneous, but consists of different substances, all, however, at the temperature T, the foregoing equation will still remain applicable, if for c we substitute the corresponding mean value. On the other hand, if different parts of the body have different temperatures, we must in the first instance apply the equation to the separate parts, and then unite the various equations by summation. If, for the sake of generality, we assume that the temperature varies continuously, so that the body must be conceived as divided into an infinite number of parts, the equation takes the following form:

$$H = \int cT\,dm \tag{7.17}$$

Applying these expressions to the integral given above for the transformation-value of the heat in the body, and denoting the initial temperature by T_0, we obtain; for the more simple case in which the temperature is uniform throughout,

$$\int \frac{dH}{T} = mc \int_{T_0}^{T} \frac{dT}{T} = mc \log \frac{T}{T_0} \tag{7.18}$$

and, as a general expression embracing all cases,

$$\int \frac{dH}{T} = \int c \log \frac{T}{T_0} dm \tag{7.19}$$

If the disgregation of a body is changed, without heat being supplied to or withdrawn from it, by an external object, the amount of heat contained in the body must be changed in consequence of the production or consumption of heat attendant on the change of disgregation, and a rise or fall of temperature must be the result; consequently the question may be raised, How great must the change of disgregation be in order to bring about a given change of temperature, it being assumed that all changes of condition take place reversibly? In this case we must apply Eq. (7.11), putting $dQ = 0$, whereby it is transformed into

$$\int \frac{dH}{T} + \int dZ = 0 \tag{7.20}$$

If we assume, for the sake of simplicity, that the temperature of the entire body varies uniformly, so that T has the same value for all parts, we may apply Eq. (7.18) to the determination of the first of the two integrals; and we thus obtain, for the required change of disgregation, the equation[3]

$$Z - Z_0 = mc \log \frac{T_0}{T} \tag{7.22}$$

If we desired to cool a body down to the absolute zero of temperature, the corresponding change of disgregation, as shown by the foregoing formula, in which we should then have $T = 0$, would be infinitely great. Hereon is based the argument by which it may be proved to be impossible practically to arrive at the absolute zero of temperature by any alteration of the condition of a body.

---

[3] (If the above simplifying hypothesis—that the temperature is the same in all parts of the body and changes in the same manner—be not made, we shall have the equation

$$Z - Z_0 = \int c \log \frac{T_0}{T} dm, \tag{7.21}$$

instead of the Eq. (7.18).—1864.)

7.3 Study Questions

QUES. 7.1. What is meant by the *disgregation* of a substance? Does the liquid, solid or gaseous state of a substance have a higher degree of disgregation? Does the disgregation of a body fully specify the arrangement of its constituent particles?

QUES. 7.2. Must all transformations in nature have positive equivalence-values?

a) When a quantity of heat is transformed into work (for instance, in heating up a gas, which in turn expands against a piston) is such a transformation positive, negative or zero? Is the change in disgregation of the gas molecules positive, negative or zero? Is the total equivalence-value of such a transformation positive, negative or zero?

b) When a quantity of work is transformed into heat (for instance, in compressing a gas, which in turn emits heat to its surroundings) is such a transformation positive, negative or zero? Is the change in disgregation of such a process positive, negative or zero? Is the total equivalence-value of such a transformation positive, negative or zero?

c) What is a *compensating* transformation? Must a negative transformation always be compensated by a positive transformation? Conversely, must a positive transformation always be compensated by a negative transformation? Upon what fundamental and well-established proposition does Clausius justify this asymmetry?

QUES. 7.3. What is the relationship between the heat added to a substance from an exterior body (dQ) the heat contained in the substance (dH) and the change in disgregation of the substance (dZ) for a reversible process? For a non-reversible process?

QUES. 7.4. What is the equivalence-value of a transformation consisting of the transfer of heat between parts of a body at different temperatures? Is this a reversible process?

QUES. 7.5. Is it possible for a substance to reach the absolute zero of temperature? Why or why not?

7.4 Exercises

EX. 7.1 (WATER EQUILIBRATION). Suppose that a bath of ice water is momentarily connected by a short copper wire to another bath of water which is 20° warmer. During this time, 1 erg of heat flows from the hot to the cold bath. Due to the immense heat capacity of the baths, their temperatures remain nearly constant during this process.

a) What is the equivalence-value of the transformation that the cold-bath undergoes? The hot bath? What is the sum of these equivalence-values? Is this a
   reversible process? (ANSWER: +0.00025 erg/K)
b) If heat flowed in the other direction (from cold to hot), how would your answers
   to the previous questions change? Would this reversed heat flow violate the
   principle of conservation of energy? Would it violate any (other) laws of physics?
c) If these tanks were not large (for example, if each contained just 1 g of water)
   then what would be the maximum amount of work which one could extract from
   the tanks in allowing them to come to thermal equilibrium—for example by
   letting heat flow from the hot to the cold bath through a reversible heat engine?
   (HINT: Heat is not conserved, but energy is, so the work being done by the heat
   engine during any time interval, dW, is just the difference between the amount
   of heat flowing out of the hot bath, dQ_h, and into the cold bath, dQ_c.)

EX. 7.2 (COPPER ROD EQUILIBRATION). Consider a cylindrical copper rod having a
length of one meter and a cross-sectional area of 2 cm^2. The two ends of the rod
are maintained at a constant temperature by keeping them in contact with boiling
water and ice water, respectively. Assuming that the thermal conductivity of the rod
does not depend on the temperature, the initial temperature distribution, $T(z)$, within
the rod is linearly decreasing from the hot to the cold end. If, now, this contact is
broken, what is the temperature distribution within the rod after its various sections
have finally come to thermal equilibrium? You may assume that once contact is
broken, no heat is gained or lost by the rod. Also: what is the equivalence-value
associated with this equilibration process? Is it reversible?

7.5   Vocabulary

1. Subsequent
2. Determinate
3. Disgregation
4. Simultaneous

Chapter 8
Language: Concepts and Conventions

*The heat which disappears during fusion or evaporation is
converted into work, and consequently exists no longer as heat;
I propose, therefore, in place of latent heat, to substitute the
term ergonized heat.*

—Rudolph Clausius

8.1 Introduction

Much of the language used to describe the effects of heat was developed at a time
when it was thought that heat was itself an invisible and indestructible fluid-like
substance. This *caloric* was able to flow into (or out of) a body, thereby raising (or
lowering) its temperature. In addition, caloric could sometimes flow into a body
without raising its temperature, as when ice melts into water of the same temperature. This water was then thought to contain a quantity of hidden, or *latent*, heat
which must be extracted in order to re-freeze the water. Perhaps unsurprisingly, the
rise of the mechanical theory of heat in the mid-nineteenth century posed a number
of conceptual and linguistic problems. First, should one even speak of *heat flow*?
Remarkably, such language is still commonplace in the twenty-first century, despite
the fact that almost nobody believes in the caloric theory of heat from which it was
derived. Second, if heat itself is not a conserved substance, then what *is* the fate
of the (so-called) heat which is added to a body—for example when a container of
gas is heated over a fire? How much of the added heat goes into motion of the gas
molecules (internal kinetic energy)? How much of it goes into changing the configuration of the molecules (internal potential energy)? How much of it goes into
moving the walls of the container which confine the gas (external work)? And how
much of it simply escapes into nearby bodies, heating them up? In the reading selection below, taken from the appendix to the sixth memoir of *The Mechanical Theory
of Heat*, Clausius attempts to introduce terminology which more accurately reflect
the new mechanical theory of heat. Has Clausius' terminology stuck?

© Springer International Publishing Switzerland 2016
K. Kuehn, *A Student's Guide Through the Great Physics Texts*,
Undergraduate Lecture Notes in Physics, DOI 10.1007/978-3-319-21828-1_8

8.2   Reading: Clausius, *The Mechanical Theory of Heat*

Clausius, R., *Mechanical Theory of Heat, with its Applications to the Steam-Engine and to the Physical Properties of Bodies*, John Van Voorst, London, 1867. Sixth Memoir, Appendix A: On Terminology.

The new conceptions which the mechanical theory of heat has introduced into science present themselves so frequently in all investigations on heat, that it has become desirable to possess simple and characteristic names for them.

I have divided into the following three parts the heat which must be imparted to a body in order to change its condition in any manner whatever: *first*, the increased amount of heat actually present in the body; *second*, the heat consumed by interior work; and *third*, the heat consumed by exterior work. Of these three quantities of heat the last can only be determined when all the changes are known which the body has suffered; for the determination of the two first quantities, however, a knowledge of the entire series of changes is not necessary, an acquaintance with the initial and final conditions of the body suffices. Given, therefore, the initial condition, proceeding from which the body arrives successively at any other conditions whatever, the first and second of the above quantities of heat may be regarded as two magnitudes which are perfectly defined by the condition of the body at the moment under consideration. The same remark applies, of course, to the sum of these two quantities which I have represented by U, and which is of great importance, inasmuch as it presents itself in the first fundamental equation of the mechanical theory of heat.

The definition I have given of this magnitude—*the sum of the increment of actually present heat, and of the heat consumed by interior work*—being for general purposes too long to serve as the name of the quantity, several more convenient ones have been proposed. As already remarked in the note on p. 83, Thomson has to this end employed the expression *"the mechanical energy of a body in a given state,"* and Kirchoff the term *"Wirkungsfunction."* Zeuner, again, in his *"Grundzüge der mechanischen Wärmetheorie,"* has called U "die innere Wärme des Körpers" (interior heat of the body).

The latter name does not appear to me to correspond quite to the signification of the magnitude U, since only a portion of the latter represents heat actually present in the body, in other words, *vis viva* of its molecular motions, the other portion having reference to heat which has been consumed by interior work, and which, therefore, no longer exists as heat. I do not for a moment imagine that Zeuner had any intention to imply, by that name, that all the heat represented by U was actually present as heat in the body; nevertheless the name might easily be interpreted in this sense.

Of the two other expressions mentioned above, the term *energy* employed by Thomson appears to me to be very appropriate; it has in its favour, too, the circumstance that it corresponds to the proposition of Rankine to include under the common name energy, both heat and everything that heat can replace. I have no hesitation, therefore, in adopting, for the quantity U, the expression *energy of the body*.

It must be here observed, however, that the total energy of a body cannot be measured, it is only the increment of energy, due to the passage of the body from any initial state to its present condition, that is susceptible of measurement. The initial condition being assumed as given, the increment of energy is a perfectly defined magnitude for every other condition of the body. The question is, are we to understand by the *energy of a body* merely the increment of energy estimated from a given initial condition, or is the energy which the body possessed at the beginning to be included in the term? In the latter case, where the *total* energy of the body is implied, we must conceive the increment of energy to be supplemented by the addition of an unknown constant having reference to the initial condition. It will not always be necessary, of course, to mention this constant expressly; we may tacitly assume that it is included.

Since the magnitude U consists of two parts which have frequently to be considered individually, it will not suffice to have an appropriate name for U merely, we must also be able to refer conveniently to these its constituent parts.

The first part presents no difficulty whatever; the heat actually present in the body may be simply called *the heat of the body*, or the *thermal content of the body* (*Wärmeinhalt des Körpers*).

In giving a name to the second part of U, however, we are at once inconvenienced by a circumstance which embarrasses the whole mechanical theory of heat,—the fact that heat and work are measured by different units. The unit of heat is the quantity of heat which is necessary to raise the temperature of a unit-weight of water from $0°$ to $1°$, and the unit of work is the quantity which is represented by the product of the unit of weight into the unit of length,—in French measure, therefore, a kilogramme-metre.

Now in the mechanical theory of heat, after admitting that heat can be transformed into work and work into heat, in other words, that either of these may replace the other, it becomes frequently necessary to form a magnitude of which heat and work are constituent parts. But heat and work being measured by different units, we cannot in such a case say, simply, the magnitude is the *sum of heat and work*; we are compelled to say either *the sum of the heat and the heat-equivalent of the work*, or *the sum of the work and the work-equivalent of the heat*.

Rankine has avoided this inconvenient mode of expression in his memoirs by assuming as his unit of heat the quantity which is equivalent to a unit of work. Nevertheless, although perfectly appropriate on theoretic grounds, it must be admitted that great difficulties oppose themselves to the general introduction of this measure of heat. On the one hand it is always difficult to change a unit when once adopted, and on the other there is here the additional circumstance that the heat-unit hitherto used is a magnitude intimately connected with ordinary calorimetric methods, and the latter being mostly based on the heating of water, necessitate only slight reductions, and these founded on very trustworthy measurements; the heat-unit adopted by Rankine, however, besides requiring the same reductions, assumes the mechanical equivalent of heat to be known,—an assumption which is only approximately correct. Accordingly, since we cannot expect the mechanical measure for heat to be universally adopted, we must always, when quantities of heat enter into an equation, first state whether these quantities are measured in the ordinary manner or by the mechanical unit, and consequently the above-mentioned inconvenience would not be removed by Rankine's procedure.

For this purpose, therefore, I will venture another proposition. Let heat and work continue to be measured each according to its most convenient unit, that is to say, heat according to the thermal unit, and work according to the mechanical one. But besides the work measured according to the mechanical unit, let another magnitude be introduced denoting *the work measured according to the thermal unit*, that is to say, *the numerical value of the work when the unit of work is that which is equivalent to the thermal unit*. For the work thus expressed a particular name is requisite. I propose to adopt for it the Greek word (εργον) *ergon*.[1]

The processes which are considered in the mechanical theory of heat may be very conveniently described by means of this new term. Heat and ergon are, in fact, two magnitudes which admit of mutual transformation and substitution, without any alteration in the numerical values of the respective quantities being thereby involved. Accordingly, heat and ergon may, without preparation, be added to, or subtracted from, one another.

When we consider the work produced during any change in the condition of a body, we must call it the *ergon produced*, if it be measured by the unit of heat, and here again we distinguish *interior ergon* and *exterior ergon*. The latter, as already stated in the memoirs, is dependent upon the entire series of successive changes, whilst the former is completely determined when the initial and final conditions, solely, are known. Assuming the initial condition to be given, therefore, the interior ergon may be regarded as a magnitude which depends solely upon the condition of the body at the moment under consideration.

Analogous to the expression *thermal content of the body*, we may introduce the expression *ergonal content of the body*. With reference to the last conception, however, the same remark applies which was previously made with reference to energy. We may understand by ergonal content, either the increment of ergon reckoned from a given initial condition, or the *total* ergonal content. In the latter case we have merely to conceive an unknown constant, having reference to the initial state, added to the increment of ergon; this is so obvious, however, that in such cases we may usually assume tacitly that the constant has been included.

The same remark also applies to the thermal content of a body. By this term we may likewise understand either the increment of heat calculated from an arbitrarily assumed initial condition, or the *total* thermal content. In the latter case a constant associated with that initial condition is to be added to the heat-increment. The only difference is that in the ease of the ergonal content, the added constant is quite unknown, whilst in the case of the thermal content, the constant may be

---

[1] The author has used the German word *Werk*, which is almost synonymous with *Arbeit*, but he proposes the term *Ergon* as more suitable for introduction into other languages. The Greek word ἔργον is so closely allied to the English word work, that both are quite well suited to designate two magnitudes which are essentially the same, but measured according to different units.—T.A.H.

approximately determined, seeing that the absolute zero of temperature is to a certain extent known.

Now the quantity U is the sum of the thermal content and ergonal content, so that in place of the word *energy*, we may use if we please the somewhat longer expression, *thermal and ergonal content*.

In connexion with these remarks on Terminology I will venture another suggestion. Hitherto the heat which disappears when a body is fused or evaporated has been termed *latent heat*. This name originated when it was thought that the heat which can no longer be detected by our senses, when a body fuses or evaporates, still exists in the body in a peculiar concealed condition. According to the mechanical theory of heat, this notion is no longer tenable. All heat actually present in a body is sensible heat; the heat which disappears during fusion or evaporation is converted into work, and consequently exists no longer as heat; I propose, therefore, in place of *latent heat*, to substitute the term *ergonized heat*.

In order to distinguish, in a similar manner, the two parts of the latent heat which I have stated to be expended, respectively, on interior and on exterior work, the expressions *interior* and *exterior ergonized heat* might be used.

It must further be observed that of the heat which must be imparted to a body in order to raise its temperature without changing its state of aggregation (all of which was formerly regarded as *free*), a great portion falls in the same category as that which has hitherto been called latent heat, and for which I now propose the term ergonized heat. For, in general, the heating of a body involves a change in the arrangement of its molecules. This change usually occasions a sensible alteration in volume, but it may occur even when the volume of the body remains the same. For every change in molecular arrangement, a certain amount of ergon is requisite, which may be partly interior and partly exterior, and in producing this ergon, heat is consumed. Only a part of the heat communicated to a body, therefore, serves to increase the heat actually present therein; the remaining part constitutes the ergonized heat.

In certain cases, such as those of evaporation and fusion, where the proposed term ergonized heat frequently presents itself, a more abbreviated form of expression may, of course, be adopted, should it be found convenient to do so. For instance, instead of using the expressions *ergonized heat of evaporation*, and *ergonized heat of fusion*, we may simply say, as I have done in my memoirs, *heat of evaporation* and *heat of fusion*.

8.3 Study Questions

QUES. 8.1. In what two ways can the internal energy of a body be changed?

a) What are the three modes by which heat may be imparted to a body so as to change its condition?
b) What is meant by the (internal) energy of a body? Can the energy of a substance be measured?

c) In what sense does the energy of a body represent both the kinetic and potential energies of the particles comprising the substance?

d) What is the unit of heat, and upon what method was this unit based? What is the unit of work, and upon what method was this unit based?

e) What difficulty was introduced by the mechanical theory of heat? What was Rankine's solution to this unit problem? And why is Clausius skeptical of this solution?

f) What is Clausius' solution to this problem? In particular, what name does he propose to denote the unit of work which is equivalent to the thermal unit?

g) What is the difference between the *ergonal content* and the *thermal content* of a body? How are these related to the energy of the body?

QUES. 8.2. Why does Clausius find the term *latent heat* to be misleading? What term does he propose instead? Why do you think the older terminology survives today? Is this a problem?

8.4 Vocabulary

1. Conception
2. Tacit
3. Constituent
4. Venture
5. Analogous
6. Arbitrary
7. Consequent
8. Requisite
9. Abbreviated
10. Fusion

Chapter 9
Energy and Entropy

The entropy of the universe tends to a maximum.

—Rudolph Clausius

9.1 Introduction

In his Sixth Memoir on *The Mechanical Theory of Heat*, Clausius explained how the internal energy of a body (U), consists of both thermal content (H) and ergonal content (Z). These represent the energy associated with the motion and the configuration of the body's particles, respectively. According to the first fundamental theorem of the mechanical theory of heat—now known as the *first law of thermodynamics*—the internal energy of a body may be changed by either adding heat (Q) to the body, or doing external work (W) on the body. But the first law of thermodynamics alone is not sufficient to explain the types of processes which tend to occur in nature. Rather, only those processes, or transformations, occur which are characterized by positive (or at best, zero) equivalence-values; any process which has a negative equivalence value must be compensated by another process having an equal or greater positive equivalence value. This is Clausius' second fundamental theorem of the mechanical theory of heat. Today, it is known as the *second law of thermodynamics*. In the reading selection below, taken from his Ninth Memoir, Clausius clarifies these ideas by writing them in a succinct mathematical form. Perhaps most notably, he introduces the concept of *entropy*, denoted by the letter S. What is meant by this term? From where was it derived? And what is the connection between entropy and the second law of thermodynamics?

9.2 Reading: Clausius, *The Mechanical Theory of Heat*

Clausius, R., *Mechanical Theory of Heat, with its Applications to the Steam-Engine and to the Physical Properties of Bodies*, John Van Voorst, London, 1867. Ninth Memoir.

© Springer International Publishing Switzerland 2016

K. Kuehn, *A Student's Guide Through the Great Physics Texts*,

Undergraduate Lecture Notes in Physics, DOI 10.1007/978-3-319-21828-1_9

In my former Memoirs on the Mechanical Theory of Heat, my chief object was to secure a firm basis for the theory, and I especially endeavoured to bring the second fundamental theorem, which is much more difficult to understand than the first, to its simplest and at the same time most general form, and to prove the necessary truth thereof. I have pursued special applications so far only as they appeared to me to be either appropriate as examples elucidating the exposition, or to be of some particular interest in practice.

The more the mechanical theory of heat is acknowledged to be correct in its principles, the more frequently endeavours are made in physical and mechanical circles to apply it to different kinds of phenomena, and as the corresponding differential equations must be somewhat differently treated from the ordinarily occurring differential equations of similar forms, difficulties of calculation are frequently encountered which retard progress and occasion errors. Under these circumstances I believe I shall render a service to physicists and mechanicians by bringing the fundamental equations of the mechanical theory of heat from their most general forms to others which, corresponding to special suppositions and being susceptible of direct application to different particular cases, are accordingly more convenient for use.

(1) The whole mechanical theory of heat rests on two fundamental theorems,— that of the equivalence of heat and work, and that of the equivalence of transformations.

In order to express the first theorem analytically, let us contemplate any body which changes its condition, and consider the quantity of heat which must be imparted to it during the change. If we denote this quantity of heat by Q, a quantity of heat given off by the body being reckoned as a negative quantity of heat absorbed, then the following equation holds for the element dQ of heat absorbed during an infinitesimal change of condition,

$$dQ = dU + A\,dW \qquad (9.1)$$

Here U denotes the magnitude which I first introduced into the theory of heat in my memoir of 1850, and defined as the sum of the free heat present in the body, and of that consumed by interior work.[1] Since then, however, W. Thomson has proposed the term *energy* of the body for this magnitude,[2] which mode of designation I have adopted as one very appropriately chosen; nevertheless, in all cases where the two elements comprised in U require to be separately indicated, we may also retain the phrase *thermal and ergonal content*, which, as already explained on p. 97, expresses my original definition of U in a rather simpler manner. W denotes the exterior work done during the change of condition of the body, and A the quantity of heat equivalent to the unit of work, or more briefly, the *thermal equivalent of work*. According to this AW is the

---

[1] Pogg. *Ann.* Bd. lxxix. S. 385.
[2] Phil. Mag. S. 4. vol. ix. p. 523.

exterior work expressed in thermal units, or according to a more convenient terminology recently proposed by me, the *exterior ergo*. (See Appendix A. to Sixth Memoir.)

If for the sake of brevity, we denote the exterior ergon by a simple letter,

$$w = AW \qquad (9.2)$$

we can write the foregoing equation as follows,

$$dQ = dU + dw. \qquad (9.3)$$

In order to express analytically the second fundamental theorem in the simplest manner, let us assume that the changes which the body suffers constitute a *cyclical process*, whereby the body returns finally to its initial condition. By dQ we will again understand an element of heat absorbed, and T shall denote the temperature, counted from the absolute zero, which the body has at the moment of absorption, or, if different parts of the body have different temperatures, the temperature of the part which absorbs the heat element dQ. If we divide the thermal element by the corresponding absolute temperature and integrate the resulting differential expression over the whole cyclical process, then for the integral so formed the relation

$$\int \frac{dQ}{T} \leq 0 \qquad (9.4)$$

holds, in which the sign of equality is to be used in cases where all changes of which the cyclical process consists are *reversible*, whilst the sign $<$ applies to cases where the changes occur in a *non-reversible* manner.[3]

(2) We will first consider more closely the magnitudes occurring in Eq. (9.3) in reference to different kinds of changes of the body.

The exterior ergon w, which is produced whilst the body passes from a given initial condition to another definite one, depends not merely on the initial and final conditions, but also on the nature of the transition.

In the first place, we have to consider the exterior forces which act on the body, and which are either overcome by, or overcome the forces of the body itself;— the exterior ergon being positive in the former, and negative in the latter case.

---

[3] In my memoir "On a Modified Form of the Second Fundamental Theorem of the Mechanical Theory of Heat" (Fourth Memoir of this collection), in which I first gave the most general expression of the Second Fundamental Theorem for a Cyclical Process, the signs of the differentials dQ were differently chosen; there a thermal element given up by a changing body to a reservoir of heat is reckoned positive, an element withdrawn from a reservoir of heat is reckoned negative. With this choice of signs, which in certain general theoretical considerations is convenient, we have to write instead of (9.4), $\int \frac{dQ}{T} \geq 0$.

In the present memoir, however, the choice mentioned in the text is everywhere retained, according to which a quantity of heat absorbed by a changing body is positive, and a quantity given off by it is negative.

The question then arises, are these exterior forces, at each moment, the same as, or different from the forces of the body? Now although we may assert that for one force to overcome another, the former must necessarily be the greater; yet since the difference between them may be as small as we please, we may consider the case where absolute equality exists as the limiting case, which, although never reached in reality, must be theoretically considered as possible. When force and counter-force are different, the mode in which the change occurs is not a reversible one.

In the second place, the change taking place in a reversible manner, the exterior ergon likewise depends upon the intermediate conditions through which the body passes when changing from the initial to the final condition, or, as it may be figuratively expressed, upon the *path* which the body pursues when passing from its initial to its final condition.

With the *energy U* of the body whose element, as well as that of the exterior ergon, enters into the Eq. (9.3), it is quite different. If the initial and final conditions of the body are given, the variation in energy is completely determined, without any knowledge of the way in which the transition from the one condition to the other took place—in fact neither the nature of the passage nor the circumstance of its being made in a reversible or non-reversible manner, has any influence on the contemporaneous change of energy. If, therefore, the initial condition and the corresponding value of the energy be supposed to be given, we may say that the energy is fully defined by the actually existing condition of the body.

Finally, since the *heat Q* which is absorbed by the body during the change of condition is the sum of the change of energy and of the exterior ergon produced, it must like the latter depend upon the way in which the transition of the body from one condition to another takes place.

Now in order to limit the field of our immediate investigation, we shall always assume, unless the contrary is expressly stated, *that we have to do with reversible changes solely.*

The Eq. (9.3) which expresses the first fundamental theorem, holds for reversible as well as for non-reversible changes; hence, in order to apply it specially to reversible changes, we have not to modify it externally in any manner, but merely to understand that by w and Q are meant the exterior ergon and quantity of heat which correspond to reversible changes.

On applying to reversible changes the relation (9.4) which expresses the second fundamental theorem, we have not only to understand by Q the quantity of heat which relates to reversible changes, but also, instead of the double sign $\leq$, we have simply to employ the sign of equality. We obtain for all reversible cyclical processes, therefore, the equation

$$\int \frac{dQ}{T} = 0 \qquad\qquad (9.5)$$

[Articles 3–13 of Clausius' memoir have here been omitted for the sake of brevity.—κ.κ.]

(14) All the foregoing considerations had reference to changes which occurred in a reversible manner. We will now also take *non-reversible* changes into consideration in order briefly to indicate at least the most important features of their treatment.

In mathematical investigations on non-reversible changes two circumstances, especially, give rise to peculiar determinations of magnitudes. In the first place, the quantities of heat which must be imparted to, or withdrawn from a changeable body are not the same, when these changes occur in a non-reversible manner, as they are when the same changes occur reversibly. In the second place, with each non-reversible change is associated an uncompensated transformation, a knowledge of which is, for certain considerations, of importance.

In order to be able to exhibit the analytical expressions corresponding to these two circumstances, I must in the first place recall a few magnitudes contained in the equations which I have previously established.

One of these is connected with the first fundamental theorem, and is the magnitude U, contained in Eq. (9.3) and discussed at the beginning of this Memoir; it represents the *thermal and ergonal content*, or the energy of the body. To determine this magnitude, we must apply the Eq. (9.3), which may be thus written,

$$dU = dQ - dw \qquad (9.6)$$

or, if we conceive it to be integrated, thus:

$$U = U_0 + Q - w \qquad (9.7)$$

Herein U_0 represents the value of the energy for an arbitrary initial condition of the body, Q denotes the quantity of heat which must be imparted to the body, and w the exterior *ergon* which is produced whilst the body passes in any manner from its initial to its present condition. As was before stated, the body can be conducted in an infinite number of ways from one condition to another, even when the changes are to be reversible, and of all these ways we may select that one which is most convenient for the calculation.

The other magnitude to be here noticed is connected with the second fundamental theorem, and is contained in Eq. (9.5). In fact if, as Eq. (9.5) asserts, the integral $\int \frac{dQ}{T}$ vanishes whenever the body, starting from any initial condition, returns thereto after its passage through any other conditions, then the expression $\frac{dQ}{T}$ under the sign of integration must be the complete differential of a magnitude which depends only on the present existing condition of the body, and not upon the way by which it reached the latter. Denoting this magnitude by S, we can write

$$dS = \frac{dQ}{T}; \qquad (9.8)$$

or, if we conceive this equation to be integrated for any reversible process whereby the body can pass from the selected initial condition to its present one, and denote at the same time by S_0 the value which the magnitude S has in that initial condition,

$$S = S_0 + \int \frac{dQ}{T} \qquad (9.9)$$

This equation is to be used in the same way for determining S as Eq. (9.7) was for defining U.

The physical meaning of the magnitude S has been already discussed in the Sixth Memoir. If in the fundamental Eq. (9.4) of the present Memoir, which holds for all changes of condition of the body that occur in a reversible manner, we make a small alteration in the notation, so that the heat taken up by the changing body, instead of the heat given off by it, is reckoned as positive, that equation will assume the form

$$\int \frac{dQ}{T} = \int \frac{dH}{T} + \int dZ \qquad (9.10)$$

The two integrals on the right are the values for the case under consideration, of two magnitudes first introduced in the Sixth Memoir.

In the first integral, H denotes the heat actually present in the body, which, as I have proved, depends solely on the temperature of the body and not on the arrangement of its parts. Hence it follows that the expression $\frac{dH}{T}$ is a complete differential, and consequently that if for the passage of the body from its initial condition to its present one we form the integral $\int \frac{dH}{T}$, we shall thereby obtain a magnitude which is perfectly defined by the present condition of the body, without the necessity of knowing in what manner the transition from one condition to the other took place. For reasons which are stated in the Sixth Memoir, I have called this magnitude the *transformation-value* of the heat present in the body.

It is natural when integrating, to take, for initial condition, that for which $H = 0$, in other words, to start from the absolute zero of temperature; for this temperature, however, the integral $\int \frac{dH}{T}$ is infinite, so that to obtain a finite value, we must take an initial condition for which the temperature has a finite value. The integral does not then represent the transformation-value of the entire quantity of heat contained in the body, but only the transformation-value of the excess of heat which the body contains in its present condition over that which it possessed in the initial condition. I have expressed this by calling the integral thus formed the *transformation-value of the body's heat, estimated from a given initial condition*. For brevity we will denote this magnitude by Y. The magnitude Z occurring in the second integral I have called the *disgregation* of the body. It depends on the arrangement of the particles of the body, and the measure of an increment of disgregation is the equivalence-value of that transformation from ergon to heat which must take place in order to cancel the increment of disgregation, and thus serve as a substitute for that increment.

Accordingly we may say that the disgregation is the transformation-value of the existing arrangement of the particles of the body. Since in determining the disgregation we must proceed from some initial condition of the body, we will assume that the initial condition selected for this purpose is the same as that which was selected for the determination of the transformation-value of the heat actually present in the body.

The sum of the two magnitudes Y and Z, just discussed, is the before-mentioned magnitude S. To show this, let us return to Eq. (9.10), and assuming, for the sake of generality, that the initial condition, to which the integrals in this equation refer, is not necessarily the same as the initial condition which was selected when determining Y and Z, but that the integrals refer to a change which originated in any manner whatever suited to any special investigation, we may then write the integrals on the right of (9.10) thus:

$$\int \frac{dH}{T} = Y - Y_0 \qquad \text{and} \qquad \int dZ = Z - Z_0, \qquad (9.11)$$

wherein Y_0 and Z_0 are the values of Y and Z which correspond to the initial condition. By these means Eq. (9.10) becomes

$$\int \frac{dQ}{T} = Y + Z - (Y_0 + Z_0) \qquad (9.12)$$

Putting herein

$$Y + Z = S \qquad (9.13)$$

and in a corresponding manner

$$Y_0 + Z_0 = S_0$$

we obtain the equation

$$\int \frac{dQ}{T} = S - S_0 \qquad (9.14)$$

which is merely a different form of the Eq. (9.9), by which S is determined.

We might call S the *transformational content* of the body, just as we termed the magnitude U its *thermal and ergonal content*. But as I hold it to be better to borrow terms for important magnitudes from the ancient languages, so that they may be adopted unchanged in all modern languages, I propose to call the magnitude S the entropy of the body, from the Greek word τροπή, *transformation*. I have intentionally formed the word entropy so as to be as similar as possible to the word *energy*; for the two magnitudes to be denoted by these words are so nearly allied in their physical meanings, that a certain similarity in designation appears to be desirable.

Before proceeding further, let us collect together, for the sake of reference, the magnitudes which have been discussed in the course of this Memoir, and which have either been introduced into science by the mechanical theory of heat, or have obtained thereby a different meaning. They are six in number, and possess in common the property of being defined by the present condition of the body, without the necessity of our knowing the mode in which the body came into this condition: (1) the *thermal content*, (2) the *ergonal content*, (3) the sum of the two foregoing, that is to say the thermal and ergonal content, or the *energy*, (4) the *transformation-value of the thermal content*, (5) the *disgregation*, which is to be considered as the transformation-value of the existing arrangement of particles, (6) the sum of the last two, that is to say, the *transformational content*, or the *entropy*.

[Article 15 of Clausius' memoir has been omitted for the sake of brevity.—K.K.]

(16) If we now assume that in one of the ways above indicated the magnitudes U and S have been determined for a body in its different conditions, the equations which hold good for *non-reversible* changes may be at once written down.
The first fundamental Eq. (9.3), and the Eq. (9.7), resulting from it by integration, which we will arrange thus,

$$Q = U - U_0 + w \qquad\qquad (9.15)$$

hold just as well for non-reversible as for reversible changes; the only difference being, that of the magnitudes standing on the right side, the exterior ergon w has a different value, in the case where a change occurs in a non-reversible manner, from that which it has in the case where the same change occurs in a reversible manner. With respect to the difference $U - U_0$ this disparity does not exist. It only depends on the initial and final condition, and not on the nature of the transition. Consequently we need only consider the nature of the transition so far as is necessary in order to determine the exterior ergon thereby performed; and on adding this exterior ergon to the difference $U - U_0$, we obtain the required quantity of heat Q which the body takes up during the transition.
The *uncompensated transformation* involved in any non-reversible change may be thus obtained:—
The expression for the uncompensated transformation which is involved in a *cyclical process*, is given in Eq. (9.11) of the Fourth Memoir.[4] If we give to the differential dQ in that equation the opposite sign, a quantity of heat given off by the body to a reservoir of heat being there reckoned positive, whilst here

---

[4] Clausius here refers to p. 127 of the Fourth Memoir in the J. Van Voorst 1867 publication, which is not included in the present volume. Here, Clausius expresses the total *transformation-value*, N, of a process whereby n bodies at temperatures T_n receive quantities of heat Q_n. He writes $N = \frac{Q_1}{T_1} + \frac{Q_2}{T_2} + \frac{Q_3}{T_3} + \ldots = \sum \frac{Q_n}{T_n}$. When a particular body's temperature changes during this process of heat transfer, then the transformation value of such a process is given by $N = \int \frac{dQ}{T}$.
—[K.K.]

we consider the heat taken up by the body to be positive, it becomes

$$N = -\int \frac{dQ}{T} \tag{9.16}$$

If the body has suffered one change or a series of changes, which do not form a cyclical process, but by which it has reached a final condition which is different from the initial condition, we may afterwards supplement this series of changes so as to form a cyclical process, by appending other changes of such a kind as to reconduct the body from its final to its initial condition. We will assume that these newly appended changes, by which the body is brought back to the initial condition, take place in a reversible manner.

On applying Eq. (9.16) to the cyclical process thus formed, we may divide the integral occurring therein into two parts, of which the first relates to the originally given passage of the body from the initial to the final condition, and the second to the supplemented return from the final to the initial condition. We will write these parts as two separate integrals, and distinguish the second, which relates to the return, by giving to its sign of integration a suffix r. Hence Eq. (9.16) becomes

$$N = -\int \frac{dQ}{T} - \int_r \frac{dQ}{T}.$$

Since by hypothesis the return takes place in a reversible manner, we can apply Eq. (9.14) to the second integral, taking care, however, to introduce the difference $S_0 - S$ instead of $S - S_0$ (where S_0 denotes the entropy in the initial condition, and S the entropy in the final condition), since the integral here in question is to be taken backwards from the final to the initial condition. We have therefore to write

$$\int_r \frac{dQ}{T} = S_0 - S.$$

By this substitution the former equation is transformed into

$$N = S - S_0 - \int \frac{dQ}{T}. \tag{9.17}$$

The magnitude N thus determined denotes the uncompensated transformation occurring in the whole cyclical process. But from the theorem, that the sum of the transformations which occur in a reversible change is null, and hence that no uncompensated transformation can arise therein, it follows that the supposed reversible return has contributed nothing to the augmentation of the uncompensated transformation, and the magnitude N represents accordingly the uncompensated transformation which has occurred in the given passage of the body from the initial to the final condition. In the deduced expression, the difference $S - S_0$ is again perfectly determined when the initial and final conditions are given, and it is only when forming the integral $\int \frac{dQ}{T}$ that the manner in which the passage from one to the other took place must be taken into consideration.

(17) In conclusion I wish to allude to a subject whose complete treatment could
      certainly not take place here, the expositions necessary for that purpose being
      of too wide a range, but relative to which even a brief statement may not be
      without interest, inasmuch as it will help to show the general importance of the
      magnitudes which I have introduced when formalizing the second fundamental
      theorem of the mechanical theory of heat.

      The second fundamental theorem, in the form which I have given to it,
      asserts that all transformations occurring in nature may take place in a certain
      direction, which I have assumed as positive, by themselves, that is, without
      compensation; but that in the opposite, and consequently negative direction,
      they can only take place in such a manner as to be compensated by simulta-
      neously occurring positive transformations. The application of this theorem to
      the Universe leads to a conclusion to which W. Thomson first drew attention,[5]
      and of which I have spoken in the Eighth Memoir. In fact, if in all the changes
      of condition occurring in the universe the transformations in one definite direc-
      tion exceed in magnitude those in the opposite direction, the entire condition
      of the universe must always continue to change in that first direction, and the
      universe must consequently approach incessantly a limiting condition.

      The question is, how simply and at the same time definitely to character-
      ize this limiting condition. This can be done by considering, as I have done,
      transformations as mathematical quantities whose equivalence-values may be
      calculated, and by algebraical addition united in one sum.

      In my former Memoirs I have performed such calculations relative to the heat
      present in bodies, and to the arrangement of the particles of the body. For every
      body two magnitudes have thereby presented themselves—the transformation-
      value of its thermal content, and its disgregation; the sum of which constitutes
      its entropy. But with this the matter is not exhausted; radiant heat must also
      be considered, in other words, the heat distributed in space in the form of
      advancing oscillations of the æther must be studied, and further, our researches
      must be extended to motions which cannot be included in the term *Heat*.

      The treatment of the last might soon be completed, at least so far as relates
      to the motions of ponderable masses, since allied considerations lead us to
      the following conclusion. When a mass which is so great that an atom in
      comparison with it may be considered as infinitely small, moves as a whole,
      the transformation-value of its motion must also be regarded as infinitesimal
      when compared with its *vis viva*; whence it follows that if such a motion by
      any passive resistance becomes converted into heat, the equivalence-value of
      the uncompensated transformation thereby occurring will be represented sim-
      ply by the transformation-value of the heat generated. Radiant heat, on the
      contrary, cannot be so briefly treated, since it requires certain special con-
      siderations in order to be able to state how its transformation-value is to be
      determined. Although I have already, in the Eighth Memoir above referred to,

---

[5] Phil. Mag. Ser. 4. vol. iv. p. 304.

spoken of radiant heat in connexion with the mechanical theory of heat, I have not alluded to the present question, my sole intention being to prove that no contradiction exists between the laws of radiant heat and an axiom assumed by me in the mechanical theory of heat. I reserve for future consideration the more special application of the mechanical theory of heat, and particularly of the theorem of the equivalence of transformations to radiant heat.

For the present I will confine myself to the statement of one result. If for the entire universe we conceive the same magnitude to be determined, consistently and with due regard to all circumstances, which for a single body I have called *entropy*, and if at the same time we introduce the other and simpler conception of energy, we may express in the following manner the fundamental laws of the universe which correspond to the two fundamental theorems of the mechanical theory of heat.

1 *The energy of the universe is constant.*
2 *The entropy of the universe tends to a maximum.*

9.3   Study Questions

QUES. 9.1. What are the fundamental equations of the mechanical theory of heat? How are they expressed mathematically? Which of these equations belong to Clausius? And what condition does Clausius' fundamental theorem place on the types of processes which can occur in nature?

QUES. 9.2. What is the *energy* of a body? How are the *thermal content* and *ergonal content* of a body related to its *energy*? Does the energy of a body in a particular state depend on how this state was achieved?

QUES. 9.3. Does the work accomplished by (or on) a body undergoing a transition between an initial and a final state depend upon the nature—the particular "path"—of the transition? What about the heat absorbed by the body? Can work and heat then be expressed mathematically as "complete differentials"? If so, under what specific conditions?

QUES. 9.4. What is the *entropy* of a body? How are the *transformation-value of the thermal content* and the *disgregation* of a body related to its *entropy*? Does the entropy of a body in a particular state depend on how this state was achieved? In what sense, then, are energy and entropy similar?

QUES. 9.5. Does the heat radiation surrounding a body have entropy? What difficulty does this present? Can the mechanical theory of heat, and Clausius' theorem in particular, be applied to radiation?

QUES. 9.6. What do the fundamental theorems of the mechanical theory of heat imply about the fate of the universe?

9.4   Exercises

EX. 9.1 (ENTROPY OF EQUILIBRATION). In this exercise, we will compute the change in entropy associated with heating up a 1-kg silver block, initially at the temperature of ice-water, using three different methods.

Method 1:   Suppose we place the silver block in direct thermal contact with a large reservoir of boiling water until its temperature rises to 100° C. What is the entropy change of the silver block as a result of this process? Of the reservoir? Of the universe as a whole? Is this method of heating the silver block a reversible process?

Method 2:   Suppose, instead, we heat up the silver block in two stages. We first let it equilibrate with a reservoir at 50° C, then we let it equilibrate with another reservoir at 100° C. What is the entropy change of the silver block as a result of this two-step process? Of the 50° reservoir? Of the 100° reservoir? Of the universe as a whole? Is this a reversible process?

Question: Which of the previous two methods of bringing the silver block to 100° results in a lower change in entropy of the universe as a whole? Why do you suppose this is? Can you imagine a process whereby you could raise the temperature of the silver block to 100° reversibly, that is, with no change in the entropy of the universe?

Method 3:   Finally, suppose we heat up the silver block to 100 °C by allowing heat to flow into it from an enormous boiling water reservoir *through a reversible heat engine*. In this case, what would be the change in entropy of the silver block? Of the reservoir? Of the universe as a whole? Is this a reversible process?

EX. 9.2 (ENERGY, ENTROPY AND THE CARNOT CYCLE). Clausius states that the energy of a substance is comprised of both *thermal* and *ergonal* content. These describe, respectively, the kinetic and potential energies of the particles comprising the substance. For an ideal (non-interacting) gas, the ergonal content must be zero. Hence the energy of an ideal gas must be determined strictly by the motion of the gas molecules. Since the temperature of a substance provides a measure of the average kinetic energy of its particles, this implies that the energy of a fixed quantity of ideal gas depends only on its temperature, regardless of its volume or pressure. With this in mind, reconsider the carnot cycle described in Ex. 4.2. Suppose that the working substance inside the cylinder acts as an ideal gas.

a) How much heat is drawn into the gas from the hot reservoir during the isothermal expansion? How much heat is ejected from the gas into the cold reservoir during the isothermal compression? What about during the other two (adiabatic) processes which make up the carnot cycle?

b) Recall that the efficiency of any engine cycle is defined by Eq. 5.4. What, then, is the efficiency of this cycle? Is your answer in agreement with Eq. 5.8? Is this what you would expect for a carnot cycle? Why?

c) What is the change in entropy of (i) the air, (ii) the hot reservoir, and (iii) the cold reservoir, during each of the four processes which make up the carnot cycle? What is the change of entropy of the entire universe during a complete cycle of this engine?

9.5   Vocabulary

1. Elucidate
2. Exposition
3. Figurative
4. Contemporaneous
5. Impart
6. Disparity
7. Ponderable
8. Infinitesimal
9. Allude
10. Incessant
11. Aether
12. Entropy

Chapter 10
The Kinetic Theory of Gases

> *The opinion that the observed properties of visible bodies*
> *apparently at rest are due to the action of invisible molecules in*
> *rapid motion is to be found in Lucretius.*
>
> —James Clerk Maxwell

10.1 Introduction

James Clerk Maxwell (1831–1879) was born in Edinburgh, Scotland. He is the
chief architect of the electromagnetic theory of light.[1] Maxwell also played a significant role in the development of both thermodynamics and the kinetic theory of
gases. Regarding the former topic (thermodynamics), Maxwell derived a set of four
equations that relate changes in the thermodynamic variables of a substance. For
example, he demonstrated that

> if the temperature is maintained constant, those substances which increase in volume as the
> temperature rises give out heat when the pressure is increased, and those which contract as
> the temperature rises absorb heat when the pressure is increased.[2]

This Maxwell relation may be expressed mathematically in terms of the partial
derivative of volume with respect to temperature (maintaining constant pressure)
and the partial derivative of entropy with respect to pressure (maintaining constant
temperature):

$$\left(\frac{\partial V}{\partial T}\right)_P = -\left(\frac{\partial S}{\partial P}\right)_T$$

He derived this equation (as well as the other three Maxwell relations) geomet-
rically, by analyzing how adjacent pairs of isothermal lines intersect adjacent
pairs of adiabatic lines on a pressure-versus-volume diagram. Maxwell's four

---

[1] For a presentation of Maxwell's electromagnetic theory of light, and also biographical notes on
Maxwell himself, refer to Chap. 30 of Vol. III.

[2] This quote is taken from Chap. IX of Maxwell, J. C., *Theory of Heat*, tenth ed., Longmans, Green,
and Co., London and New York, 1891.

© Springer International Publishing Switzerland 2016
K. Kuehn, *A Student's Guide Through the Great Physics Texts*,
Undergraduate Lecture Notes in Physics, DOI 10.1007/978-3-319-21828-1_10

thermodynamic relations are completely general in that they apply to any substance whatsoever.

Regarding the latter topic (the kinetic theory of gases) Maxwell argued that the macroscopic properties of a gas, like its pressure and temperature, may be understood in terms of the velocities of the molecules comprising the gas. The kinetic theory also provides a key to understanding phenomena such as the conduction of heat through iron, the viscosity of honey, and the diffusion of pollen through air. In the reading selection that follows, taken from his *Theory of Heat*, Maxwell provides an introduction to the kinetic theory of gases.

10.2   Reading: Maxwell, *on the Molecular Theory of the Constitution of Bodies*

Maxwell, J. C., *Theory of Heat*, tenth ed., Longmans, Green, and Co., London and New York, 1891. Chap. XXII.

We have already shown that heat is a form of energy—that when a body is hot it possesses a store of energy, part at least of which can afterwards be exhibited in the form of visible work.

Now energy is known to us in two forms. One of these is Kinetic Energy, the energy of motion. A body in motion has kinetic energy, which it must communicate to some other body during the process of bringing it to rest. This is the fundamental form of energy. When we have acquired the notion of matter in motion, and know what is meant by the energy of that motion, we are unable to conceive that any possible addition to our knowledge could explain the energy of motion, or give us a more perfect knowledge of it than we have already.

There is another form of energy which a body may have, which depends, not on its own state, but on its position with respect to other bodies. This is called Potential Energy. The leaden weight of a clock, when it is wound up, has potential energy, which it loses as it descends. It is spent in driving the clock. This energy depends, not on the piece of lead considered in itself, but on the position of the lead with respect to another body—the earth—which attracts it.

In a watch, the mainspring, when wound up, has potential energy, which it spends in driving the wheels of the watch. This energy arises from the coiling up of the spring, which alters the relative position of its parts. In both cases, until the clock or watch is set agoing, the existence of potential energy, whether in the clock-weight or in the watch-spring, is not accompanied with any visible motion. We must therefore admit that potential energy can exist in a body or system all whose parts are at rest.

It is to be observed, however, that the progress of science is continually opening up new views of the forms and relations of different kinds of potential energy, and that men of science, so far from feeling that their knowledge of potential energy is perfect in kind, and incapable of essential change, are always endeavouring to explain the different forms of potential energy; and if these explanations are in any case condemned, it is because they fail to give a sufficient reason for the fact, and not because the fact requires no explanation.

We have now to determine to which of these forms of energy heat, as it exists in hot bodies, is to be referred. Is a hot body, like a coiled-up watch-spring, devoid of motion at present, but capable of exciting motion under proper conditions? or is it like a fly-wheel, which derives all its tremendous power from the visible motion with which it is animated?

It is manifest that a body may be hot without any motion being visible, either of the body as a whole, or of its parts relatively to each other. If, therefore, the body is hot in virtue of motion, the motion must be carried on by parts of the body too minute to be seen separately, and within limits so narrow that we cannot detect the absence of any part from its original place.

The evidence for a state of motion, the velocity of which must far surpass that of a railway train, existing in bodies which we can place under the strongest microscope, and in which we can detect nothing but the most perfect repose, must be of a very cogent nature before we can admit that heat is essentially motion.

Let us therefore consider the alternative hypothesis—that the energy of a hot body is potential energy, or, in other words, that the hot body is in a state of rest, but that this state of rest depends on the antagonism of forces which are in equilibrium as long as all surrounding bodies are of the same temperature, but which as soon as this equilibrium is destroyed are capable of setting bodies in motion. With respect to a theory of this kind, it is to be observed that potential energy depends essentially on the relative position of the parts of the system in which it exists, and that potential energy cannot be transformed in any way without some change of the relative position of these parts. In every transformation of potential energy, therefore, motion of some kind is involved.

Now we know that whenever one body of a system is hotter than another, heat is transferred from the hotter to the colder body, either by conduction or by radiation. Let us suppose that the transfer takes place by radiation. Whatever theory we adopt about the kind of motion which constitutes radiation, it is manifest that radiation consists of motion of some kind, either the projection of the particles of a substance called caloric across the intervening space, or a wave-like motion propagated through a medium filling that space. In either case, during the interval between the time when the heat leaves the hot body and the time when it reaches the cold body, its energy exists in the intervening space in the form of the motion of matter.

Hence, whether we consider the radiation of heat as effected by the projection of material caloric, or by the undulations of an intervening medium, the outer surface of a hot body must be in a state of motion, provided any cold body is in its neighbourhood to receive the radiations which it emits. But we have no reason to believe that the presence of a cold body is essential to the radiation of heat by a hot one. Whatever be the mode in which the hot body shoots forth its heat, it must depend on the state of the hot body alone, and not on the existence of a cold body at a distance, so that even if all the bodies in a closed region were equally hot, every one of them would be radiating heat; and the reason why each body remains of the same temperature is, that it receives from the other bodies exactly as much heat as it emits. This, in fact, is the foundation of Prevost's Theory of Exchanges. We must therefore admit that at every part of the surface of a hot body there is a radiation of heat, and therefore a state of motion of the superficial parts of the body. Now this motion is certainly invisible to us by any direct mode of observation, and therefore the mere fact of a body appearing to be at rest cannot be taken as a demonstration that its parts may not be in a state of motion.

Hence part, at least, of the energy of a hot body must be energy arising from the motion of its parts, or kinetic energy.

The conclusion at which we shall arrive, that a very considerable part of the energy of a hot body is in the form of motion, will become more evident when we consider the thermal energy of gases.

Every hot body, therefore, is in motion. We have next to enquire into the nature of this motion. It is evidently not a motion of the whole body in one direction, for however small we make the body by mechanical processes, each visible particle remains apparently in the same place, however hot it is. The motion which we call heat must therefore be a motion of parts too small to be observed separately; the motions of different parts at the same instant must be in different directions; and the motion of any one part must, at least in solid bodies, be such that, however fast it moves, it never reaches a sensible distance from the point from which it started.

We have now arrived at the conception of a body as consisting of a great many small parts, each of which is in motion. We shall call any one of these parts a molecule of the substance. A molecule may therefore be defined as a small mass of matter the parts of which do not part company during the excursions which the molecule makes when the body to which it belongs is hot.

The doctrine that visible bodies consist of a determinate number of molecules is called the molecular theory of matter. The opposite doctrine is that, however small the parts may be into which we divide a body, each part retains all the properties of the substance. This is the theory of the infinite divisibility of bodies. We do not assert that there is an absolute limit to the divisibility of matter: what we assert is, that after we have divided a body into a certain finite number of constituent parts called molecules, then any further division of these molecules will deprive them of the properties which give rise to the phenomena observed in the substance.

The opinion that the observed properties of visible bodies apparently at rest are due to the action of invisible molecules in rapid motion is to be found in Lucretius.

Daniel Bernoulli was the first to suggest that the pressure of air is due to the impact of its particles on the sides of the vessel containing it; but he made very little progress in the theory which he suggested.

Lesage and Prevost of Geneva, and afterwards Herapath in his 'Mathematical Physics,' made several important applications of the theory.

Dr. Joule in 1848 explained the pressure of gases by the impact of their molecules, and calculated the velocity which they must have to produce the observed pressure.

Krönig also directed attention to this explanation of the phenomena of gases.

It is to Professor Clausius, however, that we owe the recent development of the dynamical theory of gases. Since he took up the subject a great advance has been made by many enquirers. I shall now endeavour to give a sketch of the present state of the theory.

All bodies consist of a finite number of small parts called molecules. Every molecule consists of a definite quantity of matter, which is exactly the same for all the molecules of the same substance. The mode in which the molecule is bound together is also the same for all molecules of the same substance. A molecule may consist of several distinct portions of matter held together by chemical bonds, and may be set in vibration, rotation, or any other kind of relative motion, but so long as the different portions do not part company, but travel together in the excursions made by the molecule, our theory calls the whole connected mass a single molecule.

The molecules of all bodies are in a state of continual agitation. The hotter a body is, the more violently are its molecules agitated. In solid bodies, a molecule, though in continual motion, never gets beyond a certain very small distance from its original position in the body. The path which it describes is confined within a very small region of space.

In fluids, on the other hand, there is no such restriction to the excursions of a molecule. It is true that the molecule generally can travel but a very small distance before its path is disturbed by an encounter with some other molecule; but after this encounter there is nothing which determines the molecule rather to return towards the place from whence it came than to push its way into new regions. Hence in fluids the path of a molecule is not confined within a limited region, as in the case of solids, but may penetrate to any part of the space occupied by the fluid.

The actual phenomena of diffusion both in liquids and in gases furnish the strongest evidence that these bodies consist of molecules in a state of continual agitation.

But when we apply the methods of dynamics to the investigation of the properties of a system consisting of a great number of small bodies in motion the resemblance of such a system to a gaseous body becomes still more apparent.

I shall endeavour to give some account of what is known of such a system, avoiding all unnecessary mathematical calculations.

10.2.1   On the Kinetic Theory of Gases

A gaseous body is supposed to consist of a great number of molecules moving with great velocity. During the greater part of their course these molecules are not acted on by any sensible force, and therefore move in straight lines with uniform velocity. When two molecules come within a certain distance of each other, a mutual action takes place between them, which may be compared to the collision of two billiard balls. Each molecule has its course changed, and starts on a new path. I have concluded from some experiments of my own that the collision between two hard spherical balls is not an accurate representation of what takes place during the encounter of two molecules. A better representation of such an encounter will be obtained by supposing the molecules to act on one another in a more gradual manner, so that the action between them goes on for a finite time, during which the centres of the molecules first approach each other and then separate.

We shall refer to this mutual action as an Encounter between two molecules, and we shall call the course of a molecule between one encounter and another the Free Path of the molecule. In ordinary gases the free motion of a molecule takes up much more time than that occupied by an encounter. As the density of the gas increases, the free path diminishes, and in liquids no part of the course of a molecule can be spoken of as its free path.

In an encounter between two molecules we know that, since the force of the impact acts between the two bodies, the motion of the centre of gravity of the two molecules remains the same after the encounter as it was before. We also know by the principle of the conservation of energy that the velocity of each molecule relatively to the centre of gravity remains the same in magnitude, and is only changed in direction.

Let us next suppose a number of molecules in motion contained in a vessel whose sides are such that if any energy is communicated to the vessel by the encounters of molecules against its sides, the vessel communicates as much energy to other molecules during their encounters with it, so as to preserve the total energy of the enclosed system. The first thing we must notice about this moving system is that even if all the molecules have the same velocity originally, their encounters will produce an inequality of velocity, and that this distribution of velocity will go on continually. Every molecule will then change both its direction and its velocity at every encounter; and, as we are not supposed to keep a record of the exact particulars of every encounter, these changes of motion must appear to us very irregular if we follow the course of a single molecule. If, however, we adopt a statistical view of the system, and distribute the molecules into groups, according to the velocity with which at a given instant they happen to be moving, we shall observe a regularity of a new kind in the proportions of the whole number of molecules which fall into each of these groups.

And here I wish to point out that, in adopting this statistical method of considering the average number of groups of molecules selected according to their velocities, we have abandoned the strict kinetic method of tracing the exact circumstances of each individual molecule in all its encounters. It is therefore possible that we may arrive at results which, though they fairly represent the facts as long as we are supposed to deal with a gas in mass, would cease to be applicable if our faculties and instruments were so sharpened that we could detect and lay hold of each molecule and trace it through all its course.

For the same reason, a theory of the effects of education deduced from a study of the returns of registrars, in which no names of individuals are given, might be found not to be applicable to the experience of a schoolmaster who is able to trace the progress of each individual pupil.

The distribution of the molecules according to their velocities is found to be of exactly the same mathematical form as the distribution of observations according to the magnitude of their errors, as described in the theory of errors of observation. The distribution of bullet-holes in a target according to their distances from the point aimed at is found to be of the same form, provided a great many shots are fired by persons of the same degree of skill.

We have already met with the same form in the case of heat diffused from a hot stratum by conduction. Whenever in physical phenomena some cause exists over which we have no control, and which produces a scattering of the particles of matter, a deviation of observations from the truth, or a diffusion of velocity or of heat, mathematical expressions of this exponential form are sure to make their appearance.

It appears then that of the molecules composing the system some are moving very slowly, a very few are moving with enormous velocities, and the greater number with intermediate velocities. To compare one such system with another, the best method is to take the mean of the squares of all the velocities. This quantity is called the Mean Square of the velocity. The square root of this quantity is called the Velocity of Mean Square.

10.2.2 Distribution of Kinetic Energy Between Two Different Sets of Molecules

If two sets of molecules whose mass is different are in motion in the same vessel, they will by their encounters exchange energy with each other till the average kinetic energy of a single molecule of either set is the same. This follows from the same investigation which determines the law of distribution of velocities in a single set of molecules.

Hence if the mass of a molecule of one kind is M_1 and that of a molecule of the other kind is M_2, and if their average velocities of agitation are V_1 and V_2, then

$$M_1 V_1^2 = M_2 V_2^2 \qquad (10.1)$$

The quantity $\frac{1}{2}mv^2$ is called the average kinetic energy of agitation of a single molecule. We shall return to this result when we come to Gay-Lussac's Law of the Volumes of Gases.

10.2.3 Internal Kinetic Energy of a Molecule

If a molecule were a mathematical point endowed with inertia and with attractive and repulsive forces, the only kinetic energy it could possess is that of translation as a whole. But if it be a body having parts and magnitude, these parts may have motions of rotation or of vibration relative to each other, independent of the motion of the centre of gravity of the molecule. We must therefore admit that part of the kinetic energy of a molecule may depend on the relative motions of its parts. We call this the Internal energy, to distinguish it from the energy due to the translation of the molecule as a whole. The ratio of the internal energy to the energy of agitation may be different in different gases.

10.2.4   Definition of the Velocity of a Gas

It is evident that if a gas consists of a great number of molecules moving about in all directions we cannot identify the velocity of any one of these molecules with what we are accustomed to consider as the velocity of the gas itself. Let us consider the case of a gas which has remained in a fixed vessel for a sufficient time to arrive at the normal distribution of velocities. This gas, according to the ordinary notions, is at rest, though the molecules of which it is composed may be flying about in all directions.

Now consider any plane area of an imaginary surface described within the vessel. This surface does not interfere with the motion of the molecules. Some molecules pass through the surface in one direction, and others in the opposite direction; but it is evident, since the gas does not tend to accumulate on one side rather than on the other, that exactly the same number of molecules pass in the one direction as in the other. If, therefore, a gas is at rest, as many molecules pass through a fixed surface in the one direction as in the other in the same time.

It is evident that if the vessel, instead of being at rest, had been in a state of uniform motion, an equal number of molecules would pass in both directions through any surface fixed with respect to the vessel. Hence we find that if a gas is in motion, and if the velocity of a surface coincides in direction and magnitude with that of the gas, the same number of molecules will pass through that surface in the positive direction as in the negative.

This leads to the following definition of the velocity of a gas:

> If we determine the motion of the centre of gravity of all the molecules within a very small region surrounding a point in a gas, then the velocity of the gas within that region is defined as the velocity of the centre of gravity of all the molecules within that region.

This is what is meant by the motion of a gas in common language. Besides this motion, there are two other kinds of motion considered in the kinetic theory of gases. The first is the motion of agitation of the molecules. This is the hitherto invisible motion of the molecule considered as a whole. Its course consists of broken portions, called free paths, interrupted by the encounters between different molecules.

The second is the internal motion of each molecule, consisting partly of rotation and partly of vibrations among the component parts of the molecule.

The velocity of the centre of gravity of a molecule is the resultant of the velocity of the gas and the velocity of agitation of the individual molecule at the given instant. The velocity of a constituent part of a molecule is the resultant of the velocity of its centre of gravity and the velocity of the constituent part relatively to the centre of gravity of the molecule.

Fig. 10.1 A plane surface
separating two regions of a
gas—[*K.K.*]

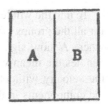

10.2.5   Theory of the Pressure of a Gas

Let us consider two portions of a gas separated by a plane surface which moves with
the same velocity as the gas. We have seen that in this case the number of molecules
which pass through the plane in opposite directions is the same.

Each molecule in crossing the plane from the region A to the region B enters the
second region in precisely the same state as it leaves the first. It therefore carries
over into the region B, not only its mass, but its momentum and its kinetic energy.
Hence, if we consider the quantity of momentum in a given direction existing at
any instant in the particles in the region B, this quantity will be altered whenever a
molecule crosses the boundary, carrying its momentum along with it (Fig. 10.1).

Now let us consider all the molecules whose velocity differs by less than a certain
quantity, c, from a given velocity the components of which are u in the direction
perpendicular to the plane from A towards B, and v and w in two other directions
parallel to the plane. Let there be N molecules whose velocity is within these limits
in every unit of volume, and let the mass of each of these be M.

Then the number of these molecules which will cross unit of area of the plane
from A to B in unit of time is

$$Nu$$

The momentum of each of these molecules resolved in the direction AB is Mu.

Hence the momentum in this direction communicated to the region B in unit of
time is

$$MNu^2.$$

Since this bombardment of the region B does not produce motion of the gas, a
pressure must be exerted on the gas by the sides of the vessel, and the amount of
this pressure for every unit of area must be MNu^2.

The region A loses positive momentum at the same rate, and in order to preserve
equilibrium there must be a pressure equal to MNu^2 on every unit of area of the
surface of the region A.

Hitherto we have considered only one group of molecules, whose velocities lie
between given limits. In every such group that which determines the pressure in the
direction AB on the surface separating A from B is a quantity of the form MNu^2,
where N is the number of molecules in the group, and u is the velocity of each
molecule resolved in the direction AB. The other components of the velocity do not
influence the pressure in this direction.

To find the whole pressure, we must find the sum of all such expressions as MNu^2 for all the groups of molecules in the system. We may write this result $p = MN\bar{u}^2$, where N now signifies the total number of molecules in unit of volume, and $\bar{u}^2$ denotes the mean value of u^2 for all these molecules. Now if V^2 is the square of the velocity without regard to direction, $V^2 = u^2 + v^2 + w^2$, where u v w are the components in three directions at right angles. Hence if $\bar{u}^2$, $\bar{v}^2$, $\bar{w}^2$ denote the mean square of these components, and $\bar{V}^2$ the mean square of the resultant, $\bar{V}^2 = \bar{u}^2 + \bar{v}^2 + \bar{w}^2$. When, as in every gas at rest, the pressure is equal in all directions, $\bar{u}^2 = \bar{v}^2 = \bar{w}^2$, and therefore $\bar{V}^2 = 3\bar{u}^2$.

Hence the pressure of a gas is

$$p = \frac{1}{3}MN\bar{V}^2 \tag{10.2}$$

where M is the mass of each molecule, N is the number of molecules in unit of volume and $\bar{V}^2$ is the mean square of the velocity.

In this expression there are two quantities which have never been directly measured—the mass of a single molecule, and the number of molecules in unit of volume. But we have here to do with the product of these quantities, which is evidently the mass of the substance in unit of volume, or in other words, its density. Hence we may write the expression

$$p = \frac{1}{3}\rho\bar{V}^2 \tag{10.3}$$

where ρ is the density of the gas.

It is easy from this expression to determine, as was first done by Joule, the mean square of the velocity of the molecules of a gas, for

$$\bar{V}^2 = 3\frac{p}{\rho} \tag{10.4}$$

where p is the pressure, and ρ the density, which must of course be expressed in terms of the same fundamental units.

For instance, under the atmospheric pressure of 2116.4 pounds weight on the square foot, and at the temperature of melting ice, the density of hydrogen is 0.005592 pounds in a cubic foot. Hence $\frac{p}{\rho} = 378816$ in gravitation units, and if the intensity of gravity where this relation was observed was 32.2, we have

$$\bar{V}^2 = 36593916,$$

or, taking the square root of this quantity,

$$\bar{V} = 6097 \text{ ft/s}.$$

This is the velocity of mean square for the molecules of hydrogen at 32 °F. and at the atmospheric pressure.

10.2.6 Law of Boyle

Two bodies are said to be of the same temperature when there is no more tendency for heat to pass from the first to the second than in the reverse direction. In the kinetic theory of heat, as we have seen, this thermal equilibrium is established when there is a certain relation between the velocities of agitation of the molecules of the two bodies. Hence the temperature of a gas must depend on the velocity of agitation of its molecules, and this velocity must be the same at the same temperature, whatever be the density.

In the expression $p = \frac{1}{3}\rho \bar{V}^2$, the quantity $\bar{V}^2$ depends only on the temperature as long as the gas remains the same. Hence when the density ρ varies, the pressure p must vary in the same proportion. This is Boyle's law, which is now raised from the rank of an experimental fact to that of a deduction from the kinetic theory of gases.

If v denotes the volume of unit of mass, we may write this expression

$$pv = \frac{1}{3}\bar{V}^2 \tag{10.5}$$

Now pv is proportional to the absolute temperature, as measured by a thermometer, of the particular gas under consideration. Hence $\bar{V}^2$, the mean square of the velocity of agitation, is proportional to the absolute temperature measured in this way.

10.2.7 Law of Gay-Lussac

Let us next consider two different gases in thermal equilibrium. We have already stated that if M_1 M_2 are the masses of individual molecules of these gases, and V_1 V_2 their respective velocities of agitation, it is necessary for thermal equilibrium that $M_1 \bar{V}_1^2 = M_2 \bar{V}_2^2$ by Eq. (10.1).

If the pressures of these gases are p_1 and p_2, and the number of molecules in unit of volume N_1 and N_2, then, by Eq. (10.2),

$$p_1 = \frac{1}{3}M_1 N_1 \bar{V}_1^2 \quad \text{and} \quad p_2 = \frac{1}{3}M_2 N_2 \bar{V}_2^2$$

If the pressures of the two gases are equal,

$$M_1 N_1 \bar{V}_1^2 = M_2 N_2 \bar{V}_2^2.$$

If their temperatures are equal,

$$M_1 \bar{V}_1^2 = M_2 \bar{V}_2^2.$$

Dividing the terms of the first of these equations by those of the second, we find

$$N_1 = N_2 \tag{10.6}$$

or *when two gases are at the same pressure and temperature, the number of molecules in unit of volume is the same in both gases.*

If we put $\rho_1 = M_1 N_1$ and $\rho_2 = M_2 N_2$ for the densities of the two gases, then, since $N_1 = N_2$, we get

$$\rho_1 : \rho_2 :: M_1 : M_2 \tag{10.7}$$

or *the densities of two gases at the same temperature and pressure are proportional to the masses of their individual molecules.*

These two equivalent propositions are the expression of a very important law established by Gay-Lussac, that the densities of gases are proportional to their molecular weights.

The proportion by weight in which different substances combine to form chemical compounds depends, according to Dalton's atomic theory, on the weights of their molecules, and it is one of the most important researches in chemistry to determine the proportions of the weights of the molecules from the proportions in which they enter into combination. Gay-Lussac discovered that in the case of gases the volumes of the combining quantities of different gases always stand in a simple ratio to each other. This law of volumes has now been raised from the rank of an empirical fact to that of a deduction from our theory, and we may now assert, as a dynamical proposition, that the weights of the molecules of gases (that is, those small portions which do not part company during their motion) are proportional to the densities of these gases at standard temperature and pressure.

10.3   Study Questions

QUES. 10.1. What is a molecule? Was Maxwell the first to assert that invisible molecules give rise to the observed properties of visible bodies? Does Maxwell believe that molecules (or perhaps atoms) are indivisible? And is there any plausible alternative to the molecular theory?

QUES. 10.2. How do the motions of molecules in solids, liquids and gases differ? What is meant by the *free path* of a molecule, and upon what does it depend?

QUES. 10.3. Do all the molecules of a gas travel at the same speed? If not, then how are they distributed? How are the average agitation velocities of molecules having different masses in a mixed gas related to one another?

QUES. 10.4. Is all of the kinetic energy of a complex molecule due to center of mass agitation? If not, then what other type of motion do molecules exhibit?

QUES. 10.5. Why does a confined gas exert pressure on the walls of its container? How is the gas pressure related to the average velocity of its constituent molecules?

QUES. 10.6. What is the definition of thermal equilibrium, and how is this stated according to the kinetic theory?

QUES. 10.7. When an ideal gas is compressed isothermally, what happens to its pressure? Is this relationship (Boyle's law) simply an empirical fact, or can it be derived from the kinetic theory?

QUES. 10.8. When a gas comprised of light molecules (say, helium) and another gas comprised of heavier molecules (say, oxygen) are held in separate chambers having the same pressure, temperature and volume, which chamber has a larger number of molecules? Which gas has the larger density? Is this (the law of Gay-Lussac) simply an empirical fact, or can it be derived from kinetic theory?

10.4 Exercises

EX. 10.1 (STATISTICAL METHODS ESSAY). What are the relative strengths and weaknesses of the kinetic and statistical methods of analysis?

EX. 10.2 (PRESSURE, TEMPERATURE AND KINETIC THEORY). A beam of hydrogen molecules (H_2) is aimed directly at a rigid wall. Each molecule in the beam has an original speed of 1860 m/s.

a) If the beam strikes the wall at a rate of 10^{23} molecules per second, and the molecules reflect perfectly elastically from the wall, then what is the force exerted by the beam on the wall? Is this a sensible force? (ANSWER: 1.2 N)
b) Is the temperature of a gas confined in a perfectly rigid chamber constant? If the wall is allowed to slowly recoil backwards as a result of the collisions, then what happens to the speed of the recoiling molecules? What does this imply about the temperature of a gas confined in a slowly expanding chamber?
c) Finally, if the wall is instead suddenly removed, so that the gas molecules no longer strike the wall at all, what happens to the speed of the molecules? What does this imply about the temperature of a gas confined in a chamber whose wall is suddenly removed, allowing it to expand into a vacuum?[3]

EX. 10.3 (ISOTOPE SPEEDS). What are the relative mean speeds of the helium-3 and helium-4 isotopes in air at standard temperature and pressure?

EX. 10.4 (INFLATED LUNGS). Assuming the air in your lungs is at atmospheric pressure, what is the average speed of the oxygen molecules inside your lungs? Based on this computed speed, about how often do the oxygen molecules collide with the interior walls of the alveoli of your lungs? (HINT: Fully inflated alveoli have a radius of about 100 microns.)

---

[3] This process is referred to as Joule-Thomson expansion.

Ex. 10.5 (MEAN FREE PATH OF HELIUM ATOMS). The *mean free path* is defined as
the average distance a particle travels before undergoing a collision with another
one. As an exercise, consider N spherical marbles of diameter d sealed in a cubical
chamber of volume V which is vigorously shaken.

a) Find a mathematical expression for the mean free path of a marble. (Hint:
   You might approach this problem by considering how much volume of space
   is "swept out", so to speak, by a moving marble when it travels an arbitrary
   distance, L, and how many other particles occupy this volume.)
b) If $d = 1$ cm, $N = 80$ and $V = 1$ l, what is the mean free path of a marble?
   At what value of N would the mean free path exceed the width of the chamber
   itself? (Answer: 4 cm)
c) Suppose that a gas of helium atoms (instead of marbles) at standard temperature
   and pressure is confined in the chamber. What is the mean free path of a helium
   atom? How about in the high-vacuum regime, when the pressure is only 1×10^{-6} atm?

Ex. 10.6 (MOLECULES ESSAY). Do you believe in molecules? If so, on what
experimental or logical grounds?

10.5   Vocabulary

1. Endeavour
2. Devoid
3. Animate
4. Manifest
5. Repose
6. Cogent
7. Antagonism
8. Superficial
9. Enquire
10. Endow

Chapter 11
Molecules and Maxwell's Demon

> *The exact equality of each molecule to all others of the same kind gives it, as Sir John Herschel has well said, the essential character of a manufactured article.*
>
> —James Clerk Maxwell

11.1 Introduction

In the previous reading selection, from the first half of Chap. 22 of his *Theory of Heat*, Maxwell explained how—according to the kinetic theory—the pressure of a gas may be understood as arising from countless collisions of molecules with the walls of the chamber which contains the gas. Most importantly, the detailed predictions of the kinetic theory (such as the relationship between the pressure, temperature and volume of a gas) agree with the empirically determined gas laws (those of Boyle and Gay-Lussac). Such agreement lends support to the molecular theory of matter. The molecular theory was controversial even at the time of Maxwell. The idea that visible matter is comprised of tiny invisible objects is, of course, not new. The doctrine of *atomism* was asserted—and debated vigorously—by the ancients. Indeed, the word *atom* itself comes from the Greek verb τεμνειν, meaning "to cut." From it is derived the term α-τομος, meaning that which is "un-cuttable."[1]

In the reading selection that follows, from the second half of Chap. 22, Maxwell develops the kinetic theory in more detail, explaining how it may be used to understand material properties (such as the viscosity and specific heat capacity of a gas) and natural processes (such as diffusion, evaporation and electrolysis). Perhaps most interestingly, near the end of this·chapter, he introduces what is now referred to as "Maxwell's Demon," a tiny intelligent being which seemingly has the ability to violate the second law of thermodynamics. Does it? Also, in the final part of this chapter, Maxwell considers the broader implications of accepting the molecular theory of matter. To what surprising conclusion is he led? Do you agree with Maxwell's conclusion?

---

[1] A very interesting and comprehensive treatment of the concept of the atom since the time of the ancient greeks is given in Melsen, A. G., *From Atomos to Atom*, Duquesne University Press, Pittsburg, 1952.

© Springer International Publishing Switzerland 2016
K. Kuehn, *A Student's Guide Through the Great Physics Texts*,
Undergraduate Lecture Notes in Physics, DOI 10.1007/978-3-319-21828-1_11

11.2   Reading: Maxwell, *on the Molecular Theory of the Constitution of Bodies*

Maxwell, J. C., *Theory of Heat*, tenth ed., Longmans, Green, and Co., London and New York, 1891. Chap. XXII.

11.2.1   Law of Charles

We must next consider the effect of changes of temperature on different gases. Since at all temperatures, when there is thermal equilibrium,

$$M_1 \bar{V_1}^2 = M_2 \bar{V_2}^2$$

and since the absolute temperature, as measured by a gas thermometer, is proportional to $\bar{V_1}^2$ when the gas is of the first kind, and to $\bar{V_2}^2$ when the gas is of the second kind; it follows, since $\bar{V_1}^2$ is itself proportional to $\bar{V_2}^2$, that the absolute temperatures, as measured by the two thermometers, are proportional, and if they agree at any one temperature (as the freezing point), they agree throughout. This is the law of the equal dilatation of gases discovered by Charles.

11.2.2   Kinetic Energy of a Molecule

The mean kinetic energy of agitation of a molecule is the product of its mass by half the mean square of its velocity, or

$$\frac{1}{2}M\bar{V}^2.$$

This is the energy due to the motion of the molecule as a whole, but its parts may be in a state of relative motion. If we assume, with Clausius, that the energy due to this internal motion of the parts of the molecule tends towards a value having a constant ratio to the energy of agitation, the whole energy will be proportional to the energy of agitation, and may be written

$$\frac{1}{2}\beta M\bar{V}^2.$$

where β is a factor, always greater than unity, and probably equal to 1.634 for air and several of the more perfect gases. For steam it may be as much as 2.19, but this is very uncertain.

To find the kinetic energy of the substance contained in unit of volume, we have only to multiply by the number of molecules, and we obtain

$$T = \frac{1}{2}\beta M N \bar{V}^2. \tag{11.1}$$

Comparing this with the Eq. (10.2) which determines the pressure, we get

$$T = \frac{3}{2}\beta p \tag{11.2}$$

or the energy in unit of volume is numerically equal to the pressure on unit of area multiplied by $\frac{3}{2}\beta$.

The energy in unit of mass is found by multiplying this by v, the volume of unit of mass:

$$T_m = \frac{3}{2}\beta p v \tag{11.3}$$

11.2.3   Specific Heat at Constant Volume

Since the product pv is proportional to the absolute temperature, the energy is proportional to the temperature.

The specific heat is measured dynamically by the increase of energy corresponding to a rise of one degree of temperature. Hence

$$K_v = \frac{3}{2}\beta\frac{pv}{\theta} \tag{11.4}$$

To express the specific heat in ordinary thermal units, we must divide this by J, the specific heat of water (Joule's equivalent). It follows from this expression that for any one gas the specific heat of unit of mass at constant volume is the same for all pressures and temperatures, because $\frac{pv}{\theta}$ remains constant. For different gases the specific heat at constant volume is inversely proportional to the specific gravity, and directly proportional to β.

Since β is nearly the same for several gases, the specific heat of these gases is inversely proportional to their specific gravity referred to air, or, since the specific gravity is proportional to their molecular weight, the specific heat multiplied by the molecular weight is the same for all these gases. This is the law of Dulong and Petit. It would be accurate for all gases if the value of β were the same in every case.

It has been shown at p. 183 that the difference of the two specific heats[2] is $\frac{pv}{\theta}$. Hence their ratio, γ, is

$$\gamma = \frac{2}{3\beta} + 1 \quad \text{and} \quad \beta = \frac{2}{3}\frac{1}{\gamma - 1}$$

---

[2] Maxwell here refers to page 183 in Chap. XI in his *Theory of Heat*, where he shows that the difference between the specific heats of a gas at constant pressure and at constant temperature is given by $c_p - c_v = R$, the ideal gas constant.

If U is the velocity of sound in a gas, we have, as at p. 228,

$$U^2 = \gamma p v \qquad (11.5)$$

The mean square of the velocity of agitation is

$$\bar{V}^2 = 3 p v \qquad (11.6)$$

Hence $U = \sqrt{\frac{\gamma}{3}}\, V$, or, if $\gamma = 1.408$, as in air and several other gases,

$$U = 0.6858\ V \qquad \text{or} \qquad V = 1.458\ U \qquad (11.7)$$

These are the relations between the velocity of sound and the velocity of mean square of agitation in any gas for which $\gamma = 1.408$.

The nature of this book admits only of a brief account of some other results of the kinetic theory of gases. Two of these are independent of the nature of the action between the molecules during their encounters.

The first of these relates to the equilibrium of a mixture of gases acted on by gravity. The result of our theory is that the final distribution of any number of kinds of gas in a vertical vessel is such that the density of each gas at a given height is the same as if all the other gases had been removed, leaving it alone in the vessel.

This is exactly the mode of distribution which Dalton supposed to exist in a mixed atmosphere in equilibrium, the law of diminution of density of each constituent gas being the same as if no other gases were present.

In our atmosphere the continual disturbances caused by winds carry portions of the mixed gases from one stratum to another, so that the proportion of oxygen and nitrogen at different heights is much more uniform than if these gases had been allowed to take their places by diffusion during a dead calm.

The second result of our theory relates to the thermal equilibrium of a vertical column. We find that if a vertical column of a gas were left to itself; till by the conduction of heat it had attained a condition of thermal equilibrium, the temperature would be the same throughout, or, in other words, gravity produces no effect in making the bottom of the column hotter or colder than the top.

This result is important in the theory of thermodynamics, for it proves that gravity has no influence in altering the conditions of thermal equilibrium in any substance, whether gaseous or not. For if two vertical columns of different substances stand on the same perfectly conducting horizontal plate, the temperature of the bottom of each column will be the same; and if each column is in thermal equilibrium of itself; the temperatures at all equal heights must be the same. In fact, if the temperatures of the tops of the two columns were different, we might drive an engine with this difference of temperature, and the refuse heat would pass down the colder column, through the conducting plate, and up the warmer column; and this would go on till all the heat was converted into work, contrary to the second law of thermodynamics.

But we know that if one of the columns is gaseous, its temperature is uniform. Hence that of the other must be uniform, whatever its material.

This result is by no means applicable to the case of our atmosphere. Setting aside the enormous direct effect of the sun's radiation in disturbing thermal equilibrium, the effect of winds in carrying large masses of air from one height to another tends to produce a distribution of temperature of a quite different kind, the temperature at any height being such that a mass of air, brought from one height to another without gaining or losing heat, would always find itself at the temperature of the surrounding air. In this condition of what Sir William Thomson has called the Convective equilibrium of heat, it is not the temperature which is constant, but the quantity ϕ, which determines the adiabatic curves.

In the convective equilibrium of temperature, the absolute temperature is proportional to the pressure raised to the power $\frac{\gamma-1}{\gamma}$, or 0.29.

The extreme slowness of the conduction of heat in air, compared with the rapidity with which large masses of air are carried from one height to another by the winds, causes the temperature of the different strata of the atmosphere to depend far more on this condition of convective equilibrium than on true thermal equilibrium.

We now proceed to those phenomena of gases which, according to the kinetic theory, depend upon the particular nature of the action which takes place when the molecules encounter each other, and on the frequency of these encounters.

There are three phenomena of this kind of which the kinetic theory takes account—the diffusion of gases, the viscosity of gases, and the conduction of heat through a gas.

We have already described the known facts about the interdiffusion of two different gases. It is only when the gases are chemically different that we can trace the process of diffusion, but on the molecular theory diffusion is always going on, even in a single gas; only it is impossible to trace the progress of the molecules, because we cannot tell one from another.

The relation between diffusion and viscosity may be explained as follows: Consider the case of motion of a mass of gas, which has already been described in Chap. XXI., in which the different horizontal layers of the gas slide over each other. In diffusion the molecules pass, some of them upwards and some of them downwards, through any horizontal plane. If the medium has different properties of any kind above and below this plane, then this interchange of molecules will tend to assimilate the properties of the two portions of the medium.

In the case of ordinary diffusion, the proportions of the two diffusing substances are different above and below, and vary in the different horizontal layers according to their height. In the case of internal friction, the mean horizontal momentum is different in the different layers, and when the molecules pass through the plane, carrying their momentum with them, this exchange of momentum between the upper and lower parts of the medium constitutes a force tending to equalize their velocity, and this is the phenomenon actually observed in the motion of viscous fluids.

The coefficient of viscosity, when measured in the kinematic way, represents the rate at which the equalization of velocity goes on by the exchange of the momentum of the molecules, just as the coefficient of diffusion represents the rate at which the equalization of chemical composition goes on by the exchange of the molecules themselves.

It appears from the kinetic theory of gases that if D is the coefficient of diffusion of the gas *into itself*, and ν the viscosity measured kinematically,

$$\nu = 0.6479 \ D \tag{11.8}$$

$$D = 1.5435 \ \nu \tag{11.9}$$

The conduction of heat in a gas, according to the kinetic theory, is simply the diffusion of the energy of the molecules by their moving about in the medium and carrying their energy with them till they encounter other molecules, when the energy is redistributed. The relation of the conductivity κ, measured thermometrically, to the viscosity ν, measured kinematically, is

$$\kappa = \frac{5}{3\gamma}\nu \tag{11.10}$$

It appears, therefore, that diffusion, viscosity, and conductivity in gases are related to each other in a very simple way, being the rate of equalization of three properties of the medium—the proportion of its ingredients, its velocity, and its temperature. The equalization is effected by the same agency in each case—namely, the agitation of the molecules. In each case, if the density remains the same, the rate of equalization is proportional to the absolute temperature; and if the temperature remains the same, the rate of equalization is inversely proportional to the density. Hence, if we consider the temperature and the pressure as defining the state of the gas, the quantities D, ν, and κ vary directly as the square of the absolute temperature and inversely as the pressure.

11.2.4   Molecular Theory of Evaporation and Condensation

The mathematical difficulties arising in the investigation of the motions of molecules are so great that it is not to be wondered at that most of the numerical results are confined to the phenomena of gases. The general character, however, of the explanation of many other phenomena by the molecular theory has been pointed out by Clausius and others.

We have seen that in the case of a gas some of the molecules have a much greater velocity than others, so that it is only to the average velocity of all the molecules that we can ascribe a definite value. It is probable that this is also true of the motions of the molecules of a liquid, so that, though the average velocity may be much smaller than in the vapour of that liquid, some of the molecules in the liquid may have velocities equal to or greater than the average velocity in the vapour. If any of the molecules at the surface of the liquid have such velocities, and if they are moving from the liquid, they will escape from those forces which retain the other molecules as constituents of the liquid, and will fly about as vapour in the space outside the liquid. This is the molecular theory of evaporation. At the same time, a molecule of the vapour striking the liquid may become entangled among the molecules of the liquid, and may thus become part of the liquid. This is the molecular explanation of condensation. The number of molecules which pass from the liquid to the vapour depends on the temperature of the liquid. The number of molecules which pass from the vapour to the liquid depends upon the density of the vapour as well as its temperature. If the temperature of the vapour is the same as that of the liquid, evaporation will take place as long as more molecules are evaporated than condensed; but when the density of the vapour has increased to such a value that as many molecules are condensed as evaporated, then the vapour has attained its maximum density. It is then said to be saturated, and it is commonly supposed that evaporation ceases. According to the molecular theory, however, evaporation is still going on as fast as ever; only, condensation is also going on at an equal rate, since the proportions of liquid and of gas remain unchanged.

A similar explanation applies to cases in which the vapour or gas is absorbed by a liquid of a different kind, as when oxygen or carbonic acid is absorbed by water or alcohol. In such cases a 'movable equilibrium' is attained when the liquid has absorbed a quantity of the gas whose volume at the density of the unabsorbed gas is a certain multiple or fraction of the volume of the liquid; or, in other words, the density of the gas in the liquid and outside the liquid stand in a certain numerical ratio to each other. This subject is treated very fully in Bunsen's 'Gasometry.'

The amount of vapour of a liquid diffused into a gas of a different kind is generally independent of the nature of the gas, except when the gas acts chemically on the vapour.

Dr. Andrews has shown ('Proc. R.S.' 1875) that by mixing nitrogen with carbonic acid, the critical temperature is lowered, and that Dalton's law of the density of mixed vapours only holds at low pressures and at temperatures greatly above their critical points.

11.2.5   Molecular Theory of Electrolysis

A very interesting part of molecular science which has not been thoroughly worked out, but which hardly belongs to a treatise on Heat, is the theory of electrolysis. Here an electromotive force acting on a liquid electrolyte causes the molecules of one of its components to be urged in one direction, while those of the other component are urged in the opposite direction. Now these components are joined together in pairs by chemical forces of great power, so that we might expect that no electrolytic effect could take place unless the electromotive force were so strong as to be able to tear these couples asunder. But, according to Clausius, in the dance of molecules which is always going on, some of the linked pairs of molecules acquire such velocities that when they have an encounter with a pair also in violent motion the molecules composing one or both of the pairs are torn asunder, and wander about seeking new partners. If the temperature is so high that the general agitation is so violent that more pairs of molecules are torn asunder than can pair again in an equal time, we have the phenomenon of Dissociation, studied by M. Ste.-Claire Deville. If, on the other hand, the separated molecules can always find partners before they are ejected from the system, the composition of the system remains apparently the same.

Now Professor Clausius considers that it is during these temporary separations that the electromotive force comes into play as a directing power, causing the molecules of one component to move on the whole one way, and those of the other the opposite way. Thus the component molecules are always changing partners, even when no electromotive force is in action, and the only effect of this force is to give direction to those movements which are already going on.

Professor Wiedemann,[3] who has also taken this view of electrolysis, compares the phenomenon with that of diffusion, and shows that the electric conductivity of an electrolyte may be considered as depending on the coefficient of diffusion of the components through each other.

11.2.6 Molecular Theory of Radiation

The phenomena already described are explained on the molecular theory by the motion of agitation of the molecules, a motion which is exceedingly irregular, the intervals between successive encounters and the velocities of a molecule during successive free paths not being subject to any law which we can express. The internal motion of a single molecule is of a very different kind. If the parts of the molecule are capable of relative motion without being altogether torn asunder, this relative motion will be some kind of vibration. The small vibrations of a connected system may be resolved into a number of simple vibrations, the law of each of which is similar to that of a pendulum. It is probable that in gases the molecules may execute many of such vibrations in the interval between successive encounters. At each encounter the whole molecule is roughly shaken. During its free path it vibrates according to its own laws, the amplitudes of the different simple vibrations being determined by the nature of the collision, but their periods depending only on the constitution of the molecule itself. If the molecule is capable of communicating these vibrations to the medium in which radiations are propagated, it will send forth radiations of certain definite kinds, and if these belong to the luminous part of the spectrum, they will be visible as light of definite refrangibility. This, then, is the explanation, or the molecular theory, of the bright lines observed in the spectra of incandescent gases. They represent the disturbance communicated to the luminiferous medium by molecules vibrating in a regular and periodic manner during their

---

[3] Gustav Heinrich Wiedemann (1826–1899) was a physicist, writer and editor of scientific periodicals. He carried out significant work on the thermal conductivity of metals and was one of the founders of the Berlin Physical Society.

free paths. If the free path is long, the molecule, by communicating its vibrations to the ether, will cease to vibrate till it encounters some other molecule.

By raising the temperature we increase the velocity of the motion of agitation and the force of each encounter. The higher the temperature the greater will be the amplitude of the internal vibrations of all kinds, and the more likelihood will there be that vibrations of short period will be excited, as well as those fundamental vibrations which are most easily produced. By increasing the density we diminish the length of the free path of each molecule, and thus allow less time for the vibrations excited at each encounter to subside, and, since each fresh encounter disturbs the regularity of the series of vibrations, the radiation will no longer be capable of complete resolution into a series of vibrations of regular periods, but will be analysed into a spectrum showing the bright bands due to the regular vibrations, along with a ground of diffused light, forming a continuous spectrum due to the irregular motion introduced at each encounter.

Hence when a gas is rare the bright lines of its spectrum are narrow and distinct, and the spaces between them are dark. As the density of the gas increases, the bright lines become broader and the spaces between them more luminous.

There is another reason for the broadening of the bright lines and the luminosity of the whole spectrum in dense gases, which we have already stated at p. 172.[4] There is this difference, however, between the effect there mentioned and that described here. At p. 172 the light from a certain stratum of incandescent gas was supposed to penetrate through other strata, which absorbed the brighter rays faster than the less luminous ones. This effect depends only on the total quantity of gas through which the rays pass, and will be the same whether it is a mile of gas at thirty inches pressure, or thirty miles at one inch pressure. The effect which we are now considering depends on the absolute density, so that it is by no means the same whether a stratum containing a given quantity of gas is 1 or 30 miles thick.

When the gas is so far condensed that it assumes the liquid or solid form, then, as the molecules have no free path, they have no regular vibrations, and no bright lines are commonly observed in incandescent liquids or solids. Mr. Huggins, however, has observed bright lines in the spectrum of incandescent erbia and lime, which appear to be due to the solid matter, and not to its vapour.

11.2.7   Limitation of the Second Law of Thermodynamics

Before I conclude, I wish to direct attention to an aspect of the molecular theory which deserves consideration.

One of the best established facts in thermodynamics is that it is impossible in a system enclosed in an envelope which permits neither change of volume nor passage of heat, and in which both the temperature and the pressure are everywhere the same,

---

[4] See Chap. XVI of Maxwell's *Theory of Heat*, included in Chap. 13 of the present volume.

to produce any inequality of temperature or of pressure without the expenditure of work. This is the second law of thermodynamics, and it is undoubtedly true as long as we can deal with bodies only in mass, and have no power of perceiving or handling the separate molecules of which they are made up. But if we conceive a being whose faculties are so sharpened that he can follow every molecule in its course, such a being, whose attributes are still as essentially finite as our own, would be able to do what is at present impossible to us. For we have seen that the molecules in a vessel full of air at uniform temperature are moving with velocities by no means uniform, though the mean velocity of any great number of them, arbitrarily selected, is almost exactly uniform. Now let us suppose that such a vessel is divided into two portions, A and B, by a division in which there is a small hole, and that a being, who can see the individual molecules, opens and closes this hole, so as to allow only the swifter molecules to pass from A to B, and only the slower ones to pass from B to A. He will thus, without expenditure of work, raise the temperature of B and lower that of A, in contradiction to the second law of thermodynamics.

This is only one of the instances in which conclusions which we have drawn from our experience of bodies consisting of an immense number of molecules may be found not to be applicable to the more delicate observations and experiments which we may suppose made by one who can perceive and handle the individual molecules which we deal with only in large masses.

In dealing with masses of matter, while we do not perceive the individual molecules, we are compelled to adopt what I have described as the statistical method of calculation, and to abandon the strict dynamical method, in which we follow every motion by the calculus.

It would be interesting to enquire how far those ideas about the nature and methods of science which have been derived from examples of scientific investigation in which the dynamical method is followed are applicable to our actual knowledge of concrete things, which, as we have seen, is of an essentially statistical nature, because no one has yet discovered any practical method of tracing the path of a molecule, or of identifying it at different times.

I do not think, however, that the perfect identity which we observe between different portions of the same kind of matter can be explained on the statistical principle of the stability of the averages of large numbers of quantities each of which may differ from the mean. For if of the molecules of some substance such as hydrogen, some were of sensibly greater mass than others, we have the means of producing a separation between molecules of different masses, and in this way we should be able to produce two kinds of hydrogen, one of which would be somewhat denser than the other. As this cannot be done, we must admit that the equality which we assert to exist between the molecules of hydrogen applies to each individual molecule, and not merely to the average of groups of millions of molecules.

11.2.8   Nature and Origin of Molecules

We have thus been led by our study of visible things to a theory that they are made up of a finite number of parts or molecules, each of which has a definite mass, and possesses other properties. The molecules of the same substance are all exactly alike, but different from those of other substances. There is not a regular gradation in the mass of molecules from that of hydrogen, which is the least of those known to us, to that of bismuth; but they all fall into a limited number of classes or species, the individuals of each species being exactly similar to each other, and no intermediate links are found to connect one species with another by a uniform gradation.

We are here reminded of certain speculations concerning the relations between the species of living things. We find that in these also the individuals are naturally grouped into species, and that intermediate links between the species are wanting. But in each species variations occur, and there is a perpetual generation and destruction of the individuals of which the species consist.

Hence it is possible to frame a theory to account for the present state of things by means of generation, variation, and discriminative destruction.

In the case of the molecules, however, each individual is permanent; there is no generation or destruction, and no variation, or rather no difference, between the individuals of each species.

Hence the kind of speculation with which we have become so familiar under the name of theories of evolution is quite inapplicable to the case of molecules.

It is true that Descartes, whose inventiveness knew no bounds, has given a theory of the evolution of molecules. He supposes that the molecules with which the heavens are nearly filled have received a spherical form from the long-continued grinding of their projecting parts, so that, like marbles in a mill, they have 'rubbed each other's angles down.' The result of this attrition forms the finest kind of molecules, with which the interstices between the globular molecules are filled. But, besides these, he describes another elongated kind of molecules, the *particula striata*, which have received their form from their often threading the interstices between three spheres in contact. They have thus acquired three longitudinal ridges, and, since some of them during their passage are rotating on their axes, these ridges are not in general parallel to the axis, but are twisted like the threads of a screw. By means of these little screws he most ingeniously attempts to explain the phenomena of magnetism.

But it is evident that his molecules are very different from ours. His seem to be produced by some general break-up of his solid space, and to be ground down in the course of ages, and, though their relative magnitude is in some degree determinate there is nothing to determine the absolute magnitude of any of them.

Our molecules, on the other hand, are unalterable by any of the processes which go on in the present state of things, and every individual of each species is of exactly the same magnitude, as though they had all been cast in the same mould, like bullets, and not merely selected and grouped according to their size, like small shot.

The individuals of each species also agree in the nature of the light which they emit—that is, in their natural periods of vibration. They are therefore like tuning-forks all tuned to concert pitch, or like watches regulated to solar time.

In speculating on the cause of this equality we are debarred from imagining any cause of equalization, on account of the immutability of each individual molecule. It is difficult, on the other hand, to conceive of selection and elimination of intermediate varieties, for where can these eliminated molecules have gone to if, as we have reason to believe, the hydrogen, &c., of the fixed stars is composed of molecules identical in all respects with our own? The time required to eliminate from the whole of the visible universe every molecule whose mass differs from that of some one of our so-called elements, by processes similar to Graham's method of dialysis, which is the only method we can conceive of at present, would exceed the utmost limits ever demanded by evolutionists as many times as these exceed the period of vibration of a molecule.

But if we suppose the molecules to be made at all, or if we suppose them to consist of something previously made, why should we expect any irregularity to exist among them? If they are, as we believe, the only material things which still remain in the precise condition in which they first began to exist, why should we not rather look for some indication of that spirit of order, our scientific confidence in which is never shaken by the difficulty which we experience in tracing it in the complex arrangements of visible things, and of which our moral estimation is shown in all our attempts to think and speak the truth, and to ascertain the exact principles of distributive justice?

11.3   Study Questions

QUES. 11.1. When an ideal gas is heated up while maintaining its pressure constant, what happens to its volume? Is this relationship (Charles' law) a strictly empirical law, or can it be derived from the kinetic theory of gases?

QUES. 11.2. What is the kinetic energy of a gas?

a) What is the mean kinetic energy of agitation of a single gas molecule?
b) Is the kinetic energy of a gas molecule due to its center of gravity motion alone? If not, what other types of motion might it undergo?
c) How is the kinetic energy associated with the internal motions of a gas molecule limited? How can this limit be expressed mathematically?

QUES. 11.3. What is the specific heat capacity of a gas?

a) How is the temperature of a gas related to its total kinetic energy? And how is the total kinetic energy related to its specific heat capacity?
b) Is the specific heat capacities of a gas larger when measured at constant pressure, or at constant volume? Why do you suppose this might be?

c) What is the relationship between the specific heat capacity of a gas measured at constant pressure and at constant volume?

d) Does the speed of sound in a gas depend on the specific heat capacity of a gas? Why might this be?

QUES. 11.4. Does gravity affect the temperature of a stationary body?

a) Is the density of a tall vertical column of gas standing in a gravitational field uniform from top to bottom? What about its pressure and its temperature?

b) Generally speaking, is the temperature distribution of a substance in thermal equilibrium affected by the presence of a gravitational field? Is the Earth's atmosphere in thermal equilibrium?

QUES. 11.5. Why does the temperature of Earth's atmosphere vary with altitude?

a) Does gravity have an effect on earth's atmosphere?

b) What is meant by convective equilibrium? How does it differ from thermal equilibrium?

c) How does the temperature of a gas which is in convective equilibrium vary with pressure? Is it hotter at high or low altitudes?

QUES. 11.6. Does the coefficient of viscosity of a gas depend on its temperature? on its pressure? How are diffusion, viscosity and heat conduction in gases related to one another?

QUES. 11.7. Why does a puddle of water on the ground evaporate if it is not boiling? Why does the same puddle of water not evaporate if it is poured into a sealed jar? Similarly, why do you feel cold when you step out of a pool? And how are all these phenomena explained by kinetic theory?

QUES. 11.8. Why does electrolysis occur even if the electromotive force applied to the molecules is much weaker than the binding forces of the molecules? In what sense is electrolysis similar to diffusion?

QUES. 11.9. Why do hot gases emit light? Does the period of vibration of a molecule depend on its velocity? What does the light emission spectrum of a hot gas look like? Does the spectrum of a hot gas depend on its temperature? on its density?

QUES. 11.10. Can the second law of thermodynamics be violated by an intelligent agent?

QUES. 11.11. What is the nature and origin of molecules?

a) How are molecules like, and unlike, species of living things?

b) Is it possible that molecules evolved by variation and natural selection? Who supported this view? Why does Maxwell reject it?

c) Can molecules suffer changes? Are they, in fact, manufactured articles?

11.4   Exercises

Ex. 11.1 (SPEED OF SOUND). Using the kinetic theory of gases, calculate the speed of sound in air at standard temperature and pressure. Do your calculations agree with measured values?

Ex. 11.2 (MOMENTUM DIFFUSION AND VISCOSITY). Consider a long train, A, having mass M and sitting stationary on an east-west railroad track. An identical train, B, is traveling at a speed V eastward down a parallel adjacent railroad track. As train B passes train A, one passenger on each train tosses a sack of grain, having mass m, across to the other train.

a) What happens to the speed of the trains as a result of this transfer of grain sacks? Do the trains speed up or slow down? Using the principle of conservation of momentum, calculate the final speed of each train in terms of M, m and V.
b) If N sacks of grain are thrown per second, then what would be the average force acting on each train? In which direction do these forces act? Do they obey Newton's third law of motion?
c) How are these trains analogous to parallel flat plates which are separated by a layer of viscous fluid and which are sliding past one another? What entities are analogous to the grain sacks being tossed back and forth?

Ex. 11.3 (INK DIFFUSION LABORATORY). Consider a tiny black particle placed in clear water. This particle is constantly bombarded by nearby water molecules: it gets kicked around in a random fashion many millions of times per second. Such agitation is referred to as "Brownian motion" after experiments performed by Robert Brown in 1827.[5] How far do we expect the particle to have moved after a given amount of time? In this laboratory exercise we will place a drop of ink into a shallow layer of water and measure the size of the ink drop as a function of elapsed time, t. If the individual ink molecules execute a random walk, then we might expect the mean square distance travelled by a collection of ink molecules to be proportional to the elapsed time:[6]

$$\overline{r^2} = 6Dt + r_0^2. \tag{11.11}$$

Here, r_0 is the initial radius of the droplet. The factor of 6 appears in Eq. 11.11 because diffusion can occur in any of three directions; this factor would be 4 in a 2-dimensional, and 2 in a 1-dimensional system. The *diffusion constant*, D, is a measure of how quickly the ink spreads into the water. According to the kinetic theory, it is related to the absolute viscosity of the fluid, μ, the radius of the water

---

[5] See Brown, R., A brief account of microsopical observations made in the months of June, July and August 1827, on the particles contained in the pollen of plants; and on the general existence of active molecules in organic and inorganic bodies, *Philosophical Magazine Series 2*, 4(21), 1828.
[6] You might with to look ahead to the random walk dice game in Ex. 15.1.

molecules, a, the temperature, T, and Boltzmann's constant k_B, by the formula[7]

$$D = \frac{k_B T}{6\pi \mu a}. \tag{11.12}$$

Therefore, if we measure D, and we known T, μ, and a, then we can calculate k_B. This, in turn, allows us to determine *Avogadro's number*. How so? The ideal gas law can be written either in terms of the number of moles n, and the gas constant, R (as $PV = nRT$). Or it can be written in terms of the number of atoms, N, and Boltzmann's constant, k_B (as $PV = Nk_B T$). It follows that

$$nR = Nk_B. \tag{11.13}$$

So if we measure Boltzmann's constant, and we know the gas constant, then we can determine Avogadro's number.

To begin your experiment, fill the bottom of a petrie dish with a shallow layer of water. Place a small drop of ink near the middle of the petrie dish and measure its radius as a function of time using a finely graded scale placed beneath the dish; it may take several tries to get a good data set. Does Eq. 11.11 fit your experimental data? What is your measured diffusion constant? Using the known absolute viscosity of water,[8] and the approximate size of a water molecule, determine Boltzmann's constant and Avogadro's number. How does your measured value of Avogadro's number compare to the accepted value? How might temperature affect your measurements of D, k_B, and N_a? Is D larger in a gas or in a liquid? Does your experiment prove that molecules exist?

EX. 11.4 (MAXWELL'S DEMON ESSAY). Consider a tiny clever being who is nearly the size of a molecule. Let us refer to this being as *Maxwell's demon*.[9] He guards a pinhole between two chambers by means of a little gate, which he can open and close at will with minimal effort. One chamber contains a high-temperature gas, the other a low-temperature gas. How might Maxwell's demon allow heat to flow from the cold into the hot chamber? Would this process violate of the second law of thermodynamics? If so, what does this imply? If not, why not?

EX. 11.5 (MOLECULES ESSAY). What do you think: is Maxwell correct in saying that molecules have the "essential character of a manufactured article"? If so, what does this imply? If not, then what *are* the marks of a manufactured article?

---

[7] Equation. 11.12 is derived in Sect. 15.6 of Reif, F., *Fundamentals of statistical and thermal physics*, McGraw-Hill, 1965.

[8] The viscosity of various liquids can be found in Haynes, W. M. (Ed.), *The CRC Handbook of Chemistry and Physics*, 95 ed., The Chemical Rubber Company, 2014.

[9] Sir William Thomson (Lord Kelvin) first referred to Maxwell's intelligent molecule-sorting agent as "Maxwell's demon." See Thomson, S. W., Kinetic Theory of the Dissipation of Energy, *Nature*, pp. 441–444, 1874. and also Thomson, S. W., The Sorting Demon of Maxwell, *Proceedings of the Royal Institution*, ix, 113, 1879.

11.5   Vocabulary

1. Diffusion
2. Viscosity
3. Assimilate
4. Saturate
5. Electrolyte
6. Asunder
7. Dissociation
8. Incandescent
9. Refrangibility
10. Luminiferous
11. Spectrum
12. Evolution
13. Attrition
14. Interstices

Chapter 12
The Diffusion Equation

> *If no change has occurred in the order of things, it cannot have been more than 200,000,000 years since the earth was in the condition of a mass of molten matter, on which a solid crust was just beginning to form.*
>
> —James Clerk Maxwell

12.1   Introduction

In his *Analytical Theory of Heat*, Joseph Fourier developed a mathematical equation which governs the diffusion of heat—and the resulting distribution of temperatures—in bodies of various shapes, sizes and compositions.[1] Essential to Fourier's analysis is the precise stipulation of the thermal boundary conditions of the body, such as the initial temperature within the body and the rate of heat flow through the exterior surfaces of the body. Fourier's treatment of thermal diffusion was both exhaustive and mathematically sophisticated. On the other hand, in Chap. 18 of his popular *Theory of Heat*, Maxwell presents Fourier's method in a simpler and more intuitive manner. This is the text included below. Maxwell begins by defining the *thermal conductivity* of a substance. This includes a discussion of *dimensional analysis* (another topic which was pioneered by Fourier). He then explains how Fourier's diffusion equation can be used to calculate the time-evolution of the temperature distribution within a heated substance. In so doing he recalls Fourier's method of describing the temperature within a body as a series of harmonic temperature distributions of various spatial frequencies.[2] The high-frequency (short wave-length) terms in the series die out most rapidly as heat diffuses through the substance, leaving the low-frequency (long wave-length) components as the most long-lived. Maxwell then goes on to explain how the diffusion equation can be used to solve geophysical problems such as seasonal variations of

---

[1] The introduction to Fourier's *Analytical Theory of Heat* is included in Chap. 1 of the present volume.

[2] A derivation of the one-dimensional diffusion equation is presented in Ex. 12.3 at the end of the present chapter.

© Springer International Publishing Switzerland 2016
K. Kuehn, *A Student's Guide Through the Great Physics Texts*,
Undergraduate Lecture Notes in Physics, DOI 10.1007/978-3-319-21828-1_12

Fig. 12.1 A rectangular section of a boiler plate whose
bottom surface is heated by
fire.—[K.K.]

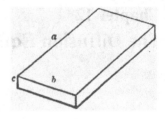

Fig. 12.1 A rectangular section of a boiler plate whose
bottom surface is heated by
fire.—[K.K.]

Earth's subsurface temperature, and perhaps even to determine the maximum age of
Earth's crust.

12.2   Reading: Maxwell, *On the Diffusion of Heat by Conduction*

Maxwell, J. C., *Theory of Heat*, tenth ed., Longmans, Green, and Co., London and
New York, 1891. Chap. XVIII.

Whenever different parts of a body are at different temperatures, heat flows from the
hotter parts to the neighbouring colder parts. To obtain an exact notion of conduction, let us consider a large boiler with a flat bottom, whose thickness is c. The fire
maintains the lower surface at the temperature T, and heat c flows upwards through
the boiler plate to the upper surface, which is in contact with the water at the lower
temperature, S.

Let us now restrict ourselves to the consideration of a rectangular portion of the
boiler plate, whose length is a, its breadth b, and its thickness c (see Fig. 12.1).

The things to be considered are the dimensions of this portion of the body, and
the nature of the material of which it is made, the temperatures of its upper and
lower surfaces, and the flow of heat through it as determined by these conditions. In
the first place it is found that when the difference of the temperature S and T is not
so great as to make a sensible difference between the properties of the substance at
these two temperatures, the flow of heat is exactly proportional to the difference of
temperatures, other things being the same.

Let us suppose that when a, b, and c are each equal to the unit of length, and
when T is 1° above s, the steady flow of heat is such that the quantity which enters
the lower surface or leaves the upper surface in the unit of time is k, then k is defined
as the specific thermal conductivity of the substance. To find H, the quantity of heat
which flows in a time t through the portion of boiler plate whose area is ab, and
whose thickness is c, when the lower surface is kept at a temperature T, and the
upper at a temperature S, till the flow has become steady, divide the plate into c
horizontal layers, the thickness of each layer being unity, and divide each layer into
ab cubes, the sides of each cube being unity.

Since the flow of heat is steady, the difference of temperature of the upper and
lower faces of each cube will be $\frac{1}{c}(T - S)$. The flow of heat through each cube will

be $\frac{k}{c}(T - S)$ in unit of time. Now, in each layer there are ab such cubes, and the flow goes on for t units of time, so that we obtain for the whole heat conducted in time t

$$H = \frac{abtk}{c}(T - S)$$

where ab is the area and c the thickness of the plate, t the time, $T - S$ the difference of temperature which causes the flow, and k the specific thermal conductivity of the substance of the plate.

It appears, therefore, that the heat conducted is directly proportional to the area of the plate, to the time, to the difference of temperature, and to the conductivity, and inversely proportional to the thickness of the plate.

12.2.1   On the Dimensions of k, the Specific Thermal Conductivity

From the equation we find

$$k = \frac{cH}{abt(T - S)}.$$

Hence if $[L]$ be the unit of length, $[T]$ the unit of time, $[H]$ the unit of heat, and $[\Theta]$ the unit of temperature, the dimensions of k will be $\frac{[H]}{[LT\Theta]}$.

The further discussion of the dimensions of k will depend on the mode of measuring heat and temperature.

1. If heat is measured as energy, its dimensions are $[\frac{L^2 M}{T^2}]$ and those of k become $[\frac{LM}{T^3\Theta}]$. This may be called the *dynamical* measure of the conductivity.
2. If heat is measured in thermal units, such that each thermal unit is capable of raising unit of mass of a standard substance through $1°$ of temperature, the dimensions of H are $[M\Theta]$ and those of k will be $[\frac{M}{LT}]$. This may be called the *calorimetric* measure of the conductivity.
3. If we take as the unit of heat that which will raise unit of *volume* of the substance itself $1°$, the dimensions of H are $[L^3\Theta]$ and those of k are $[\frac{L^2}{T}]$. This may be called the *thermometric* measure of the conductivity.

In order to obtain a distinct conception of the flow of heat through a solid body, let us suppose that at a given instant we know the temperature of every point of the body. If we now suppose a surface or interface to be described within the body such that at every point of this interface the temperature has a given value $T°$, we may call this interface the isothermal interface of $T°$. (Of course, when we suppose this interface to exist in the body, we do not conceive the body to be altered in any way by this supposition, as if the body were really cut in two by it.) This isothermal interface separates those parts of the body which are hotter than the temperature To from those which are colder than this temperature.

Let us now suppose the isothermal interfaces drawn for every exact degree of temperature, from that of the hottest part of the body to that of the coldest part.

These interfaces may be curved in any way, but no two different interface can meet each other, because no part of the body can at the same time have two different temperatures. The body will therefore be divided into layers or shells by these interfaces, and the space between two isothermal surfaces differing by 1° of temperature will be in the form of a thin shell, whose thickness may vary from one part to another.

At every point of this shell there is a flow of heat from the hotter surface to the colder surface through the substance of the shell.

The direction of this flow is perpendicular to the surface of the shell, and the rate of flow is greater the thinner the shell is at the place, and the greater its conductivity.

If we draw a line perpendicular to the surface of the shell, and of length unity, then if c is the thickness of the shell, and if the neighbouring shells are of nearly the same thickness, this line will cut a number of shells equal to $\frac{1}{c}$. This, then, is the difference of temperature between two points in the body at unit of distance, measured in the direction of the flow of heat, and therefore the flow of heat along this line is measured by $\frac{k}{c}$, where k is the conductivity.

We can now imagine, with the help of the isothermal interfaces, the state of the body at a given instant. Wherever there is inequality of temperature between neighbouring parts of the body a flow of heat is going on. This flow is everywhere perpendicular to the isothermal interfaces, and the flow through unit of area of one of these interfaces in unit of time is equal to the conductivity divided by the distance between two consecutive isothermal interfaces.

The knowledge of the actual thermal state of the body, and of the law of conduction of heat, thus enables us to determine the flow of heat at every part of the body. If the flow of heat is such that the amount of heat which flows into any portion of the body is exactly equal to that which flows out of it, then the thermal state of this portion of the body will remain the same as long as the flow of heat fulfills this condition.

If this condition is fulfilled for every part of the body, the temperature at any point will not alter with the time, the system of isothermal interfaces will continue the same, and the flow of heat will go on without alteration, being always the same at the same part of the body.

This state of things is referred to as the *state of steady flow of heat*. It cannot exist unless heat is steadily supplied to the hotter parts of the surface of the body, from some source external to the body, and an equal quantity removed from the colder parts of the surface by some cooling medium, or by radiation.

The state of steady flow of heat requires the fulfillment at every part of the body of a certain condition, similar to that which is fulfilled in the flow of an incompressible fluid.

When this condition is not fulfilled, the quantity of heat which enters any portion of the body may be greater or less than that which escapes from it. In the one case heat will accumulate, and the portion of the body will rise in temperature. In the other case the heat of the portion will diminish, and it will fall in temperature. The amount of this rise or fall of temperature will be measured numerically by the gain or loss of heat, divided by the capacity for heat of the portion considered.

If the portion considered is unit of volume, and if we measure heat as in the third method given at p. 145 by the quantity required to raise unit of volume of the substance, in its actual state, 1°, then the rise of temperature of this portion will be numerically equal to the total flow of heat into it.

We are now able, by means of a thorough knowledge of the thermal state of the body at a given instant, to determine the rate at which the temperature of every part must be changing, and therefore we are able to predict its state in the succeeding instant. Knowing this, we can predict its state in the next instant following, and so on.

The only parts of the body to which this method does not apply are those parts of its surface to which heat is supplied, or from which heat is abstracted, by agencies external to the body. If we know either the rate at which heat is supplied or abstracted at every part of the surface, or the actual temperature of every part of the surface during the whole time, either of these conditions, together with the original thermal state of the body, will afford sufficient data for calculating the temperature of every point during all time to come.

The discussion of this problem is the subject of the great work of Joseph Fourier, *Théorie de la Chaleur*. It is not possible in a treatise of the size and scope of this book to reproduce, or even to explain, the powerful analytical methods employed by Fourier to express the varied conditions, as to the form of its surface and its original thermal state, to which the body may be subjected. These methods belong, rather, to the general theory of the application of mathematics to physics; for in every branch of physics, when the investigation turns upon the expression of arbitrary conditions, we have to follow the method which Fourier first pointed out in his 'Theory of Heat.'

I shall only mention one or two of the results given by Fourier in which the intricacies arising from the arbitrary conditions of the problem are avoided.

The first of these is the case in which the solid is supposed of infinite extent, and of the same conductivity in every part.

The temperature of every point of this body at a given time is supposed to be known, and it is required to determine the temperature of any given point P after a time t has elapsed.

Fourier has given a complete solution of this problem, of which we may obtain some idea by means of the following considerations. Let k be the conductivity, measured by the third method, in which the unit of heat adopted is that which will raise unit of volume of the substance 1°; then if we make

$$kt = \alpha^2, \tag{12.1}$$

α will be a line the length of which will be proportional to the square root of the time.

Let Q be any point in the body, and lets its distance from P be r. Let the original temperature of Q be θ. Now take a quantity of matter proportional to $e^{-\frac{r^2}{4kt}}$ and of the temperature θ, and mix it with portions of matter taken from every other part of the body, the temperature of each portion being the original temperature of that point, and the quantity of each portion being proportional to $e^{-\frac{r^2}{4kt}}$. The mean temperature of all such portions will be the temperature of the point P after a time t.

In other words, the temperature of P after a time t may be regarded as in some sense the mean of the original temperatures of all parts of the body. In taking this mean, however, different parts are allowed different weights, depending on their distance from P, the parts near P having more influence on the result than those at a greater distance.

The mathematical formula which indicates the weight to be given to the temperature of each part in taking the mean is a very important one. It occurs in several branches of physics, particularly in the theory of errors and in that of the motions of systems of molecules.

It follows from this result that, in calculating the temperature of the point P, we must take into account the temperature of every other point Q, however distant, and however short the time may be during which the propagation of heat has been going on. Hence, in a strict sense, the influence of a heated part of the body extends to the most distant parts of the body in an incalculably short time, so that it is impossible to assign to the propagation of heat a definite velocity. The velocity of propagation of thermal effects depends entirely on the magnitude of the effect which we are able to recognise; and if there were no limit to the sensibility of our instruments, there would be no limit to the rapidity with which we could detect the influence of heat applied to distant parts of the body. But while this influence on distant points can be expressed mathematically from the first instant, its numerical value is excessively small until, by the lapse of time, the line α has grown so as to be comparable with r, the distance of P from Q. If we take this into consideration, and remember that it is only when the changes of temperature are comparable with the original differences of temperature that we can detect them with our instruments, we shall see that the sensible propagation of heat, so far from being instantaneous, is an excessively slow process, and that the time required to produce a similar change of temperature in two similar systems of different dimensions is proportional to the square of the linear dimensions. For instance, if a red-hot ball of 4 in. diameter fired into a sandbank has in an hour raised the temperature of the sand 6 in. from its centre $10\,°F$, then a red-hot ball of 8 in. diameter would take 4 h to raise the temperature of the sand 12 in. from its centre by the same number of degrees. This result, which is very important in practical questions about the time of cooling or heating of bodies of any form, may be deduced directly from the consideration of the dimensions of the quantity k—namely, the square of a length divided by a time. It follows from this that if in two unequally heated systems of similar form but different dimensions the conductivity and the temperature are the same at corresponding points at first, then the process of diffusion of heat will go on at different rates in the two systems, so that if for each system the time be taken proportional to the square of the linear dimensions, the temperatures of corresponding points will still be the same in both systems.

The method just described affords a complete determination of the temperature of any point of a homogeneous infinite solid at any future time, the temperature of every point of the solid being given at the instant from which we begin to count the time. But when we attempt to deduce from a knowledge of the present thermal state of the body what must have been its state at some past time, we find that the method ceases to be applicable.

To make this attempt, we have only to make t, the symbol of the time, a negative quantity in the expressions given by Fourier. If we adopt the method of taking the mean of the temperatures of all the particles of the solid, each particle having a certain weight assigned to it in taking the mean, we find that this weight, according to the formula, is greater for the distant particles than for the neighbouring ones, a result sufficiently startling in itself. But when we find that, in order to obtain the mean, after taking the sum of the temperatures multiplied by their proper factors, we have to divide by a quantity involving the square root of t, the time, we are assured that when t is taken negative the operation is simply impossible, and devoid of any physical meaning, for the square root of a negative quantity, though it may be interpreted with reference to some geometrical operations, is absolutely without meaning with reference to time.

It appears, therefore, that Fourier's solution of this problem, though complete considered with reference to future time, fails when we attempt to discover the state of the body in past time.

In the diagram Fig. 12.2 the curves show the distribution of temperature in an infinite mass at different times, after the sudden introduction of a hot horizontal stratum in the midst of the infinite solid. The temperature is indicated by the horizontal distance to the right of the vertical line, and the hot stratum is supposed to have been introduced at the middle of the figure.

The curves indicate the temperatures of the various strata 1, 4, and 16 h after the introduction of the hot stratum. The gradual diffusion of the heat is evident, and also the diminishing rate of diffusion as its extent increases.

The problem of the diffusion of heat in an infinite solid does not present those difficulties which occur in problems relating to a solid of definite shape. These difficulties arise from the conditions to which the surface of the solid may be subjected, as, for instance, the temperature may be given over part of the surface, the quantity of heat supplied to another part may be given, or we may only know that the surface is exposed to air of a certain temperature.

The method by which Fourier was enabled to solve many questions of this kind depends on the discovery of harmonic distributions of heat.

Suppose the temperatures of the different parts of the body to be so adjusted that when the body is left to itself under the given conditions relating to the surface, the temperatures of all the parts converge to the final temperature, their differences from the final temperature always preserving the same proportion during the process; then this distribution of temperature is called an harmonic distribution. If we suppose the final temperature to be taken as zero, then the temperatures in the harmonic distribution diminish in a geometrical progression as the times increase in arithmetical progression, the ratio of cooling being the same for all parts of the body.

In each of the cases investigated by Fourier there may be an infinite series of harmonic distributions. One of these, which has the slowest rate of diminution,

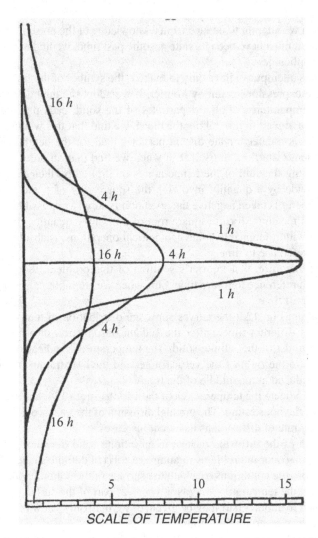

Fig. 12.2 As time evolves, the temperature of a hot stratum of material in the middle of an infinite body cools as heat diffuses into the surrounding material.—[K.K.]

may be called the fundamental harmonic; the rates of diminution of the others are proportional to the squares of the natural numbers.

If the body is originally heated in any arbitrary manner, Fourier shows how to express the original temperature as the sum of a series of harmonic distributions. When the body is left to itself the part depending on the higher harmonics rapidly dies away, so that after a certain time the distribution of heat continually approximates to that due to the fundamental harmonic, which therefore represents the law of cooling of a body after the process of diffusion of heat has gone on for a long time.

Sir William Thomson has shown, in a paper published in the 'Cambridge and Dublin Mathematical Journal' in 1844, how to deduce, in certain cases, the thermal state of a body in past time from its observed condition at present.

For this purpose, the present distribution of temperature must be expressed (as it always may be) as the sum of a series of harmonic distributions. Each of these harmonic distributions is such that the difference of the temperature of any point from the final temperature diminishes in a geometrical progression as the time increases in arithmetical progression, the ratio of the geometrical progression being the greater the higher the degree of the harmonic.

If we now make t negative, and trace the history of the distribution of temperature up the stream of time, we shall find each harmonic increasing as we go backwards, and the higher harmonics increasing faster than the lower ones.

If the present distribution of temperature is such that it may be expressed in a finite series of harmonics, the distribution of temperature at any previous time maybe calculated; but if (as is generally the case) the series of harmonics is infinite, then the temperature can be calculated only when this series is convergent. For present and future time it is always convergent, but for past time it becomes ultimately divergent when the time is taken at a sufficiently remote epoch. The negative value of t, for which the series becomes ultimately divergent, indicates a certain date in past time such that the present state of things cannot be deduced from any distribution of temperature occurring previously to that date, and becoming diffused by ordinary conduction. Some other event besides ordinary conduction must have occurred since that date in order to produce the present state of things. This is only one of the cases in which a consideration of the dissipation of energy leads to the determination of a superior limit to the antiquity of the observed order of things.

A very important class of problems is that in which there is a steady flow of heat into the body at one point of its surface, and out of it at another part. There is a certain distribution of temperature in all such cases, which if once established will not afterwards change: this is called the permanent distribution. If the original distribution differs from this, the effect of the diffusion of heat will be to cause the distribution of temperature to approximate without limit to this permanent distribution. Questions relating to the permanent distribution of temperature and the steady flow of heat are in general less difficult than those in which this state is not established.

Another important class of problems is that in which heat is supplied to a portion of the surface in a periodic manner, as in the case of the surface of the earth, which receives and emits heat according to the periods of day and night, and the longer periods of summer and winter.

The effect of such periodic changes of temperature at the surface is to produce waves of heat, which descend into the earth and gradually die away. The length of these waves is proportional to the square root of the periodic time. If we examine the wave at a depth such that the greatest heat occurs when it is coldest at the surface, then the extent of the variation of temperature at this depth is only $\frac{1}{23}$ of its value at the surface. In the rocks of this country this depth is about 25 ft for the annual variations.

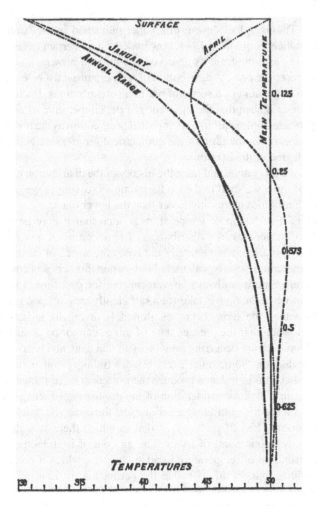

Fig. 12.3 Subsurface temperature variations during different seasons.—[*K.K.*]

In the diagram Fig. 12.3 the distribution of temperature in the different strata is represented at two different times. If we suppose the figure to represent the diurnal variation of temperature, then the curves indicate the temperatures at 2 A.M. and 8 A.M. If we suppose it to represent the annual variation, then the curves correspond to January and April.

Since the depth of the wave varies as the square root of the periodic time, the wave-length of the annual variation of temperature will be about 19 times the depth of those of the diurnal variation. At a depth of about 50 ft the variation of annual temperature is about a year in arrear.

The actual variation of temperature at the surface does not follow the law which gives a simple harmonic wave, but, however complicated the actual variation may be, Fourier shows how to decompose it into a number of harmonic waves of which it is the sum. As we descend into the earth these waves die away, the shortest most rapidly, so that we lose the irregularities of the diurnal variation in a few inches, and the diurnal variation itself in a few feet. The annual variation can be traced to a much greater depth; but at depths of 50 ft and upwards the temperature is sensibly constant throughout the year, the variation being less than the 500th part of that at the surface.

But if we compare the mean temperatures at different depths, we find that as we descend the mean temperature rises, and that after we have passed through the upper strata, in which the periodic variations of temperature are observed, this increase of temperature goes on as we descend to the greatest depths known to man. In this country the rate of increase of temperature appears to be about 1 °F for 50 ft of descent.

The fact that the strata of the earth are hotter below than above shows that heat must be flowing through them from below upwards. The amount of heat which thus flows upwards in a year through a square foot of the surface can easily be found if we know the conductivity of the substance through which it passes. For several kinds of rock the conductivity has been ascertained by means of experiments made upon detached portions of the rock in the laboratory. But a still more satisfactory method, where it can be employed, is to make a register of the temperature at different depths throughout the year, and from this to determine the length of the annual wave of temperature, or its rate of decay. From either of these data the conductivity of the substance of the earth may be found without removing the rocks from their bed.

By observations of this kind made at different points of the earth's surface we might determine the quantity of heat which flows out of the earth in a year. This can be done only roughly at present, on account of the small number of places at which such observations have been made, but we know enough to be certain that a great quantity of heat escapes from the earth every year. It is not probable that any great proportion of this heat is generated by chemical action within the earth. We must therefore conclude that there is less heat in the earth now than in former periods of its existence, and that its internal parts were formerly very much hotter than they are now.

In this way Sir W. Thomson has calculated that, if no change has occurred in the order of things, it cannot have been more than 200,000,000 years since the earth was in the condition of a mass of molten matter, on which a solid crust was just beginning to form.

12.2.2   On the Determination of the Thermal Conductivity of Bodies

The most obvious method of determining the conductivity of a substance is to form it into a plate of uniform thickness, to bring one of its surfaces to a known temperature and the other to a known lower temperature, and to determine the quantity of heat which passes through the plate in a given time.

For instance, if we could bring one surface to the temperature of boiling water by a current of steam, and keep the other at the freezing temperature by means of ice, we might measure the heat transmitted either by the quantity of steam condensed, or by the quantity of ice melted.

The chief difficulty in this method is that the surface of the plate does not acquire the temperature of the steam or the ice with which it is in contact, and that it is difficult to ascertain its real temperature with the accuracy necessary for a determination of this kind.

Most of the actual determinations of conductivity have been made in a more indirect way—by observing the permanent distribution of temperature in a bar, one end of which is maintained at a high temperature, while the rest of its surface is exposed to the cooling effects of the atmosphere.

The temperatures of a series of points in the bar are ascertained by means of thermometers inserted into holes drilled in it, and brought into thermal connexion with its substance by means of fluid metal surrounding the bulbs.

In this way the rate of diminution of temperature with the distance can be ascertained at various points on the bar.

To determine the conductivity, we must compare the rate of variation of temperature with the flow of heat which is due to it. It is in the determination of this flow of heat that the indirectness of the method consists. The most trustworthy method of determining the flow of heat is that employed by Principal Forbes in his experiments on the conduction of heat in an iron bar.[3] He took a bar of exactly the same section and material as the experimental bar, and, after heating it uniformly, allowed it to cool in air of the same temperature as that surrounding the experimental bar. By observing the temperature of the cooling bar at frequent intervals of time, he ascertained the quantity of heat which escaped from the sides of the bar, this heat being measured in terms of the quantity of heat required to raise unit of volume *of the bar* 1°. This loss of heat depended of course on the temperature of the bar at the time, and a table was formed showing the loss from a linear foot of the bar in a minute at any temperature.

Now, in the experimental bar the temperature of every part was known, and therefore the loss of heat from any given portion of the bar could be found by making use of the table. To determine the flow of heat across any particular section, it was necessary to sum up the loss of heat from all parts of the bar beyond this section, and when this was done, by comparing the flow of heat across the section with the rate of diminution of temperature per linear foot in the curve of temperature, the conductivity of the bar for the temperature of the section was ascertained. Principal Forbes found that the thermal conductivity of iron decreases as the temperature increases.

---

[3] *Trans. Roy. Soc. Edinb.* 1861–1862.

The conductivity thus determined is expressed in terms of the quantity of heat required to raise unit of volume *of the substance* 1°. If we wish to express it in the ordinary way in terms of the thermal unit as defined with reference to water at its maximum density, we must multiply our result by the specific heat of the substance, and by its density; for the quantity of heat required to raise unit of mass of the substance 1° is its specific heat, and the number of units of mass in unit of volume is the density of the substance.

As long as we are occupied with questions relating to the diffusion of heat and the waves of temperature in a single substance, the quantity on which the phenomena depend is the thermometric conductivity expressed in terms of the substance itself; but whenever we have to do with the effects of the flow of heat upon other bodies, as in the case of boiler plates, steam-condensers, &c., we must use a definite thermal unit, and express the calorimetric conductivity in terms of it. It has been shown by Professor Tyndall that the wave of temperature travels faster in bismuth than in iron, though the conductivity of bismuth is much less than that of iron. The reason is that the thermal capacity of the iron is much greater than that of an equal volume of bismuth.

Forbes was the first to remark that the order in which the metals follow one another in respect of thermal conductivity is nearly the same as their order as regards electric conductivity. This remark is an important one as regards certain metals, but it must not be pushed too far; for there are substances which are almost perfect insulators of electricity, whereas it is impossible to find a substance which will not transmit heat.

The electric conductivity of metals diminishes as the temperature rises. The thermal conductivity of iron also diminishes, but in a smaller ratio, as the temperature rises.

Professor Tait has given reasons for believing that the thermal conductivity of metals may be inversely proportional to their absolute temperature.

The electric conductivity of most non-metallic substances, and of all electrolytes and dielectrics, *increases* as the temperature rises. We have not sufficient data to determine whether this is the case as regards their thermal conductivity. According to the molecular theory of Chap. 22. the thermal conductivity of gases increases as the temperature rises.

12.2.3   On the Conductivity of Fluids

It is very difficult to determine the thermal conductivity of fluids, because the variation of temperature which is part of the phenomenon produces a variation of density, and unless the surfaces of equal temperature are horizontal, and the upper strata are the warmest, currents will be produced in the fluid which will entirely mask the phenomena of true conduction.

Another difficulty arises from the fact that most fluids have a very small conductivity compared with solid bodies. Hence the sides of the vessel containing the fluid are often the principal channel for the conduction of heat.

In the case of gaseous fluids the difficulty is increased by the greater mobility of their parts, and by the great variation of density with change of temperature. Their conductivity is extremely small, and the mass of the gas is generally small compared with that of the vessel in which it is contained. Besides this, the effect of direct radiation from the source of heat through the gas on the thermometer produces a heating effect which may, in some cases, completely mask the effect of true conduction. For all these reasons, the determination of the thermal conductivity of a gas is an investigation of extreme difficulty.

12.2.4   Applications of the Theory

The great thermal conductivity of the metals, especially of copper, furnishes the means of producing many thermal effects in a convenient manner. For instance, in order to maintain a body at a high temperature by means of a source of heat at some distance from it, a thick rod of copper may be used to conduct the heat from the source to the body we wish to heat; and when it is desired to warm the air of a room by means of a hot pipe of small dimensions, the effect may be greatly increased by attaching copper plates to the pipe, which become hot by conduction, and expose a great heating surface to the air.

To ensure an exact equality of temperature in all the parts of a body, it may be placed in a closed chamber formed of thick sheet copper. If the temperature is not quite uniform outside this chamber, any difference of temperature between one part of the outer surface and another will produce such a flow of heat in the substance of the copper that the temperature of the inner surface will be very nearly uniform. To maintain the chamber at a uniform high temperature by means of a flame, as is sometimes necessary, it may be placed in a larger copper chamber, and so suspended by strings or supported on legs that very little heat can pass by direct conduction from the outer to the inner wall. Thus we have first an outer highly conducting shell of copper; next a slowly conducting shell of air, which, however, tends to equalize the temperature by convection; then another highly conducting shell of copper; and lastly the inner chamber. The whole arrangement facilitates the flow of heat parallel to the walls of the chambers, and checks its flow perpendicular to the walls. Now differences of temperature within the chamber must arise from the passage of heat from without to within, or in the reverse direction, and the flow of heat along the successive envelopes tends only to equalize the temperature. Hence, by the arrangement of successive shells, alternately of highly conducting and slowly conducting matter, and still more if the slowly conducting matter is fluid, an almost complete uniformity of temperature may be maintained within the inner chamber, even when the outer chamber has all the heat applied to it at one point.

This arrangement was employed by M. Fizeau in his researches on the dilatation of bodies by heat.

12.3 Study Questions

QUES. 12.1. How can the rate of steady heat flow from the hot to the cold surface of a rectangular boilerplate be maximized? Consider: what properties of the plate, and of its surroundings, increase or decrease the rate of heat flow? And how can the rate of heat flow be expressed mathematically?

QUES. 12.2. What are the dimensions of the specific thermal conductivity of a substance? What is the difference between *dynamical, calorimetric*, and *thermometric* measures of the thermal conductivity?

QUES. 12.3. Can the past as well as the future isothermal surfaces within a body be calculated from a complete knowledge of the present isothermal surfaces?

a) What are isothermal surfaces? May different isothermal surfaces intersect one another? And how are the shape and spacing of isothermal surfaces related to the direction and rate of heat flow within a substance?
b) Can the rate of heat flow into and out of a region of a body be unequal? If so, what happens? And what is it called when the isothermal surfaces within a body are time-independent?
c) What specific information is required to determine the shape of the isothermal surfaces within a body at any particular instant?
d) At what velocity does heat propagate through a substance? What makes the assignment of such a velocity difficult? And why does the assignment of a particular velocity to heat propagation depend on the sensitivity of the measuring device?
e) How does the rate of thermal diffusion within a body scale with its size? For example, if its size is doubled, by what factor does the thermal diffusion time increase?
f) If the original temperature distribution of body is mathematically modeled as a series of harmonic temperature distributions having different spatial frequencies, then which frequencies tend to die away most rapidly? Why does this make it difficult to extrapolate the diffusion of heat to previous times?

QUES. 12.4. What is the difference between steady heat flow and periodic heat flow? Does Earth's surface experience steady heat flow? To what depth are Earth's seasonal temperature variations felt? Are these subsurface temperature variations contemporaneous with the surface temperature variations?

QUES. 12.5. Are there any limits to the age of the earth's crust? Upon what assumptions is Maxwell's analysis based?

QUES. 12.6. Why is it difficult to measure the thermal conductivity of a body by maintaining opposite boundaries of the substance at known temperatures (using,

say, steam and ice-water) and measuring the heat flow? What happens at the surface
of the body? What alternative method is typically employed?

QUES. 12.7. If the thermal conductivity of iron is greater than that of bismuth, then
why do periodic temperature variations travel faster in bismuth than in iron?

QUES. 12.8. Are the thermal and electrical conductivities of materials correlated?
And are these properties temperature dependent?

QUES. 12.9. Why is the thermal conductivity of fluids and gases particularly difficult
to measure?

QUES. 12.10. Why do radiators make use of high-conductivity metal fins to heat
your house and to cool computer chips? And how can a chamber be constructed
whose interior temperature is practically uniform?

12.4 Exercises

EX. 12.1 (THERMAL CONDUCTIVITY). Two 1 cm thick boilerplates (as in Fig. 12.1)
are stacked on top of each other. The top one is made of iron, the bottom one of
copper. The bottom surface of the cooper boilerplate is subjected to a continuous
flow of steam at 100 °C; the top surface of the iron boilerplate is subjected to a
constant flow of ice water at 0 °C. What is the temperature at the interface between
the iron and copper boilerplates? How much heat is transported across one square
centimeter of these boilerplates each second? (ANSWER: 83 ° C.)

EX. 12.2 (SCALING AND HEAT FLOW). Suppose a red-hot ball 6 in. in diameter is
fired into a sandbank. After 1 h, it has raised the temperature of the sand 9 in. from
its center by 5 °C. Now if, instead, a red-hot ball 12 in. in diameter was used, how
long would it take the sand 18 in. from its center be raised by 5 °C?

EX. 12.3 (DERIVING THE DIFFUSION EQUATION). In this exercise, we will derive the
diffusion equation for heat flow in one-dimension. We will then obtain a solution to
this equation for the special case of an infinitely long rod subjected to a short heat
pulse at one of its ends. The rate at which heat energy flows into (or out of) a body
is called the heating (or cooling) power:

$$P = \frac{dQ}{dt},$$ (12.2)

The dimensions of P are $[\frac{ML^2}{T^3}]$. Now consider a rod of length L and cross sectional
area A whose left end is held at a higher temperature than its right end. The amount
of heat deposited in a tiny segment of length dz, in time dt, is simply the difference
between the amount of heat coming in from its left boundary and going out of its
right boundary:

$$dQ = [P(z) - P(z + dz)]dt.$$ (12.3)

a) Taylor expand Eq. 12.3 about z and linearize to obtain

$$dQ = \left[P(z) - P(z) - \left(\frac{\partial P}{\partial z} \right) dz \right] dt. \tag{12.4}$$

b) Rearrange Eq. 12.4 to show that the rate at which heat accumulates in this segment is

$$\frac{dQ}{dt} = -\left(\frac{\partial P}{\partial z} \right) dz. \tag{12.5}$$

c) Show that the rate of temperature rise for a rod segment of length dz and density ρ can be written in terms of the specific heat capacity, c, as

$$\frac{\partial T}{\partial t} = \frac{1}{\rho A c \, dz} \left(\frac{dQ}{dt} \right). \tag{12.6}$$

d) Combine Eqs. 12.5 and 12.6, and show that

$$\frac{\partial T}{\partial t} = -\frac{1}{\rho A c} \left(\frac{\partial P}{\partial z} \right). \tag{12.7}$$

e) If the heat flow down the rod is directly proportional to the temperature gradient, then

$$P = -kA \left(\frac{\partial T}{\partial z} \right). \tag{12.8}$$

Eq. 12.8 provides a definition of k, the *thermal conductivity* of a substance. Based on the dimensions of k, does it correspond to what Maxwell calls the *dynamical, calorimetric*, or *thermometric* thermal conductivity?

f) Now combine Eqs. 12.7 and 12.8 to show that, if the thermal conductivity of the rod is uniform, then we obtain the *one-dimensional diffusion equation*:

$$\frac{\partial T}{\partial t} = D \frac{\partial^2 T}{\partial z^2}, \tag{12.9}$$

where $D \equiv k/\rho c$ is the *thermal diffusivity* of the substance. Is Eq. 12.9 correct if k is itself temperature-dependent?

g) What are the dimensions of the thermal diffusivity? Does it correspond to what Maxwell calls the *dynamical, calorimetric*, or *thermometric* thermal conductivity?

h) Finally, what is the thermal diffusivity of copper at room temperature?

EX. 12.4 (HEAT DIFFUSION IN AN INFINITE ROD). In Ex. 12.3, we derived the one-dimensional diffusion equation. In this exercise, we will solve the diffusion equation analytically for the special case of an infinitely long one-dimensional rod. We will assume that all of the heat is deposited at one location at $t = 0$.

a) Verify, by direct substitution into Eq. 12.9, that the solution to the diffusion equation may be written as

$$T' = B_1 + B_2 \frac{1}{\sqrt{t}} e^{\frac{-z^2}{4Dt}}. \tag{12.10}$$

Here, B_1 and B_2 are constants. The reason for the T-prime notation on the left-hand-side of Eq. 12.10 will become apparent in a moment; it does *not* indicate a derivative.

b) What is the temperature, T', after a very long time? Prove that the difference between the rod temperature and ambient temperature can be written as

$$T = B_2 \frac{1}{\sqrt{t}} e^{\frac{-z^2}{4Dt}}. \tag{12.11}$$

c) The total heat added to the rod, Q, may be found by summing the heats, dq, added to each segment of the rod.[4] These dq's may be expressed in terms of the temperature change of each segment as $dq = dm\, cT$. Thus

$$Q = \int_0^\infty dq = \int_0^\infty dz\, \rho A\, cT \tag{12.12}$$

Integrate Eq. 12.12 to show that

$$B_2 = \frac{Q}{A\sqrt{\pi \rho c k}}. \tag{12.13}$$

and thus that the solution to the one-dimensional diffusion equation is

$$T = \frac{Q}{A\sqrt{\pi \rho c k t}} e^{-z^2/4Dt}. \tag{12.14}$$

d) Eq. 12.14 is a bit complicated. To clarify its meaning, let us define a characteristic time scale, τ, and a characteristic temperature scale, Θ, associated with a particular position along the rod:

$$\tau = \frac{z^2}{4D} \qquad \text{and} \qquad \Theta = \frac{2Q}{A\rho c z \sqrt{\pi}}. \tag{12.15}$$

Show that the solution to the diffusion equation can now be expressed in dimensionless form:

$$\frac{T}{\Theta} = \sqrt{\frac{\tau}{t}} e^{-\tau/t} \tag{12.16}$$

---

[4] See Clausius' discussion of the diffusion of heat, especially the text surrounding Eq. 7.15 in Chap. 7 of the present volume.

e) Show that the maximum of T/Θ occurs when $t/\tau = 2$. What is T/Θ at this instant?

f) Suppose that the temperature at a distance z_0 from the end of a long copper rod is measured as a function of time. From the amount of heat added to the end of the rod and the maximum temperature at z_0 (*i.e.* T_{max}), one can determine the value of the product ρc. Show that

$$\rho c = \frac{0.86\, Q}{\sqrt{\pi}\, A z_0\, T_{max}} \qquad (12.17)$$

g) Furthermore, from measuring the time at which T_{max} occurs, one can determine τ (and hence the thermal diffusivity, D). Finally, from D and ρc, one can determine the thermal conductivity, k. Find an appropriate mathematical expression for the thermal conductivity of copper in terms of these measurable quantities.

EX. 12.5 (HEAT DIFFUSION—FOURIER METHOD). In his famous *Analytical Theory of Heat*, Fourier claimed that an arbitrary periodic function may be written as a (perhaps very complicated) sum of sine and cosine functions, each having a different amplitude and frequency.[5] Such a *linear combination* of trigonometric functions is thus called a Fourier series. In this exercise, we will very briefly introduce Fourier's method of solving the one-dimensional diffusion equation. Suppose that at a particular time, t, the temperature at a position, x, inside a body is given by $T(x, t)$. Now instead of representing this temperature distribution as a sum of trigonometric functions (as Fourier would have), let us appeal to Euler's formula[6] and write the temperature distribution as a sum of N exponential functions which are each periodic in x:

$$T(x, t) = \sum_{k=0}^{N} \widetilde{T}_k(t)\, e^{ikx} \qquad (12.18)$$

Here, $\widetilde{T}_k(t)$ represents the complex amplitude (at time t) of the term in the sum having wave-number k. Note that k here is *not* the thermal conductivity; it is related to the wavelength, λ, according to $k = 2\pi/\lambda$. The initial temperature distribution in the body will determine each of the N values of $\widetilde{T}_k(0)$. Show that Eq. 12.18 is indeed a solution to the one-dimensional diffusion equation, provided that each of the amplitudes, $\widetilde{T}_k(t)$, satisfy

$$\widetilde{T}_k(t) = \widetilde{T}_k(0)\, e^{-Dk^2 t} \qquad (12.19)$$

What does this imply regarding the time-evolution of an arbitrary non-uniform temperature distribution within a body? In particular, do the long-wavelength or the short-wavelength terms in Eq. 12.18 die out sooner? Is this consistent with what Maxwell claims on p. 150, above?

---

[5] For an introduction to Joseph Fourier's *Analytical Theory of Heat*, see Chaps. 1 and 2 of the present volume.

[6] Euler's formula says that $e^{i\theta} = \cos\theta + i\sin\theta$.

12.5   Vocabulary

1. Conception
2. Arbitrary
3. Propagation
4. Deduce
5. Stratum
6. Harmonic
7. Diurnal
8. Annual
9. Ascertain
10. Intricacy
11. Conduction
12. Diminution
13. Isothermal
14. Convection
15. Conduction
16. Supposition

Chapter 13
Radiant Heat

> The brighter the surface of a silver teapot, the longer will it
> retain the heat of the tea.
>
> —James Clerk Maxwell

13.1 Introduction

In the previous reading, from Chap. 18 of Maxwell's *Theory of Heat*, we focused on
the transport of heat by *conduction* and (to a lesser extent) *convection*. In the former
case, heat is transported between stationary bodies which are in thermal contact—
like a hot lead ball buried in cool sand. In the latter case, a warm substance moves
into a cooler region, bringing heat with it—like an ascending hot-air balloon. A third
mode of heat transport, which has only briefly been mentioned, consists of bodies
exchanging heat across empty space in the form of *radiation*. This phenomenon is
illustrated by the heating of a cool thermometer suspended in a sealed warm vacuum chamber, or by the heating of the earth across vast space by the sun. In fact,
all bodies emit heat radiation, whether a red-hot rod of iron, a warm human body or
a cold dust cloud in a distant nebula. In the reading selection that follows Maxwell
describes the nature of radiant heat, and how it depends not only on the temperature of the body, but also on the characteristics of the body's surface. Although
these phenomena had been studied extensively by many scientists up to the time
of Maxwell, no satisfactory theory had been provided to accurately explain them.
Within 30 years of Maxwell's publication, Max Planck's careful examination of the
color and intensity of radiation emitted by black-bodies—those which appear black
when cold—led to the birth of modern quantum theory.

13.2 Reading: Maxwell, *On Radiation*

Maxwell, J. C., *Theory of Heat*, tenth ed., Longmans, Green, and Co., London and
New York, 1891. Chapter XVI.

© Springer International Publishing Switzerland 2016   163
K. Kuehn, *A Student's Guide Through the Great Physics Texts,*
Undergraduate Lecture Notes in Physics, DOI 10.1007/978-3-319-21828-1_13

We have already noticed some of the phenomena of radiation, and have shown that they do not properly belong to the science of Heat, and that they should rather be treated, along with sound and light, as a branch of the great science of Radiation.

The phenomenon of radiation consists in the transmission of energy from one body to another by propagation through the intervening medium, in such a way that the progress of the radiation may be traced, after it has left the first body and before it reaches the second, travelling through the medium with a certain velocity, and leaving the medium behind it in the condition in which it found it.

We have already considered one instance of radiation in the case of waves of sound. In this case the energy communicated to the air by a vibrating body is propagated through the air, and may finally set some other body, as the drum of the ear, in motion. During the propagation of the sound this energy exists in the portion of air through which it is travelling, partly in the form of motion of the air to and fro, and partly in the form of condensation and rarefaction. The energy due to sound in the air is distinct from heat, because it is propagated in a definite direction, so that in a certain time it will have entirely left the portion of air under consideration, and will be found in another portion of air to which it has travelled. Now heat never passes out of a hot body except to enter a colder body, so that the energy of sound-waves, or any other form of energy which is propagated so as to pass wholly out of one portion of the medium and into another, cannot be called heat.

There are, however, important thermal effects produced by radiation, so that we cannot understand the science of heat without studying some of the phenomena of radiation.

When a body is raised to a very high temperature it becomes visible in the dark, and is said to shine, or to emit light. The velocity of propagation of the light emitted by the sun and by very hot bodies has been approximately measured, and is estimated to be between 180,000 and 192,000 miles/s, or about 900,000 times faster than sound in air.

The time taken by the light in passing from one place to another within the limited range which we have at our command in a laboratory is exceedingly short, and it is only by means of the most refined experimental methods that it has been measured. It is certain, however, that there is an interval of time between the emission of light by one body and its reception by another, and that during this time the energy transmitted from the one body to the other has existed in some form in the intervening medium. The opinions with regard to the relation between light and heat have suffered several alternations, according as these agents were regarded as substances or as accidents. At one time light was regarded as a substance projected from the luminous body, which, if the luminous body were hot, might itself become hot like any other substance. Heat was thus regarded as an accident of the substance light.

When the progress of science had rendered the measurement of quantities of heat as accurate as the measurement of quantities of gases, heat, under the name of caloric, was placed in the list of substances. Afterwards, the independent progress of optics led to the rejection of the corpuscular theory of light, and the establishment of the undulatory theory, according to which light is a wave-like motion of a medium already existing. The caloric theory of heat, however, still prevailed even after the corpuscular theory of light was rejected, so that heat and light seemed almost to have exchanged places.

When the caloric theory of heat was at length demonstrated to be false, the grounds of the argument were quite independent of those which had been used in the case of light.

We shall therefore consider the nature of radiation, whether of light or heat, in an independent manner, and show why we believe that what is called radiant heat is the same thing as what is called light, only perceived by us through a different channel. The same radiation which when we become aware of it by the eye we call light, when we detect it by a thermometer or by the sensation of heat we call radiant heat.

In the first place, radiant heat agrees with light in always moving in straight lines through any uniform medium. It is not, therefore, propagated by diffusion, as in the case of the conduction of heat, where the heat always travels from hotter to colder parts of the medium in whatever direction this condition may lead it.

The medium through which radiant heat passes is not heated if perfectly diathermanous, any more than a perfectly transparent medium through which light passes is rendered luminous. But if any impurity or defect of transparency causes the medium to become visible when light passes through it, it will also cause it to become hot and to stop part of the heat when traversed by radiant heat.

In the next place, radiant heat is reflected from the polished surfaces of bodies according to the same laws as light. A concave mirror collects the rays of the sun into a brilliantly luminous focus. If these collected rays fall on a piece of wood, they will set it on fire. If the luminous rays are collected by means of a convex lens, similar heating effects are produced, showing that radiant heat is refracted when it passes from one transparent medium to another.

When light is refracted through a prism, so as to change its direction through a considerable angle of deviation, it is separated into a series of kinds of light which are easily distinguished from each other by their various colours. The radiant heat which is refracted through the prism is also spread out through a considerable angular range, which shows that it also consists of radiations of various kinds. The luminosity of the different radiations is evidently not in the same proportion as their heating effects. For the blue and green rays have very little heating power compared with the extreme red, which are much less luminous, and the heating rays are found far beyond the end of the red, where no light at all is visible.

There are other methods of separating the different kinds of light, which are sometimes more convenient than the use of a prism. Many substances are more transparent to one kind of light than another, and are therefore called coloured media. Such media absorb certain rays and transmit others. If the light transmitted by a stratum of a coloured medium afterwards passes through another stratum of the same medium, it will be much less diminished in intensity than at first. For the kind of light which is most absorbed by the medium has been already removed, and what is transmitted by the first stratum is that which can pass most readily through the second. Thus a very thin stratum of a solution of bichromate of potash cuts off the whole of the spectrum from the middle of the green to the violet, but the remainder of the light, consisting of the red, orange, yellow, and part of the green, is very slightly diminished in intensity by passing through another stratum of the same medium.

If, however, the second stratum be of a different medium, which absorbs most of the rays which the first transmits, It will cut off nearly the whole light, though it may be itself very transparent for other rays absorbed by the first medium. Thus a stratum of sulphate of copper absorbs nearly all the rays transmitted by the bichromate of potash, except a few of the green rays.

Melloni found that different substances absorb different kinds of radiant heat, and that the heat sifted by a screen of any substance will pass in greater proportion through a screen of the same substance than unsifted heat, while it may be stopped in greater proportion than unsifted heat by a screen of a different substance.

These remarks may illustrate the general similarity between light and radiant heat. We must next consider the reasons which induce us to regard light as depending on a particular kind of motion in the medium through which it is propagated. These reasons are principally derived from the phenomena of the interference of light. They are explained more at large in treatises on light, because it is much easier to observe these phenomena by the eye than by any kind of thermometer. We shall therefore be as brief as possible.

There are various methods by which a beam of light from a small luminous object may be divided into two portions, which, after travelling by slightly different paths, finally fall on a white screen. Where the two portions of light overlap each other on the screen, a series of long narrow stripes may be seen, alternately lighter and darker than the average brightness of the screen near them, and when white light is used, these stripes are bordered with colours. By using light of one kind only, such as that obtained from the salted wick of a spirit-lamp, a greater number of bands or fringes may be seen, and a greater difference of brightness between the light and the dark bands. If we stop either of the portions of light into which the original beam was divided, the whole system of bands disappears, showing that they are due, not to either of the portions alone, but to both united.

If we now fix our attention on one of the dark bands, and then cut off one of the partial beams of light, we shall observe that instead of appearing darker it becomes actually brighter, and if we again allow the light to fall on the screen it becomes dark again. Hence it is possible to produce darkness by the addition of two portions of light. If light is a substance, there cannot be another substance which when added to it shall produce darkness. We are therefore compelled to admit that light is not a substance.

Now is there any other instance in which the addition of two apparently similar things diminishes the result? We know by experiments with musical instruments that a combination of two sounds may produce less audible effect than either separately, and it can be shown that this takes place when the one is half a wave-length in advance of the other. Here the mutual annihilation of the sounds arises from the fact that a motion of the air towards the ear is the exact opposite of a motion away from the ear, and if the two instruments are so arranged that the motions which they tend to produce in the air near the ear are in opposite directions and of equal magnitude, the result will be no motion at all. Now there is nothing absurd in one motion being the exact opposite of another, though the supposition that one substance is the exact opposite of another substance, as in some forms of the Two-Fluid theory of Electricity, is an absurdity.

We may show the interference of waves in a visible manner by dipping a two-pronged fork into water or mercury. The waves which diverge from the two centres where the prongs enter or leave the fluid are seen to produce a greater disturbance when they exactly coincide than, when one gets ahead of the other.

Now it is found, by measuring the positions of the bright and dark bands on the screen, that the difference of the distances travelled by the two portions of light is for the bright bands always an exact multiple of a certain very small distance which we shall call a wave-length, whereas for the dark bands it is intermediate between two multiples of the wave-length, being ½, 1½, 2½, &c., times that length.

We therefore conclude that whatever exists or takes place at a certain point in a ray of light, then, at the same instant, at a point at ½ or 1½ of the wave-length in advance, something exactly the opposite exists or takes place, so that in going along a ray we find an alternation of conditions which we may call positive and negative.

In the ordinary statement of the theory of undulations these conditions are described as motion of the medium in opposite directions. The essential character of the theory would remain the same if we were to substitute for ordinary motion to and fro any other succession of oppositely directed conditions. Professor Rankine has suggested opposite rotations of molecules about their axes, and I have suggested oppositely directed magnetizations and electromotive forces; but the adoption of either of these hypotheses would in no way alter the essential character of the undulation theory.

Now it is found that if a very narrow thermo-electric pile be placed in the position of the screen, and moved so that sometimes a bright band and sometimes a dark one falls on the pile, the galvanometer indicates that the pile receives more heat when in the bright than when in the dark band, and that when one portion of the beam is cut off the heat in the dark band is increased. Hence in the interference of radiations the heating effect obeys the same laws as the luminous effect.

Indeed it has been found that even when the source of radiation is a hot body which emits no luminous rays, the phenomena of interference can be traced, showing that two rays of dark heat can interfere no less than two rays of light. Hence all that we have said about the waves of light is applicable to the heat-radiation, which is therefore a series of waves.

It is also known in the case of light that after passing through a plate cut from a crystal of tourmaline parallel to its axis the transmitted beam cannot pass through a second similarly cut plate of tourmaline whose axis is perpendicular to that of the first, though it can pass through it when the axis is in any other position. Such a beam of light, which has different properties according as the second plate is turned into different positions round the beam as an axis is called a polarized beam. There are many other ways of polarizing a beam of light, but the result is always of the same kind. Now this property of polarized light shows that the motion which constitutes light cannot be in the direction of the ray, for then there could be no difference between different sides of the ray. The motion must be transverse to the direction of the ray, so that we may now describe a ray of polarized light as a condition of disturbance in a direction at right angles to the ray propagated through a medium so that the disturbance is in opposite directions at every half wave-length measured along the ray. Since Principal J.D. Forbes showed that a ray of dark heat can be polarized we can make the same assertion about the heat radiation.

Let us now consider the consequences of admitting that what we call radiation, whether of heat, light, or invisible rays which act on chemical preparations, is of the nature of a transverse undulation in a medium.

A transverse undulation is completely defined when we know—

1. Its wave-length, or the distance between two places in which the disturbance is in the same phase.
2. Its amplitude, or the greatest extent of the disturbance.
3. The plane in which the direction of the disturbance lies.
4. The phase of the wave at a particular point.
5. The velocity of propagation through the medium.

When we know these particulars about an undulation, it is completely defined, and cannot be altered in any way without changing some of these specifications.

Now by passing a beam consisting of any assemblage of undulations through a prism, we can separate it into portions according to their wave lengths, and we can select rays of a particular wave-length for examination. Of these we may, by means of a plate of tourmaline, select those whose plane of polarization is the principal plane of the tourmaline, but this is unnecessary for our purpose. We have now got rays of a definite wave-length. Their velocity of propagation depends only on the nature of the ray and of the medium, so that we cannot alter it at pleasure, and the phase changes so rapidly (billions of times in a second) that it cannot be directly observed. Hence the only variable quantity remaining is the amplitude of the disturbance, or, in other words, the intensity of the ray.

Now the ray may be observed in various ways. We may, if it excites the sensation of sight, receive it in to our eye. If it affects chemical compounds, we may observe its effect on them, or we may receive the ray on a thermo-electric pile and determine its heating effect.

But all these effects, being effects of one and the same thing, must rise and fall together. A ray of specified wavelength and specified plane of polarization cannot be a combination of several different things, such as a light-ray, a heat-ray, and an actinic ray. It must be one and the same thing, which has luminous, thermal, and actinic effects, and everything which increases one of these effects must increase the others also.

The chief reason why so much that has been written on this subject is tainted with the notion that heat is one thing and light is another seems to be that the arrangements for operating on radiations of a selected wave-length are troublesome, and when mixed radiations are employed, in which the luminous and the thermal effects are in different proportions, anything which alters the proportion of the different radiations in the mixture alters also the proportion of the resulting thermal and luminous effect, as indeed it generally alters the colour of the mixed light.

We have seen that the existence of these radiations may be detected in various ways—by photographic preparations, by the eye, and by the thermometer. There can be no doubt, however, as to which of these methods gives the true measure of the energy transmitted by the radiation. This is exactly measured by the heating effect of the ray when completely absorbed by any substance.

When the wave-length is greater than 812 millionths of a millimetre no luminous effect is produced on the eye, though the effect on the thermometer may be very great. When the wave-length is 650 millionths of a millimetre the ray is visible as a red light, and a considerable heating effect is observed. But when the wave-length is 500 millionths of a millimetre, the ray, which is seen as a brilliant green, has much less heating effect than the dark or the red rays, and it is difficult to obtain strong thermal effects with rays of smaller wave-lengths, even when concentrated.

But, on the other hand, the photographic effect of the radiation on salts of silver, which is very feeble in the red rays, and even in the green rays, becomes more powerful the smaller the wave-length, till for rays whose wave-length is 400, which have a feeble violet luminosity and a still feebler thermal effect, the photographic effect is very powerfull and even far beyond the visible spectrum, for wavelengths of less than 200 millionths of a millimetre, which are quite invisible to our eyes and quite undiscoverable by our thermometers, the photographic effect is still observed. This shows that neither the luminous nor the photographic effect is in any way proportional to the energy of the radiation when different kinds of radiation are concerned. It is probable that when the radiation produces the photographic effect it is not by its energy doing work on the chemical compound, but rather by a well-timed vibration of the molecules dislodging them from the position of almost indifferent equilibrium into which they had been thrown by previous chemical manipulations, and enabling them to rush together according to their more permanent affinities, so as to form stabler compounds. In cases of this kind the effect is no more a dynamical measure of the cause than the effect of the fall of a tree is a measure of the energy of the wind which uprooted it.

It is true that in many cases the amount of the radiation may be very accurately estimated by means of its chemical effects even when these chemical effects tend to diminish the intrinsic energy of the system. But by estimating the heating effect of a radiation which is entirely absorbed by the heated body we obtain a true measure of the energy of the radiation. It is found that a surface thickly coated with lampblack absorbs nearly the whole of every kind of radiation which falls on it. Hence surfaces of this kind are of great value in the thermal study of radiation.

We have now to consider the conditions which determine the amount and quality of the radiation from a heated body. We must bear in mind that temperature is a property of hot bodies and not of radiations, and that qualities such as wave-lengths, &c., belong to radiations, but not to the heat which produces them or is produced by them.

On Prevost's Theory of Exchanges

When a system of bodies at different temperatures is left to itself, the transfer of heat which takes place always has the effect of rendering the temperatures of the different bodies more nearly equal, and this character of the transfer of heat, that it passes from hotter to colder bodies, is the same whether it is by radiation or by conduction that the transfer takes place.

Let us consider a number of bodies, all at the same temperature, placed in a chamber the walls of which are maintained at that temperature, and through which no heat can pass by radiation (suppose the walls of metal, for instance). No change of temperature will occur in any of these bodies. They will be in thermal equilibrium with each other and with the walls of the chamber. This is a consequence of the definition of equal temperature at Chap. 2.[1]

Now if any one of these bodies had been taken out of the chamber and placed among colder bodies there would be a transfer of heat by radiation from the hot body to the colder ones; or if a colder body had been introduced into the chamber it would immediately begin to receive heat by radiation from the hotter bodies round it. But the cold body has no power of acting directly on the hot bodies at a distance, so as to cause them to begin to emit radiations, nor has the hot chamber any power to stop the radiation of any one of the hot bodies placed within it. We therefore conclude with Prevost that a hot body is always emitting radiations, even when no colder body is there to receive them, and that the reason why there is no change of temperature when a body is placed in a chamber of the same temperature is that it receives from the radiation of the walls of the chamber exactly as much heat as it loses by radiation towards these walls.

If this is the true explanation of the thermal equilibrium of radiation, it follows that if two bodies have the same temperature the radiation emitted by the first and absorbed by the second is equal in amount to the radiation emitted by the second and absorbed by the first during the same time.

The higher the temperature of a body, the greater its radiation is found to be, so that when the temperatures of the bodies are unequal the hotter bodies will emit more radiation than they receive from the colder bodies, and therefore, on the whole, heat will be lost by the hotter and gained by the colder bodies till thermal equilibrium is attained. We shall return to the comparison of the radiation at different temperatures after we have examined the relations between the radiation of different bodies at the same temperature.

The application of the theory of exchanges has at various times been extended to the phenomena of heat as they were successively investigated. Fourier has considered the law of radiation as depending on the angle which the ray makes with the surface, and Leslie has investigated its relation to the state of polish of the surface; but it is in recent times, and chiefly by the researches of B. Stewart, Kirchhoff, and

---

[1] Maxwell refers here to Chap. 2 of Maxwell, J. C., *Theory of Heat*, tenth ed., Longmans, Green, and Co., London and New York, 1891.

De la Provostaye, that the theory of exchanges has been shown to be applicable, not only to the total amount of the radiation, but to every distinction in quality of which the radiation is capable.

For, by placing between two bodies of the same temperature a contrivance such as that already noticed at p. 168, so that only radiations of a determinate wave-length and in a determinate plane can pass from the one body to the other, we reduce the general proposition about thermal equilibrium to a proposition about this particular kind of radiation. We may therefore transform it into the following more definite proposition.

If two bodies are at the same temperature, the radiation emitted by the first and absorbed by the second agrees with the radiation emitted by the second and absorbed by the first, not only in its total heating effect, but in the intensity, wave-length, and plane of polarization of every component part of either radiation. And the law that the amount of radiation increases with the temperature must be true, not only for the whole radiation, but for all the component parts of it when analysed according to their wave-lengths and planes of polarization.

The consequences of these two propositions, applying as they do to every kind of radiation, whether detected by its thermal or by its luminous effects, are so numerous and varied that we cannot attempt any full enumeration of them in this treatise. We must confine ourselves to a few examples.

When a radiation falls on a body, part of it is reflected and part enters the body. The latter part again may either be wholly absorbed by the body or partly absorbed and partly transmitted.

Now lampblack reflects hardly any of the radiation which falls on it, and it transmits none. Nearly the whole is absorbed.

Polished silver reflects nearly the whole of the radiation which falls upon it, absorbing only about a fortieth part, and transmitting none.

Rock salt reflects less than a twelfth part of the radiation which falls on it; it absorbs hardly any, and transmits 92 %.

These three substances, therefore, may be taken as types of absorption, reflexion, and transmission respectively.

Let us suppose that these properties have been observed in these substances at the temperature, say, of 212 °F, and let them be placed at this temperature within a chamber whose walls are at the same temperature. Then the amount of the radiation from the lampblack which is absorbed by the other two substances is, as we have seen, very small. Now the lampblack absorbs the whole of the radiation from the silver or the salt. Hence the radiation from these substances must also be small, or, more precisely—

> *The radiation of a substance at a given temperature is to the radiation of lampblack at that temperature as the amount of radiation absorbed by the substance at that temperature is to the whole radiation which falls upon it.*

Hence a body whose surface is made of polished silver will emit a much smaller amount of radiation than one whose surface is of lampblack. The brighter the surface of a silver teapot, the longer will it retain the heat of the tea; and if on the surface of a metal plate some parts are polished, others rough, and others blackened, when the plate is made red hot the blackened parts will appear brightest, the rough parts not so bright, and the polished parts darkest. This is well seen when melted lead is made red hot. When part of the dross is removed, the polished surface of the melted metal, though really hotter than the dross, appears of a less brilliant red.

A piece of glass when taken red hot out of the fire appears of a very faint red compared with a piece of iron taken from the same part of the fire, though the glass is really hotter than the iron, because it does not throw off its heat so fast. Air or any other transparent gas, even when raised to a heat at which opaque bodies appear white hot, emits so little light that its luminosity can hardly be observed in the dark, at least when the thickness of the heated air is not very great.

Again, when a substance at a given temperature absorbs certain kinds of radiation and transmits others, it emits at that temperature only those kinds of radiation which it absorbs. A very remarkable instance of this is observed in the vapour of sodium. This substance when heated emits rays of two definite kinds, whose wave-lengths are 0.00059053 and 0.00058989 mm respectively. These rays are visible, and may be seen in the form of two bright lines by directing a spectroscope upon a flame in which any compound of sodium is present.

Now if the light emitted from an intensely heated solid body, such as a piece of lime in the oxyhydrogen light, be transmitted through sodium-vapour at a temperature lower than that of the lime, and then analysed by the spectroscope, two dark lines are seen, corresponding to the two bright ones formerly observed, showing that sodium-vapour absorbs the same definite kinds of light which it radiates.

If the temperature of the sodium-vapour is raised, say by using a Bunsen's burner instead of a spirit-lamp to produce it, or if the temperature of the lime is lowered till it is the same as that of the vapour, the dark lines disappear, because the sodium-vapour now radiates exactly as much light as it absorbs from the light of the lime-ball at the same temperature. If the sodium-flame is hotter than the lime-ball the lines appear bright.

This is an illustration of Kirchhoff's principle, that the radiation of every kind increases as the temperature rises.

In performing this experiment we suppose the light from the lime-ball to pass through the sodium-flame before it reaches the slit of the spectroscope. If, however, the flame is interposed between the slit and the eye, or the screen on which the spectrum is projected, the dark lines may be seen distinctly, even when the temperature of the sodium-flame is higher than that of the lime-ball. For in the parts of the spectrum near the lines the light is now compounded of the analysed light of the lime-ball and the direct light of the sodium-flame, while at the lines themselves the light of the spectrum of the lime-ball is cut off, and only the direct light of the sodium-flame remains, so that the lines appear darker than the rest of the field.

It does not belong to the scope of this treatise to attempt to go over the immense field of research which has been opened up by the application of the spectroscope to distinguish different incandescent vapours, and which has led to a great increase of our knowledge of the heavenly bodies.

If the thickness of a medium, such as sodium-vapour, which radiates and absorbs definite kinds of light, be very great, the whole being at a high temperature, the light emitted will be of exactly the same composition as that emitted from lampblack at the same temperature. For, though some kinds of radiation are much more feebly emitted by the substance than others, these are also so feebly absorbed that they can reach the surface from immense depths, whereas the rays which are so copiously radiated are also so rapidly absorbed that it is only from places very near the surface that they can escape out of the medium. Hence both the depth and the density of an incandescent gas cause its radiation to assume more and more of the character of a continuous spectrum.

When the temperature of a substance is gradually raised, not only does the intensity of every particular kind of radiation increase, but new kinds of radiation are produced. Bodies of low temperature emit only rays of great wavelength. As the temperature rises these rays are more copiously emitted, but at the same time other rays of smaller wave-length make their appearance. When the temperature has risen to a certain point, part of the radiation is luminous and of a red colour, the luminous rays of greatest wave-length being red. As the temperature rises, the other luminous rays appear in the order of the spectrum, but every rise of temperature increases the intensity of all the rays which have already made their appearance. A white-hot body emits more red rays than a red-hot body, and more non-luminous rays than any non-luminous body.

The total thermal value of the radiation at any temperature, depending as it does upon the amount of all the different kinds of rays of which it is composed, is not likely to be a simple function of the temperature. Nevertheless, Dulong and Petit succeeded in obtaining a formula which expresses the facts observed by them with tolerable exactness. It is of the form

$$R = ma^\theta,$$

where R is the total loss of heat in unit of time by radiation from unit of area of the surface of the substance at the temperature θ, m is a constant quantity depending only on the substance and the nature of its surface, and a is a numerical quantity which, when θ expresses the temperature on the Centigrade scale, is 1.0077.

If the body is placed in a chamber devoid of air, whose walls are at the temperature t, then the heat radiated from the walls to the body and absorbed by it will be

$$r = ma^t,$$

so that the actual loss of heat will be

$$R - r = ma^\theta - ma^t$$

The constancy of the amount of radiation between the same surfaces at the same temperatures affords a very convenient method of comparing quantities of heat. This method was referred to in our chapter on Calorimetry (Chap. 3),[2] under the name of the Method of Cooling.

The substance to be examined is heated and put into a thin copper vessel, the outer surface of which is blackened, or at least is preserved in the same state of roughness or of polish throughout the experiments. This vessel is placed in a larger copper vessel so as not to touch it, and the outer vessel is placed in a bath of water kept at a constant temperature. The temperature of the substance in the smaller vessel is observed from time to time, or, still better, the times are observed at which the reading of a thermometer immersed in the substance is an exact number of degrees. In this way the time of cooling, say from 100° to 90°, from 90° to 80°, is registered, the temperature of the outer vessel being kept always the same.

Suppose that this observation of the time of cooling is made first when the vessel is filled with water, and then when some other substance is put into it. The rate at which heat escapes by radiation is the same for the same temperature in both experiments. The quantity of heat which escapes during the cooling, say from 100° to 90°, in the two experiments, is proportional to the time of cooling. Hence the capacity of the vessel and its contents in the first experiment is to its capacity in the second experiment as the time of cooling from 100° to 90° in the first experiment is to the time of cooling from 100° to 90° in the second experiment.

The method of cooling is very convenient in certain cases, but it is necessary to keep the temperature of the whole of the substance in the inner vessel as nearly uniform as possible, so that the method must be restricted to liquids which we can stir, and to solids whose conductivity is great, and which may be cut in pieces and immersed in a liquid.

The method of cooling has been found very applicable to the measurement of the quantity of heat conducted through a substance. (See the chapter on Conduction[3])

13.2.1   *Effect of Radiation on Thermometers*

On account of the radiation passing in all directions through the atmosphere, it is a very difficult thing to determine the true temperature of the air in any place out of doors by means of a thermometer.

If the sun shines on the thermometer, the reading is of course too high; but if we put it in the shade, it may be too low, because the thermometer may be emitting more radiation than it receives from the clear sky. The ground, walls of houses, clouds, and the various devices for shielding the thermometer from radiation, may all become sources of error, by causing an unknown amount of radiation on the bulb. For rough purposes the effects of radiation may be greatly removed by giving the

---

[2] Chap. 3 of Maxwell, J. C., *Theory of Heat*, tenth ed., Longmans, Green, and Co., London and New York, 1891.—[*K.K.*]

[3] Included in Chap. 12 of the present volume.—[*K.K.*]

bulb a surface of polished silver, of which, as we have seen, the absorption is only a fortieth of that of lampblack.

A method described by Dr. Joule in a communication to the Philosophical Society of Manchester, November 26, 1867, seems the only one free from all objections. The thermometer is placed in a long vertical copper tube open at both ends, but with a cap to close the lower end, which may be removed or put on without warming it by the hand. Whatever radiation affects the thermometer must be between it and the inside of the tube, and if these are of the same temperature, the radiation will have no effect on the observed reading of the thermometer. Hence, if we can be sure that the copper tube and the air within it are at the temperature of the atmosphere, and that the thermometer is in thermal equilibrium, the thermometer reading will be the true temperature.

Now, if the air within the tube is of the same temperature as the air outside, it will be of the same density, and it will therefore be in statical equilibrium with it. If it is warmer it will be lighter, and an upward current will be formed in the tube when the cap is removed. If it is colder, a downward current will be formed.

To detect these currents a spiral wire is suspended in the tube by a fine fibre, so that an upward or downward current causes the spiral to twist the fibre, and any motion of the spiral is made apparent by means of a small mirror attached to it.

To vary the temperature of the copper tube it is enclosed in a wider tube, so that water may be placed in the space between the tubes, and by pouring in warmer or cooler water the temperature may be adjusted till there is no current.

We then know that the air is of the same temperature within the tube as it is without. But we know that the tube is also of the same temperature as the air, for if it were not it would heat or cool the air and produce a current. Finally, we know that the thermometer, if stationary, is at the temperature of the atmosphere; for the air in contact with it, and the sides of the tube, which alone can exchange radiations with it, have the same temperature as the atmosphere.

13.3 Study Questions

QUES. 13.1. In what sense is wave energy different than the heat energy within a substance? What evidence supports the idea that energy may reside in space which is devoid of matter?

QUES. 13.2. What is radiant heat? In particular, (a) does radiant heat travel in straight lines? (b) Can it be reflected and refracted? (c) Does it heat up transparent substances through which it travels? (d) Can it be filtered, or sifted, by transparent media? (e) Does radiant heat exhibit interference patterns? (f) Can it be polarized?

QUES. 13.3. What is the strongest evidence that light is not itself a substance? And if it is not a substance, then what is it? What are two hypotheses which attempt to account for the undulation (wave) theory of light? Does the undulation theory require either of these to be true?

QUES. 13.4. How are heat, light, and actinic rays alike and different? (a) What are the properties or quantities which define a transverse wave? (b) How are heat rays, light rays, and actinic rays experimentally detected? (c) Why were heat and light thought to be fundamentally different phenomena for many years? (d) How can the energy of a ray be accurately measured?

QUES. 13.5. Which properties of a body affect the amount of heat radiation it emits and absorbs?

a) Do bodies in a vacuum exchange heat? What if they are at identical temperatures?
b) What two propositions, or laws, govern the emission and absorption of radiant heat? What are some of the consequences of these laws?
c) Why should you polish your silver teapot? Consider: when heated to the same temperature, which glows brightest—a polished mirror, a dull mirror, or a black-painted mirror?
d) Does the relationship between a substance's ability to absorb and emit heat radiation apply to each particular frequency of heat radiation, or merely to the total radiant power?
e) Does the emission spectrum of a substance depend on the size and density of the substance, or merely on the composition of the substance? For example, would a star composed entirely of hot sodium gas emit the same color of light as a small sodium-vapor lamp at the same temperature?
f) What is Kirchoff's principle? Does the emission spectrum of a medium depend on its temperature? If so, how?
g) How can an understanding of radiant heat be used to accurately measure the heat capacity of a solid or liquid substance?

QUES. 13.6. What difficulty arises when using a thermometer to measure air temperature? How can this difficulty be minimized by coating the thermometer bulb with polished silver? And what is Dr. Joule's more sophisticated method of alleviating the aforementioned difficulty?

13.4 Exercises

EX. 13.1 (LESLIE'S CUBE LABORATORY). A Leslie's Cube can be used to illustrate how the characteristics of a body's surface affects the amount of radiant heat it emits. It consists of a small cubical container that can be filled with very hot water through a sealable hole in the top and whose four sides are coated with (i) white paint, (ii) black paint, (iii) rough brass, and (iv) polished brass, respectively. Fill a Leslie's cube with very hot water (it doesn't need to be boiling) and use a radiometer to explore the intensity of the radiation emitted from each of the surfaces.[4] Does the radiation intensity change as the cube cools down?

Ex. 13.2 (STEFAN-BOLTZMANN LAW LABORATORY). The *Stefan-Boltzmann law* relates the intensity of radiation by a black body to its absolute temperature. It expresses the empirical fact that the total energy radiated per unit time from a unit area of a black body's surface is proportional to the fourth power of its temperature:

$$P = \sigma T^4. \tag{13.1}$$

The Stefan-Boltzmann constant, σ, has a value of 5.67×10^{-8} W/m^2-K^4; from a theoretical perspective, it may be expressed in terms of Boltzmann's constant, k_B, the speed of light, c, and Planck's constant, h:

$$\sigma = \frac{2\pi^4 k_B^4}{15\, c^2 h^3}. \tag{13.2}$$

In this laboratory exercise, we will use a radiometer to measure how the energy radiated from a hot tungsten filament depends on the temperature of the filament.[5] The temperature of the tungsten wire may be obtained from its resistance. In particular, the ratio of the resistance at temperature T to its resistance at temperature 293 K, is given by

$$\frac{R_T}{R_{293}} = 1 + \alpha(T - 293) \tag{13.3}$$

The temperature coefficient, α, for tungsten is about 0.0045.

Place the radiometer about 10 cm from the tungsten filament bulb. Attach the bulb to a power supply and turn it on. Use a resistor to limit the electrical current through the filament so you don't burn it out. Simultaneously measure the electrical current through the filament and the voltage drop across the filament. Then you can use Ohm's law to calculate the resistance, and hence the temperature, of the filament. How does the radiated power depend on the temperature of the filament? Does it agree with the Stefan-Boltzmann law?

Ex. 13.3 (RADIANT COOLING OF A COPPER SPHERE). It has been found that good absorbers of radiation are also good emitters of radiation. This is why polishing a tea-pot minimizes radiative heat loss; it is also why emergency thermal blankets are coated with reflective material. The *emissivity*, ε, describes the fraction of radiation a surface absorbs. For a black body $\varepsilon = 1$, for a mirror $\varepsilon = 0$, and for a grey body $1 < \varepsilon < 0$. Generally speaking, the Stefan-Boltzmann law (Eq 13.1) must be modified to account for the emissivity of the body under consideration:

$$P = \varepsilon \sigma T^4. \tag{13.4}$$

---

[4] A Leslie's Cube (Model EH-10) and a Radiometer (Model EG-45) are available from Daedelon Corp., Salem, MA.

[5] A Stefan-Boltzmann source (Model EH-15) and a Radiometer (Model EG-45) are available from Daedelon Corp., Salem, MA.

As an exercise, suppose that a copper sphere of radius $r = 1$ cm with an emissivity $\varepsilon = 0.5$ (its surface is oxidized) is suspended in an evacuated cryogenic chamber. The chamber is thermally regulated at $T_0 = 3$ K; the initial temperature of the copper sphere is $T = 10$ K.

a) At what rate does the copper sphere emit radiant energy from its surface? At what rate does it absorb radiant energy from the surrounding chamber? Is the copper sphere in thermal equilibrium?

b) Show that the rate of change of temperature of the sphere is given by

$$\frac{dT}{dt} = \frac{3\varepsilon\sigma}{r\rho c}(T^4 - T_0^4). \tag{13.5}$$

Here, c and ρ are the specific heat capacity and density of the sphere, respectively.

c) At sufficiently low-temperatures—below the so-called *Debye temperature*—the heat capacity of many solid materials is proportional to the third power of its absolute temperature:

$$C = \frac{12\pi^4}{5} N k_B \left(\frac{T}{\Theta_D}\right)^3. \tag{13.6}$$

For copper, the Debye temperature $\Theta_D = 344$ K. In Eq. 13.6, N is the number of atoms in the solid and k_B is Boltzmann's constant. How much time, τ, does it take for the copper's temperature to drop by a factor of e^{-1}? What if the sphere was polished so that $\varepsilon = 0.05$?

13.5   Vocabulary

1. Transmission
2. Intervene
3. Rarefaction
4. Corpuscular
5. Caloric
6. Diathermanous
7. Luminous
8. Refract
9. Stratum
10. Potash
11. Undulation
12. Interference
13. Polarize
14. Transverse
15. Assemblage
16. Tourmaline

17. Actinic
18. Wavelength
19. Taint
20. Affinity
21. Intrinsic
22. Lampblack
23. Enumerate
24. Treatise
25. Spectroscope
26. Copious
27. Incandescent
28. Spectrum
29. Calorimetry
30. Equilibrium

Chapter 14
From Positivism to Objectivity

> *The system of physics is still suffering from a strong dose of anthropomorphism.*
>
> —Max Planck

14.1 Introduction

Max Planck (1858–1947) was born in the city of Kiel in the Duchy of Holstein—now part of Germany.[1] As a youth, he attended the classical Königliche Maximilian Gymnasium, where he was introduced to the principle of the conservation of energy by one of his teachers, Hermann Müller. Although the young Planck was gifted in music—he excelled at performing on both the piano and the organ—he opted to study physics and mathematics when he enrolled at the University of Munich in 1874. Three years later, he traveled to Berlin for a year to study under Hermann von Helmholtz, Gustav Kirchoff and the mathematician Karl Weierstrass. Inspired by a careful reading of the theoretical papers of Rudolph Clausius during his time in Berlin, Planck wrote his 1879 doctoral dissertation on Clausius' second law of thermodynamics.[2] After completing his habilitation thesis on *The equilibrium states of isotropic bodies at different temperatures*, Planck returned to Munich to work as a lecturer (*Privatdozent*) for a few years before being appointed Associate Professor of Theoretical Physics at the University of Kiel in 1885. His subsequent work began to combine Clausius' theory of thermodynamics with Maxwell's electromagnetic theory of light. An outstanding problem, identified by Kirchoff, was obtaining a proper theoretical account of the electromagnetic radiation emitted by a heated body which was itself coated with a completely absorbing substance—the so-called "black-body radiation." Existing theories, which treated the body as a collection of tiny oscillating electrical charges, failed to account for the observed black-body emission spectrum. After completing his *Treatise on Thermodynamics* in 1897, he

---

[1] Much of this biographical information was gleaned from "Planck, Max Karl Ernst Ludwig." Complete Dictionary of Scientific Biography. 2008. *Encyclopedia.com*. (December 1, 2014). http://www.encyclopedia.com/doc/1G2-2830903438.html.

[2] See Chaps. 6–9 of the present volume.

© Springer International Publishing Switzerland 2016
K. Kuehn, *A Student's Guide Through the Great Physics Texts*,
Undergraduate Lecture Notes in Physics, DOI 10.1007/978-3-319-21828-1_14

moved to the University of Berlin where he succeeded Kirchoff as assistant professor and director of the new Institute of Theoretical Physics. It was during his time in Berlin that Planck solved the problem of black-body radiation. He did so by making the radical assumption that electromagnetic energy can only exist in discrete units, or *quanta*. The publication of Planck's quantum hypothesis in 1900 is now regarded as marking the birth of modern quantum theory. In the reading selection below, Planck presents some of his mature views on the relationship between heat, energy and probability.

Planck was elected a member of the Prussian Academy of Sciences in 1894, and he served as president of the German Physical Society from 1905–1909. For his work on black-body radiation, Planck was awarded the Nobel Prize for physics in 1918. And after retiring in 1926, he served as president of the Kaiser Wilhelm Society. But while he enjoyed great success and recognition in his professional life, his personal life was marked by profound tragedy. His wife died in 1909; his son Karl was killed in action during the first world war; his daughters Margarete and Emma both died in childbirth; and his son Erwin was executed in 1945 for his suspected role in the failed attempt to assassinate Adolf Hitler. Most of his personal books and manuscripts were destroyed during the Allied bombing of Berlin.

The two reading selections that follow are drawn from a series of lectures delivered by Max Planck in 1909 at Columbia University. In his first lecture, Planck begins with a review of how physical science has been historically organized anthropomorphically—according to the particular sensitivities of man. He then describes how the advances in thermodynamics—specifically Rudolph Clausius's development of the second law—has inspired a reorganization of the whole of physical science. Along what particular lines does Planck propose to divide physics henceforth? Would such a division attain the goal of complete objectivity?

14.2   Reading: Planck, *Reversibility and Irreversibility*

Planck, M., *Eight Lectures on Theoretical Physics*, Columbia University Press, New York, 1915. First Lecture.

Colleagues, ladies and gentlemen: The cordial invitation, which the President of Columbia University extended to me to deliver at this prominent center of American science some lectures in the domain of theoretical physics, has inspired in me a sense of the high honor and distinction thus conferred upon me and, in no less degree, a consciousness of the special obligations which, through its acceptance, would be imposed upon me. If I am to count upon meeting in some measure your just expectations, I can succeed only through directing your attention to the branches of my science with which I myself have been specially and deeply concerned, thus exposing myself to the danger that my report in certain respects shall thereby have somewhat too subjective a coloring. From those points of view which appear to me the most striking, it is my desire to depict for you in these lectures the present status of the system of theoretical physics. I do not say: the present status of theoretical physics; for to cover this far broader subject, even approximately, the number of lecture hours at my disposal would by no means suffice. Time limitations forbid the extensive consideration of the details of this great field of learning; but it will be quite possible to develop for you, in bold outline, a representation of the system as a whole, that is, to give a sketch of the fundamental laws which rule in the physics of today, of the most important hypotheses employed, and of the great ideas which have recently forced themselves into the subject. I will often gladly endeavor to go into details, but not in the sense of a thorough treatment of the subject, and only with the object of making the general laws more clear, through appropriate specially chosen examples. I shall select these examples from the most varied branches of physics.

If we wish to obtain a correct understanding of the achievements of theoretical physics, we must guard in equal measure against the mistake of overestimating these achievements, and on the other hand, against the corresponding mistake of underestimating them. That the second mistake is actually often made, is shown by the circumstance that quite recently voices have been loudly raised maintaining the bankruptcy and, débâcle of the whole of natural science. But I think such assertions may easily be refuted by reference to the simple fact that with each decade the number and the significance of the means increase, whereby mankind learns directly through the aid of theoretical physics to make nature useful for its own purposes. The technology of today would be impossible without the aid of theoretical physics. The development of the whole of electro-technics from galvanoplasty to wireless telegraphy is a striking proof of this, not to mention aerial navigation. On the other hand, the mistake of overestimating the achievements of theoretical physics appears to me to be much more dangerous, and this danger is particularly threatened by those who have penetrated comparatively little into the heart of the subject. They maintain that some time, through a proper improvement of our science, it will be possible, not only to represent completely through physical formulae the inner constitution of the atoms, but also the laws of mental life. I think that there is nothing in the world entitling us to the one or the other of these expectations. On the other hand, I believe that there is much which directly opposes them. Let us endeavor then to follow the middle course and not to deviate appreciably toward the one side or the other.

When we seek for a solid immovable foundation which is able to carry the whole structure of theoretical physics, we meet with the questions: What lies at the bottom of physics? What is the material with which it operates? Fortunately, there is a complete answer to this question. The material with which theoretical physics operates is measurements, and mathematics is the chief tool with which this material is worked. All physical ideas depend upon measurements, more or less exactly carried out, and each physical definition, each physical law, possesses a more definite significance the nearer it can be brought into accord with the results of measurements. Now measurements are made with the aid of the senses; before all with that of sight, with hearing and with feeling. Thus far, one can say that the origin and the foundation of all physical research are seated in our sense perceptions. Through sense perceptions only do we experience anything of nature; they are the highest court of appeal in questions under dispute. This view is completely confirmed by a glance at the historical development of physical science. Physics grows upon the ground of sensations. The first physical ideas derived were from the individual perceptions of man, and, accordingly, physics was subdivided into: physics of the eye (optics), physics of the car (acoustics), and physics of heat sensation (theory of heat). It may well be said that so far as there was a domain of sense, so far extended originally the domain of physics. Therefore it appears that in the beginning the division of physics was based upon the peculiarities of man. It possessed, in short, an anthropomorphic character. This appears also, in that physical research, when not occupied with special sense perceptions, is concerned with practical life, and particularly with the practical needs of men. Thus, the art of geodesy led to geometry, the study of machinery to mechanics, and the conclusion lies near that physics in the last analysis had only to do with the sense perceptions and needs of mankind.

In accordance with this view, the sense perceptions are the essential elements of the world; to construct an object as opposed to sense perceptions is more or less an arbitrary matter of will. In fact, when I speak of a tree, I really mean only a complex of sense perceptions: I can see it, I can hear the rustling of its branches, I can smell its fragrance, I experience pain if I knock my head against it, but disregarding all of these sensations, there remains nothing to be made the object of a measurement, wherewith, therefore, natural science can occupy itself. This is certainly true. In accordance with this view, the problem of physics consists only in the relating of sense perceptions, in accordance with experience, to fixed laws; or, as one may express it, in the greatest possible economic accommodation of our ideas to our sensations, an operation which we undertake solely because it is of use to us in the general battle of existence.

All this appears extraordinarily simple and clear and, in accordance with it, the fact may readily be explained that this positivist view is quite widely spread in scientific circles today. It permits, so far as it is limited to the standpoint here depicted (not always done consistently by the exponents of positivism), no hypothesis, no metaphysics; all is clear and plain. I will go still further; this conception never leads to an actual contradiction. I may even say, it can lead to no contradiction. But, ladies and gentlemen, this view has never contributed to any advance in physics. If physics is to advance, in a certain sense its problem must be stated in quite the inverse way, on account of the fact that this conception is inadequate and at bottom possesses only a formal meaning.

The proof of the correctness of this assertion is to be found directly from a consideration of the process of development which theoretical physics has actually undergone, and which one certainly cannot fail to designate as essential. Let us compare the system of physics of today with the earlier and more primitive system which I have depicted above. At the first glance we encounter the most striking difference of all, that in the present system, as well in the division of the various physical domains as in all physical definitions, the historical element plays a much smaller role than in the earlier system. While originally, as I have shown above, the fundamental ideas of physics were taken from the specific sense perceptions of man, the latter are today in large measure excluded from physical acoustics, optics, and the theory of heat. The physical definitions of tone, color, and of temperature are today in no wise derived from perception through the corresponding senses; but tone and color are defined through a vibration number or wave length, and the temperature through the volume change of a thermometric substance, or through a temperature scale based on the second law of thermodynamics; but heat sensation is in no wise mentioned in connection with the temperature. With the idea of force it has not been otherwise. Without doubt, the word force originally meant bodily force, corresponding to the circumstance that the oldest tools, the ax, hammer, and mallet, were swung by man's hands, and that the first machines, the lever, roller, and screw, were operated by men or animals. This shows that the idea of force was originally derived from the sense of force, or muscular sense, and was, therefore, a specific sense perception. Consequently, I regard it today as quite essential in a lecture on mechanics to refer, at any rate in the introduction, to the original meaning of the force idea. But in the modern exact definition of force the specific notion of sense perception is eliminated, as in the case of color sense, and we may say, quite in general, that in modern theoretical physics the specific sense perceptions play a much smaller rôle in all physical definitions than formerly. In fact, the crowding into the background of the specific sense elements goes so far that the branches of physics which were originally completely and uniquely characterized by an arrangement in accordance with definite sense perceptions have fallen apart, in consequence of the loosening of the bonds between different and widely separated branches, on account of the general advance towards simplification and coordination. The best example of this is furnished by the theory of heat. Earlier, heat formed a separate and unified domain of physics, characterized through the perceptions of heat sensation. Today one finds in well nigh all physics textbooks dealing with heat a whole domain, that of radiant heat, separated and treated under optics. The significance of heat perception no longer suffices to bring together the heterogeneous parts.

In short, we may say that the characteristic feature of the entire previous development of theoretical physics is a definite elimination from all physical ideas of the anthropomorphic elements, particularly those of specific sense perceptions. On the other hand, as we have seen above, if one reflects that the perceptions form the point of departure in all physical research, and that it is impossible to contemplate their absolute exclusion, because we cannot close the source of all our knowledge, then this conscious departure from the original conceptions must always appear astonishing or even paradoxical. There is scarcely a fact in the history of physics which today stands out so clearly as this. Now, what are the great advantages to be gained through such a real obliteration of personality? What is the result for the sake of whose achievement are sacrificed the directness and succinctness such as only the special sense perceptions vouchsafe to physical ideas?

The result is nothing more than the attainment of unity and compactness in our system of theoretical physics, and, in fact, the unity of the system, not only in relation to all of its details, but also in relation to physicists of all places, all times, all peoples, all cultures. Certainly, the system of theoretical physics should be adequate, not only for the inhabitants of this earth, but also for the inhabitants of other heavenly bodies. Whether the inhabitants of Mars, in case such actually exist, have eyes and ears like our own, we do not know,—it is quite improbable; but that they, in so far as they possess the necessary intelligence, recognize the law of gravitation

and the principle of energy, most physicists would hold as self evident: and anyone
to whom this is not evident had better not appeal to the physicists, for it will always
remain for him an unsolvable riddle that the same physics is made in the United
States as in Germany.

To sum up, we may say that the characteristic feature of the actual development
of the system of theoretical physics is an ever extending emancipation from the
anthropomorphic elements, which has for its object the most complete separation
possible of the system of physics and the individual personality of the physicist.
One may call this the objectiveness of the system of physics. In order to exclude
the possibility of any misunderstanding, I wish to emphasize particularly that we
have here to do, not with an absolute separation of physics from the physicist—for a
physics without the physicist is unthinkable,—but with the elimination of the individuality of the particular physicist and therefore with the production of a common
system of physics for all physicists.

Now, how does this principle agree with the positivist conceptions mentioned
above? Separation of the system of physics from the individual personality of the
physicist? Opposed to this principle, in accordance with those conceptions, each
particular physicist must have his special system of physics, in case that complete
elimination of all metaphysical elements is effected; for physics occupies itself only
with the facts discovered through perceptions, and only the individual perceptions
are directly involved. That other living beings have sensations is, strictly speaking, but a very probable: though arbitrary, conclusion from analogy. The system of
physics is therefore primarily an individual matter and, if two physicists accept the
same system, it is a very happy circumstance in connection with their personal relationship, but it is not essentially necessary. One can regard this view-point however
he will; in physics it is certainly quite fruitless, and this is all that I care to maintain here. Certainly, I might add, each great physical idea means a further advance
toward the emancipation from anthropomorphic ideas. This was true in the passage
from the Ptolemaic to the Copernican cosmical system, just as it is true at the present
time for the apparently impending passage from the so-called classical mechanics
of mass points to the general dynamics originating in the principle of relativity. In
accordance with this, man and the earth upon which he dwells are removed from
the centre of the world. It may be predicted that in this century the idea of time
will be divested of the absolute character with which men have been accustomed
to endow it (*cf.* the final lecture). Certainly, the sacrifices demanded by every such
revolution in the intuitive point of view are enormous; consequently, the resistance
against such a change is very great. But the development of science is not to be
permanently halted thereby; on the contrary, its strongest impetus is experienced
through precisely those forces which attain success in the struggle against the old
points of view, and to this extent such a struggle is constantly necessary and useful.

Now, how far have we advanced today toward the unification of our system of
physics? The numerous independent domains of the earlier physics now appear
reduced to two; mechanics and electrodynamics, or, as one may say: the physics
of material bodies and the physics of the ether. The former comprehends acoustics, phenomena in material bodies, and chemical phenomena; the latter, magnetism,

optics, and radiant heat. But is this division a fundamental one? Will it prove final? This is a question of great consequence for the future development of physics. For myself, I believe it must be answered in the negative, and upon the following grounds: mechanics and electrodynamics cannot be permanently sharply differentiated from each other. Does the process of light emission, for example, belong to mechanics or to electrodynamics? To which domain shall be assigned the laws of motion of electrons? At first glance, one may perhaps say: to electrodynamics, since with the electrons ponderable matter docs not play any rule. But let one direct his attention to the motion of free electrons in metals. There he will find, in the study of the classical researches of H. A. Lorentz, for example, that the laws obeyed by the electrons belong rather to the kinetic theory of gases than to electrodynamics. In general, it appears to me that the original differences between processes in the ether and processes in material bodies are to be considered as disappearing. Electrodynamics and mechanics are not so remarkably far apart, as is considered to be the case by many people, who already speak of a conflict between the mechanical and the electrodynamic views of the world. Mechanics requires for its foundation essentially nothing more than the ideas of space, of time, and of that which is moving, whether one considers this as a substance or a state. The same ideas are also involved in electrodynamics. A sufficiently generalized conception of mechanics can therefore also well include electrodynamics, and, in fact, there are many indications pointing toward the ultimate amalgamation of these two subjects, the domains of which already overlap in some measure.

If, therefore, the gulf between ether and matter be once bridged, what is the point of view which in the last analysis will best serve in the subdivision of the system of physics? The answer to this question will characterize the whole nature of the further development of our science. It is, therefore, the most important among all those which I propose to treat today. But for the purposes of a closer investigation it is necessary that we go somewhat more deeply into the peculiarities of physical principles.

We shall best begin at that point from which the first step was made toward the actual realization of the unified system of physics previously postulated by the philosophers only; at the principle of conservation of energy. For the idea of energy is the only one besides those of space and time which is common to all the various domains of physics. In accordance with what I have stated above, it will be apparent and quite self evident to you that the principle of energy, before its general formularization by Mayer, Joule, and Helmholtz, also bore an anthropomorphic character. The roots of this principle lay already in the recognition of the fact that no one is able to obtain useful work from nothing; and this recognition had originated essentially in the experiences which were gathered in attempts at the solution of a technical problem: the discovery of perpetual motion. To this extent, perpetual motion has come to have for physics a far reaching significance, similar to that of alchemy for the chemist, although it was not the positive, but rather the negative results of these experiments, through which science was advanced. Today we speak of the principle of energy quite without reference to the technical viewpoint or to that of man. We say that the total amount of energy of an isolated system of bodies is a quantity whose amount can be neither increased nor diminished through any kind of process within the system, and we no longer consider the accuracy with which this law holds as dependent upon the refinement of the methods, which we at present possess, of testing experimentally the question of the realization of perpetual motion. In this, strictly speaking, unprovable generalization, impressed upon us with elemental force, lies the emancipation from the anthropomorphic elements mentioned above.

While the principle of energy stands before us as a complete independent structure, freed from and independent of the accidents appertaining to its historical development, this is by no means true in equal measure in the case of that principle which R. Clausius introduced into physics; namely, the second law of thermodynamics. This law plays a very peculiar role in the development of physical science, to the extent that one is not able to assert today that for it a generally recognized, and therefore objective formularization, has been found. In our present consideration it is therefore a matter of particular interest to examine more closely its significance.

In contrast to the first law of thermodynamics, or the energy principle, the second law may be characterized as follows. While the first law permits in all processes of nature neither the creation nor destruction of energy, but permits of transformations only, the second law goes still further into the limitation of the possible processes of nature, in that it permits, not all kinds of transformations, but only certain types, subject to certain conditions. The second law occupies itself, therefore, with the question of the kind and, in particular, with the direction of any natural process.

At this point a mistake has frequently been made, which has hindered in a very pronounced manner the advance of science up to the present day. In the endeavor to give to the second law of thermodynamics the most general character possible, it has been proclaimed by followers of W. Ostwald as the second law of energetics, and the attempt made so to formulate it that it shall determine quite generally the direction of every process occurring in nature. Some weeks ago I read in a public academic address of an esteemed colleague the statement that the import of the second law consists in this, that a stone falls downwards, that water flows not up hill, but down, that electricity flows from a higher to a lower potential, and so on. This is a mistake which at present is altogether too prevalent not to warrant mention here.

The truth is, these statements are false. A stone can just as well rise in the air as fall downwards; water can likewise flow upwards, as, for example, in a spring; electricity can flow very well from a lower to a higher potential, as in the case of oscillating discharge of a condenser. The statements are obviously quite correct, if one applies them to a stone originally at rest, to water at rest, to electricity at rest; but then they follow immediately from the energy principle, and one does not need to add a special second law. For, in accordance with the energy principle, the kinetic energy of the stone or of the water can only originate at the cost of gravitational energy, *i.e.*, the center of mass must descend. If, therefore, motion is to take place at all, it is necessary that the gravitational energy shall decrease. That is, the center of mass must descend. In like manner, an electric current between two condenser plates can originate only at the cost of electrical energy already present; the electricity must therefore pass to a lower potential. If, however, motion and current be already present, then one is not able to say, a priori, anything in regard to the direction of the change; it can take place just as well in one direction as the other. Therefore, there is no new insight into nature to be obtained from this point of view.

Upon an equally inadequate basis rests another conception of the second law, which I shall now mention. In considering the circumstance that mechanical work may very easily be transformed into heat, as by friction, while on the other hand heat can only with difficulty be transformed into work, the attempt has been made so to characterize the second law, that in nature the transformation of work into heat can take place completely, while that of heat into work, on the other hand, only incompletely and in such manner that every time a quantity of heat is transformed into work another corresponding quantity of energy must necessarily undergo at the same time a compensating transformation, as, *e.g.*, the passage of heat from a higher to a lower temperature. This assertion is in certain special cases correct, but does not strike in general at the true import of the matter, as I shall show by a simple example.

One of the most important laws of thermodynamics is, that the total energy of an ideal gas depends only upon its temperature, and not upon its volume. If an ideal gas be allowed to expand while doing work, and if the cooling of the gas be prevented through the simultaneous addition of heat from a heat reservoir at higher temperature, the gas remains unchanged in temperature and energy content, and one may say that the heat furnished by the heat reservoir is completely transformed into work without exchange of energy. Not the least objection can be urged against this assertion. The law of incomplete transformation of heat into work is retained only through the adoption of a different point of view, but which has nothing to do with the status of the physical facts and only modifies the way of looking at the matter, and therefore can neither be supported nor contradicted through facts; namely, through the introduction ad hoc of new particular kinds of energy, in that one divides the energy of the gas into numerous parts which individually can depend upon the volume. But it is a priori evident that one can never derive from so artificial a definition a new physical law, and it is with such that we have to do when we pass from the first law, the principle of conservation of energy, to the second law.

I desire now to introduce such a new physical law: "It is not possible to construct a periodically functioning motor which in principle does not involve more than the raising of a load and the cooling of a heat reservoir." It is to be understood, that in one cycle of the motor quite arbitrary complicated processes may take place, but that after the completion of one cycle there shall remain no other changes in the surroundings than that the heat reservoir is cooled and that the load is raised a corresponding distance, which may be calculated from the first law. Such a motor could of course be used at the same time as a refrigerating machine also, without any further expenditure of energy and materials. Such a motor would moreover be the most efficient in the world, since it would involve no cost to run it; for the earth, the atmosphere, or the ocean could be utilized as the heat reservoir. We shall call this, in accordance with the proposal of W. Ostwald, perpetual motion of the second kind. Whether in nature such a motion is actually possible cannot be inferred from the energy principle, and may only be determined by special experiments.

Just as the impossibility of perpetual motion of the first kind leads to the principle of the conservation of energy, the quite independent principle of the impossibility of perpetual motion of the second kind leads to the second law of thermodynamics, and, if we assume this impossibility as proven experimentally, the general law follows immediately: *there are processes in nature which is no possible way can be made completely reversible.* For consider, *e.g.*, a frictional process through which mechanical work is transformed into heat with the aid of suitable apparatus, if it were actually possible to make in some way such complicated apparatus completely reversible, so that everywhere in nature exactly the same conditions be reestablished as existed at the beginning of the frictional process, then the apparatus considered would be nothing more than the motor described above, furnishing a perpetual motion of the second kind. This appears evident immediately, if one clearly perceives what the apparatus would accomplish: transformation of heat into work without any further outstanding change.

We call such a process, which in no wise can be made completely reversible, an irreversible process, and all other processes reversible processes; and thus we strike the kernel of the second law of thermodynamics when we say that irreversible processes occur in nature. In accordance with this, the changes in nature have a unidirectional tendency. With each irreversible process the world takes a step forward, the traces of which under no circumstances can be completely obliterated. Besides friction, examples of irreversible processes are: heat conduction, diffusion, conduction of electricity in conductors of finite resistance, emission of light and heat radiation, disintegration of the atom in radioactive substances, and so on. On the other hand, examples of reversible processes are: motion of the planets, free fall in empty space, the undamped motion of a pendulum, the frictionless flow of liquids, the propagation of light and sound waves without absorption and refraction, undamped electrical vibrations, and so on. For all these processes are already periodic or may be made completely reversible through suitable contrivances, so that there remains no outstanding change in nature; for example, the free fall of a body whereby the acquired velocity is utilized to raise the body again to its original height; a light or sound wave which is allowed in a suitable manner to be totally reflected from a perfect mirror.

What now are the general properties and criteria of irreversible processes, and what is the general quantitative measure of irreversibility? This question has been examined and answered in the most widely different ways, and it is evident here again how difficult it is to reach a correct formularization of a problem. Just as originally we came upon the trail of the energy principle through the technical problem of perpetual motion, so again a technical problem, namely, that of the steam engine, led to the differentiation between reversible and irreversible processes. Long ago Sadi Carnot recognized, although he utilized an incorrect conception of the nature of heat, that irreversible processes are less economical than reversible, or that in an irreversible process a certain opportunity to derive mechanical work from heat is lost. What then could have been simpler than the thought of making, quite in general, the measure of the irreversibility of a process the quantity of mechanical work which is unavoidably lost in the process. For a reversible process then, the unavoidably lost work is naturally to be set equal to zero. This view, in accordance with which the import of the second law consists in a dissipation of useful energy, has in fact, in certain special cases, *e.g.*, in isothermal processes, proved itself useful. It has persisted, therefore, in certain of its aspects up to the present day; but for the general case, however, it has shown itself as fruitless and, in fact, misleading. The reason for this lies in the fact that the question concerning the lost work in a given irreversible process is by no means to be answered in a determinate manner, so long as nothing further is specified with regard to the source of energy from which the work considered shall be obtained.

An example will make this clear. Heat conduction is an irreversible process, or as Clausius expresses it: Heat cannot without compensation pass from a colder to a warmer body. What now is the work which in accordance with definition is lost when the quantity of heat Q passes through direct conduction from a warmer body at the temperature T_1 to a colder body at the temperature T_2? In order to answer this question, we make use of the heat transfer in carrying out a reversible Carnot cyclical process between the two bodies employed as heat reservoirs. In this process a certain amount of work would be obtained, and it is just the amount sought, since it is that which would be lost in the direct passage by conduction; but this has no definite value so long as we do not know whence the work originates, whether, *e.g.*, in the warmer body or in the colder body, or from somewhere else. Let one reflect that the heat given up by the warmer body in the reversible process is certainly not equal to the heat absorbed by the colder body, because a certain amount of heat is transformed into work, and that we can identify, with exactly the same right, the quantity of heat Q transferred by the direct process of conduction with that which in the cyclical process is given up by the warmer body, or with that absorbed by the colder body. As one does the former or the latter, he accordingly obtains for the quantity of lost work in the process of conduction:

$$Q \cdot \frac{T_1 - T_2}{T_1} \quad \text{or} \quad Q \cdot \frac{T_1 - T_2}{T_2}. \tag{14.1}$$

We see, therefore, that the proposed method of expressing mathematically the irreversibility of a process does not in general effect its object, and at the same time we recognize the peculiar reason which prevents its doing so. The statement of the question is too anthropomorphic. It is primarily too much concerned with the needs of mankind, in that it refers directly to the acquirement of useful work. If one require from nature a determinate answer, he must take a more general point of view, more disinterested, less economic. We shall now seek to do this.

Let us consider any typical process occurring in nature. This will carry all bodies concerned in it from a determinate initial state, which I designate as state A, into a determinate final state B. The process is either reversible or irreversible. A third possibility is excluded. But whether it is reversible or irreversible depends solely upon the nature of the two states A and B, and not at all upon the way in which the process has been carried out; for we are only concerned with the answer to the question as to whether or not, when the state B is once reached, a complete return to A in any conceivable manner may be accomplished. If now, the complete return from B to A is not possible, and the process therefore irreversible, it is obvious that the state B may be distinguished in nature through a certain property from state A. Several years ago I ventured to express this as follows: that nature possesses a greater "preference" for state B than for state A. In accordance with this mode of expression, all those processes of nature are impossible for whose final state nature possesses a smaller preference than for the original state. Reversible processes constitute a limiting case; for such, nature possesses an equal preference for the initial and for the final state, and the passage between them takes place as well in one direction as the other.

We have now to seek a physical quantity whose magnitude shall serve as a general measure of the preference of nature for a given state. This quantity must be one which is directly determined by the state of the system considered, without reference to the previous history of the system, as is the case with the energy, with the volume, and with other properties of the system. It should possess the peculiarity of increasing in all irreversible processes and of remaining unchanged in all reversible processes, and the amount of change which it experiences in a process would furnish a general measure for the irreversibility of the process.

R. Clausius actually found this quantity and called it "entropy." Every system of bodies possesses in each of its states a definite entropy, and this entropy expresses the preference of nature for the state in question. It can, in all the processes which take place within the system, only increase and never decrease. If it be desired to consider a process in which external actions upon the system are present, it is necessary to consider those bodies in which these actions originate as constituting part of the system; then the law as stated in the above form is valid. In accordance with it, the entropy of a system of bodies is simply equal to the sum of the entropies of the individual bodies, and the entropy of a single body is, in accordance with Clausius, found by the aid of a certain reversible process. Conduction of heat to a body increases its entropy, and, in fact, by an amount equal to the ratio of the quantity of heat given the body to its temperature. Simple compression, on the other hand, does not change the entropy.

Returning to the example mentioned above, in which the quantity of heat Q is conducted from a warmer body at the temperature T_1 to a colder body at the temperature T_2, in accordance with what precedes, the entropy of the warmer body decreases in this process, while, on the other hand, that of the colder increases, and the sum of both changes, that is, the change of the total entropy of both bodies, is:

$$-\frac{Q}{T_1} + \frac{Q}{T_2} > 0. \tag{14.2}$$

This positive quantity furnishes, in a manner free from all arbitrary assumptions, the measure of the irreversibility of the process of heat conduction. Such examples may be cited indefinitely. Every chemical process furnishes an increase of entropy.

We shall here consider only the most general case treated by Clausius: an arbitrary reversible or irreversible cyclical process, carried out with any physico-chemical arrangement, utilizing an arbitrary number of heat reservoirs. Since the arrangement at the conclusion of the cyclical process is the same as that at the beginning, the final state of the process is to be distinguished from the initial state solely through the different heat content of the heat reservoirs, and in that a certain amount of mechanical work has been furnished or consumed. Let Q be the heat given up in the course of the process by a heat reservoir at the temperature T_1, and let A be the total work yielded (consisting, *e.g.* in the raising of weights); then, in accordance with the first law of thermodynamics:

$$\sum Q = A \qquad (14.3)$$

In accordance with the second law, the sum of the changes in entropy of all the heat reservoirs is positive, or zero. It follows, therefore, since the entropy of a reservoir is decreased by the amount Q/T through the loss of heat Q that:

$$\sum \frac{Q}{T} \leq 0. \qquad (14.4)$$

This is the well-known inequality of Clausius.

In an isothermal cyclical process, T is the same for all reservoirs. Therefore:

$$\sum Q \leq 0, \qquad \text{hence:} \qquad A \leq 0. \qquad (14.5)$$

That is: in an isothermal cyclical process, heat is produced and work is consumed. In the limiting case, a reversible isothermal cyclical process, the sign of equality holds, and therefore the work consumed is zero, and also the heat produced. This law plays a leading role in the application of thermodynamics to physical chemistry.

The second law of thermodynamics including all of its consequences has thus led to the principle of increase of entropy. You will now readily understand, having regard to the questions mentioned above, why I express it as my opinion that in the theoretical physics of the future the first and most important differentiation of all physical processes will be into reversible and irreversible processes.

In fact, all reversible processes, whether they take place in material bodies, in the ether, or in both together, show a much greater similarity among themselves than to any irreversible process. In the differential equations of reversible processes the time differential enters only as an even power, corresponding to the circumstance that the sign of time can be reversed. This holds equally well for vibrations of the pendulum, electrical vibrations, acoustic and optical waves, and for motions of mass points or of electrons, if we only exclude every kind of damping. But to such processes also belong those infinitely slow processes of thermodynamics which consist of states of equilibrium in which the time in general plays no role, or, as one may also say, occurs with the zero power, which is to be reckoned as an even power. As Helmholtz has pointed out, all these reversible processes have the common property that they may be completely represented by the principle of least action, which gives a definite answer to all questions concerning any such measurable process, and, to this extent, theory of reversible processes may be regarded as completely established. Reversible processes have, however, the disadvantage that singly and collectively they are only ideal: in actual nature there is no such thing as a reversible process. Every natural process involves in greater or less degree friction or conduction of heat. But in the domain of irreversible processes the principle of least action is no longer sufficient; for the principle of increase of entropy brings into the system of physics a wholly new element, foreign to the action principle, and which demands special mathematical treatment. The unidirectional course of a process in the attainment of a fixed final state is related to it.

I hope the foregoing considerations have sufficed to make clear to you that the distinction between reversible and irreversible processes is much broader than that between mechanical and electrical processes and that, therefore, this difference, with better right than any other, may be taken advantage of in classifying all physical processes, and that it may eventually play in the theoretical physics of the future the principal rôle.

However, the classification mentioned is in need of quite an essential improvement, for it cannot be denied that in the form set forth, the system of physics is still suffering from a strong dose of anthropomorphism. In the definition of irreversibility, as well as in that of entropy, reference is made to the possibility of carrying out in nature certain changes, and this means, fundamentally, nothing more than that the division of physical processes is made dependent upon the manipulative skill of man in the art of experimentation, which certainly does not always remain at a fixed stage, but is continually being more and more perfected. If, therefore, the distinction between reversible and irreversible processes is actually to have a lasting significance for all times, it must be essentially broadened and made independent of any reference to the capacities of mankind. How this may happen, I desire to state 1 week from tomorrow. The lecture of tomorrow will be devoted to the problem of bringing before you some of the most important of the great number of practical consequences following from the entropy principle.

14.3 Study Questions

QUES. 14.1. What are the dangers of underestimating, or overestimating, the value of theoretical physics? Which does Planck think is worse? Which do you think is worse?

QUES. 14.2. Is positivism compatible with the goal of a "common system of physics for all physicists"?

a) What, according to Planck, is the material (so to speak) upon which theoretical physics operates? With what tool does it operate? And what is the foundation of all physical research?

b) How was physics historically divided? In what sense was this anthropomorphic? How has physics been more recently reorganized? What specific benefit has this conferred?

c) How does the goal of reducing anthropomorphism conflict with the positivistic philosophy of science? In particular, if all physical discovery is based on individual perception, then is a common system of physics even possible?

d) How far has science advanced in the goal of complete unification? Is further unification possible?

QUES. 14.3. What is the root of the law of conservation of energy? From what technical problem was this law derived? Is the law of conservation of energy, strictly speaking, provable?

QUES. 14.4. How is the second law of thermodynamics different than the first law? In particular, on what independent principle is it founded? And from what technical problem was it derived?

QUES. 14.5. What common mistakes are made regarding the second law of thermodynamics?

a) Does the second law dictate the direction of every process in nature—for example the direction of flow of water or electricity?
b) Does the second law imply that whenever a quantity of heat is transformed into work, an equal quantity of energy must always be transformed into heat (so as to compensate this transformation)?

QUES. 14.6. What is an irreversible process? What are some examples of irreversible, and reversible, processes?

QUES. 14.7. How did Sadi Carnot define an irreversible process? Is this a good definition? In particular, is mechanical work actually lost in an irreversible process?

QUES. 14.8. How, alternatively, did Clausius define an irreversible process?

a) What does it mean for nature to possess a greater "preference" for state A than state B? What does this have to do with the notion of irreversibility, and the second law?
b) What physical quantity of states A and B express this preference? Upon what can this quantity depend, and upon what can it not depend?
c) By how much does the entropy of a body at temperature T increase when a quantity of heat Q is added to the body?
d) How is the change in entropy computed when heat is conducted between two bodies? Is entropy an additive quantity?
e) How, then, is Clausius' statement of the second law expressed in terms of changes in entropy?

QUES. 14.9. What is the essential difference between differential equations which govern reversible and irreversible physical processes? What underlying principle represents all reversible processes? Does this also represent irreversible processes?

QUES. 14.10. If physical science was historically organized according to the sense perceptions of man, then what (according to Planck) is the best way to organize it henceforth? Can this be done objectively, that is, without resorting to anthropomorphism?

14.4   Exercises

Ex. 14.1 (POSITIVISM, ANTHROPOMORPHISM AND OBJECTIVITY ESSAY). Do you agree
with the positivist view of physics? In particular: does physics consist only in the
relating of sense perceptions? What are the strengths and weaknesses of this view?
Do you believe that physics should eschew any kind of anthropomorphism? If so, to
what extent is this even possible?[3]

14.5   Vocabulary

1. Cordial
2. Confer
3. Refute
4. Anthropomorphic
5. Geodesy
6. Positivism
7. Metaphysics
8. Perception
9. Heterogeneous
10. Succinct
11. Vouchsafe
12. Emancipation
13. Intuitive
14. Electrodynamics
15. Ponderable
16. Amalgamation
17. Endeavor
18. Prevalent
19. *Ad hoc*
20. Assertion
21. *A priori*
22. Isothermal

---

[3] There is an enormous body of scholarship which deals with the philosophy of science from
the perspective of empiricist/positivist/modernist philosophers, relativist/postmodernist philosophers and rationalist/objectivist philosophers. One recent critique of the positivist position from
the perspective of rationalist philosophy can be found in Hoppe, H.-H., In Defense of Extreme
Rationalism, *Rev. Austrian Econ.*, *3*(1), 179–214, 1989.

Chapter 15
Entropy, Probability and Atomism

> *Nature prefers the more probable states to the less probable,*
> *because in nature processes take place in the direction of*
> *greater probability.*
>
> —Max Planck

15.1 Introduction

In Max Planck's first lecture delivered at Columbia University in 1909, he explained how physical science was historically organized *anthropomorphically*—that is, according to man's particular senses. Thus we had the sciences of optics (what the eye can perceive), acoustics (what the ear can hear) and heat (what the skin can feel). With new apparatus and measurement techniques, however, physical science has become less concerned with subjective sensory experience and more concerned with objective quantification. For example, radio antennae and bolometers have enabled scientists to measure the properties of electromagnetic radiation that lies far outside of the spectrum of visible light. Moreover, the laws of thermodynamics—and especially Clausius' second law—has inspired a reorganization of physical science into just two classes of phenomena: those which are *reversible* and those which are *irreversible*. Unfortunately, according to Planck, this new classification scheme still has the (undesirable) mark of anthropomorphism. For the definition of irreversibility—and the associated concept of entropy—still relies on the skill of an experimenter in devising an efficient heat engine for the purpose of accomplishing useful work. In other words, the purely thermodynamic definition of entropy is based on the inclinations and limitations of man. Is there a better definition of entropy—one which is more objective? This is the question to which Planck now turns in his third lecture...

15.2 Reading: Planck, *The Atomic Theory of Matter*

Planck, M., *Eight Lectures on Theoretical Physics*, Columbia University Press, New York, 1915. Third Lecture.

© Springer International Publishing Switzerland 2016   197
K. Kuehn, *A Student's Guide Through the Great Physics Texts,*
Undergraduate Lecture Notes in Physics, DOI 10.1007/978-3-319-21828-1_15

The problem with which we shall be occupied in the present lecture is that of a closer investigation of the atomic theory of matter. It is, however, not my intention to introduce this theory with nothing further, and to set it up as something apart and disconnected with other physical theories, but I intend above all to bring out the peculiar significance of the atomic theory as related to the present general system of theoretical physics; for in this way only will it be possible to regard the whole system as one containing within itself the essential compact unity, and thereby to realize the principal object of these lectures.

Consequently it is self evident that we must rely on that sort of treatment which we have recognized in last week's lecture as fundamental. That is, the division of all physical processes into reversible and irreversible processes. Furthermore, we shall be convinced that the accomplishment of this division is only possible through the atomic theory of matter, or, in other words, that irreversibility leads of necessity to atomistics.

I have already referred at the close of the first lecture to the fact that in pure thermodynamics, which knows nothing of an atomic structure and which regards all substances as absolutely continuous, the difference between reversible and irreversible processes can only be defined in one way, which a priori carries a provisional character and does not withstand penetrating analysis. This appears immediately evident when one reflects that the purely thermodynamic definition of irreversibility which proceeds from the impossibility of the realization of certain changes in nature, as, *e.g.* the transformation of heat into work without compensation, has at the outset assumed a definite limit to man's mental capacity, while, however, such a limit is not indicated in reality. On the contrary: mankind is making every endeavor to press beyond the present boundaries of its capacity, and we hope that later on many things will be attained which, perhaps, many regard at present as impossible of accomplishment. Can it not happen then that a process, which up to the present has been regarded as irreversible, may be proved, through a new discovery or invention, to be reversible? In this case the whole structure of the second law would undeniably collapse, for the irreversibility of a single process conditions that of all the others.

It is evident then that the only means to assure to the second law real meaning consists in this, that the idea of irreversibility be made independent of any relationship to man and especially of all technical relations.

Now the idea of irreversibility harks back to the idea of entropy; for a process is irreversible when it is connected with an increase of entropy. The problem is hereby referred back to a proper improvement of the definition of entropy. In accordance with the original definition of Clausius, the entropy is measured by means of a certain reversible process, and the weakness of this definition rests upon the fact that many such reversible processes, strictly speaking all, are not capable of being carried out in practice. With some reason it may be objected that we have here to do, not with an actual process and an actual physicist, but only with ideal processes, so-called thought experiments, and with an ideal physicist who operates with all the experimental methods with absolute accuracy. But at this point the difficulty is encountered: How far do the physicist's ideal measurements of this sort suffice? It may be understood, by passing to the limit, that a gas is compressed by a pressure which is equal to the pressure of the gas, and is heated by a heat reservoir which possesses the same temperature as the gas, but, for example, that a saturated vapor shall be transformed through isothermal compression in a reversible manner to a liquid without at any time a part of the vapor being condensed, as in certain thermodynamic considerations is supposed, must certainly appear doubtful. Still more striking, hover, is the liberty as regards thought experiments, which in physical chemistry, is granted the theorist. With his semi-permeable membranes, which in reality are only realizable under certain special conditions and then only with a certain approximation, he separates in a reversible manner, not only all possible varieties of molecules, whether or not they are in stable or unstable conditions, but he also separates the oppositely charged ions from one another and from the undissociated molecules, and he is disturbed, neither by the enormous electrostatic forces which resist such a separation, nor by the circumstance that in reality, from the beginning of the separation, the molecules become in part dissociated while the ions in part again combine. But such ideal processes are necessary throughout in order to make possible the comparison of the entropy of the undissociated molecules with the entropy of the dissociated molecules; for the law of thermodynamic equilibrium does not permit in general of derivation in any other way, in case one wishes to retain pure thermodynamics as a basis. It must be considered remarkable that all these ingenious thought processes have so well found confirmation of their results in experience, as is shown by the examples considered by us in the last lecture.

If now, on the other hand, one reflects that in all these results every reference to the possibility of actually carrying out each ideal process has disappeared—there are certainly left relations between directly measurable quantities only, such as temperature, heat effect, concentration, *etc.*—the presumption forces itself upon one that perhaps the introduction as above of such ideal processes is at bottom a round-about method, and that the peculiar import of the principle of increase of entropy with all its consequences can be evolved from the original idea of irreversibility or, just as well, from the impossibility of perpetual motion of the second kind, just as the principle of conservation of energy has been evolved from the law of impossibility of perpetual motion of the first kind.

This step: to have completed the emancipation of the entropy idea from the experimental art of man and the elevation of the second law thereby to a real principle, was the scientific life's work of Ludwig Boltzmann. Briefly stated, it consisted in general of referring back the idea of entropy to the idea of probability. Thereby is also explained, at the same time, the significance of the above (p. 192) auxiliary term used by me; "preference" of nature for a definite state. Nature prefers the more probable states to the less probable, because in nature processes take place in the direction of greater probability. Heat goes from a body at higher temperature to a body at lower temperature because the state of equal temperature distribution is more probable than a state of unequal temperature distribution.

Through this conception the second law of thermodynamics is removed at one stroke from its isolated position, the mystery concerning the preference of nature vanishes, and the entropy principle reduces to a well understood law of the calculus of probability.

The enormous fruitfulness of so "objective" a definition of entropy for all domains of physics I shall seek to demonstrate in the following lectures. But today we have principally to do with the proof of its admissibility; for on closer consideration we shall immediately perceive that the new conception of entropy at once introduces a great number of questions, new requirements and difficult problems. The first requirement is the introduction of the atomic hypothesis into the system of physics. For, if one wishes to speak of the probability of a physical state, *i.e.*, if he wishes to introduce the probability for a given state as a definite quantity into the calculation, this can only be brought about, as in cases of all probability calculations, by referring the state back to a variety of possibilities; *i.e.*, by considering a finite number of a priori equally likely configurations (complexions) through each of which the state considered may be realized. The greater the number of complexions, the greater is the probability of the state. Thus, *e.g.*, the probability of throwing a total of four with two ordinary six-sided dice is found through counting the complexions by which the throw with a total of four may be realized. Of these there are three complexions:

> with the first die, 1, with the second die, 3,
> with the first die, 2, with the second die, 2,
> with the first die, 3, with the second die, 1.

On the other hand, the throw of two is only realized through a single complexion. Therefore, the probability of throwing a total of four is three times as great as the probability of throwing a total of two.

Now, in connection with the physical state under consideration, in order to be able to differentiate completely from one another the complexions realizing it, and to associate it with a definite reckonable number, there is obviously no other means than to regard it as made up of numerous discrete homogeneous elements—for in perfectly continuous systems there exist no reckonable elements—and hereby the atomistic view is made a fundamental requirement. We have, therefore, to regard all bodies in nature, in so far as they possess an entropy, as constituted of atoms, and we therefore arrive in physics at the same conception of matter as that which obtained in chemistry for so long previously.

But we can immediately go a step further yet. The conclusions reached hold, not only for thermodynamics of material bodies, but also possess complete validity for the processes of heat radiation, which are thus referred hack to the second law of thermodynamics. That radiant heat also possesses an entropy follows from the fact that a body which emits radiation into a surrounding diathermanous medium experiences a loss of heat and, therefore, a decrease of entropy. Since the total entropy of a physical system can only increase, it follows that one part of the entropy of the whole system, consisting of the body and the diathermanous medium, must be contained in the radiated heat. If the entropy of the radiant heat is to be referred back to the notion of probability, we are forced, in a similar way as above, to the conclusion that for radiant heat the atomic conception possesses a definite meaning. But, since radiant heat is not directly connected with matter, it follows that this atomistic conception relates, not to matter, but only to energy, and hence, that in heat radiation certain energy elements play an essential role. Even though this conclusion appears so singular and even though in many circles today vigorous objection is strongly urged against it, in the long run physical research will not be able to withhold its sanction from it, and the less, since it is confirmed by experience in quite a satisfactory manner. We shall return to this point in the lectures on heat radiation. I desire here only to mention that the novelty involved by the introduction of atomistic conceptions into the theory of heat radiation is by no means so revolutionary as, perhaps, might appear at the first glance. For there is, in my opinion at least, nothing which makes necessary the consideration of the heat processes in a complete vacuum as atomic, and it suffices to seek the atomistic features at the source of radiation, *i.e.* in those processes which have their play in the centres of emission and absorption of radiation. Then the Maxwellian electrodynamic differential equations can retain completely their validity for the vacuum, and, besides, the discrete elements of heat radiation are relegated exclusively to a domain which is still very mysterious and where there is still present plenty of room for all sorts of hypotheses.

Returning to more general considerations, the most important question comes up as to whether, with the introduction of atomistic conceptions and with the reference of entropy to probability, the content of the principle of increase of entropy is exhaustively comprehended, or whether still further physical hypotheses are required in order to secure the full import of that principle. If this important question had been settled at the time of the introduction of the atomic theory into thermodynamics, then the atomistic views would surely have been spared a large number of conceivable misunderstandings and justifiable attacks. For it turns out, in fact—and our further considerations will confirm this conclusion—that there has as yet nothing been done with atomistics which in itself requires much more than an essential generalization, in order to guarantee the validity of the second law.

We must first reflect that, in accordance with the central idea laid down in the first lecture (p. 185), the second law must possess validity as an objective physical law, independently of the individuality of the physicist. There is nothing to hinder us from imagining a physicist—we shall designate him a "microscopic" observer— whose senses are so sharpened that he is able to recognize each individual atom and to follow it in its motion. For this observer each atom moves exactly in accordance with the elementary laws which general dynamics lays down for it, and these laws allow, so far as we know, of an inverse performance of every process. Accordingly, here again the question is neither one of probability nor of entropy and its increase. Let us imagine, on the other hand, another observer, designated a "macroscopic" observer, who regards an ensemble of atoms as a homogeneous gas, say, and consequently applies the laws of thermodynamics to the mechanical and thermal processes within it. Then, for such an observer, in accordance with the second law, the process in general is an irreversible process. Would not now the first observer be justified in saying: "The reference of the entropy to probability has its origin in the fact that irreversible processes ought to be explained through reversible processes. At any rate, this procedure appears to me in the highest degree dubious. In any case, I declare each change of state which takes place in the ensemble of atoms designated a gas, as reversible, in opposition to the macroscopic observer." There is not the slightest thing, so far as I know, that one can urge against the validity of these statements. But do we not thereby place ourselves in the painful position of the judge who declared in a trial the correctness of the position of each separately of two contending parties and then, when a third contends that only one of the parties could emerge from the process victorious, was obliged to declare him also correct? Fortunately we find ourselves in a more favorable position. We can certainly mediate between the two parties without its being necessary for one or the other to give up his principal point of view. For closer consideration shows that the whole controversy rests upon a misunderstanding—a new proof of how necessary it is before one begins a controversy to come to an understanding with his opponent concerning the subject of the quarrel. Certainly, a given change of state cannot be both reversible and irreversible. But the one observer connects a wholly different idea with the phrase "change of state" than the other. What is then, in general, a change of state? The state of a physical system cannot well be otherwise defined than as the aggregate of all those physical quantities, through whose instantaneous values the time changes of the quantities, with given boundary conditions, are uniquely determined. If we inquire now, in accordance with the import of this definition, of the two observers as to what they understand by the state of the collection of atoms or the gas considered, they will give quite different answers. The microscopic observer will mention those quantities which determine the position and the velocities of all the individual atoms. There are present in the simplest case, namely, that in which the atoms may be considered as material points, six times as many quantities as atoms, namely, for each atom the three coordinates and the three velocity components, and in the case of combined molecules, still more quantities. For him the state and the progress of a process is then first determined when all these various quantities are individually given. We shall designate the state defined in this way the "micro-state," The macroscopic observer, on the other hand, requires fewer data. He will say that the state of the homogeneous gas considered by him is determined by the density, the visible velocity and the temperature at each point of the gas, and he will expect that, when these quantities arc given, their time variations and, therefore, the progress of the process, to be completely determined in accordance with the two laws of thermo-dynamics, and therefore accompanied by an increase in entropy. In this collection he can call upon all the experience at his disposal, which will fully confirm his expectation. If we call this state the "macro-state," it is clear that the two laws: "the micro-changes of state are reversible" and "the macro-changes of state are irreversible," lie in wholly different domains and, at any rate, are not contradictory.

But now how can we succeed in bringing the two observers to an understanding? This is a question whose answer is obviously of fundamental significance for the atomic theory. First of all, it is easy to see that the macro-observer reckons only with mean values; for what he calls density, visible velocity and temperature of the gas are, for the micro-observer, certain mean values, statistical data, which are derived from the space distribution and from the velocities of the atoms in an appropriate manner. But the micro-observer cannot operate with these mean values alone, for, if these are given at one instant of time, the progress of the process is not determined throughout; on the contrary: he can easily find with given mean values an enormously large number of individual values for the positions and the velocities of the atoms, all of which correspond with the same mean values and which, in spite of this, lead to quite different processes with regard to the mean values. It follows from this of necessity that the micro-observer must either give up the attempt to understand the unique progress, in accordance with experience, of the macroscopic changes of state—and this would be the end of the atomic theory—or that he, through the introduction of a special physical hypothesis, restrict in a suitable manner the manifold of micro-states considered by him. There is certainly nothing to prevent him from assuming that not all conceivable micro-states are realizable in nature, and that certain of them are in fact thinkable, but never actually realized. In the formularization of such a hypothesis, there is of course no point of departure to be found from the principles of dynamics alone; for pure dynamics leaves this case undetermined. But on just this account any dynamical hypothesis, which involves nothing further than a closer specification of the micro-states realized in nature, is certainly permissible. Which hypothesis is to be given the preference can only be decided through comparison of the results to which the different possible hypotheses lead in the course of experience.

In order to limit the investigation in this way, we must obviously fix our attention only upon all imaginable configurations and velocities of the individual atoms which are compatible with determinate values of the density, the velocity and the temperature of the gas, or in other words: we must consider all the micro-states which belong to a determinate macro-state, and must investigate the various kinds of processes which follow in accordance with the fixed laws of dynamics from the different micro-states. Now, precise calculation has in every case always led to the important result that an enormously large number of these different micro-processes relate to one and the same macro-process, and that only proportionately few of the same, which are distinguished by quite special exceptional conditions concerning the positions and the velocities of neighboring atoms, furnish exceptions. Furthermore, it has also shown that one of the resulting macro-processes is that which the macroscopic observer recognizes, so that it is compatible with the second law of thermodynamics.

Here, manifestly, the bridge of understanding is supplied. The micro-observer needs only to assimilate in his theory the physical hypothesis that all those special cases in which special exceptional conditions exist among the neighboring configurations of interacting atoms do not occur in nature, or, in other words, that the micro-states are in elementary disorder. Then the uniqueness of the macroscopic process is assured and with it, also, the fulfillment of the principle of increase of entropy in all directions.

Therefore, it is not the atomic distribution, but rather the hypothesis of elementary disorder, which forms the real kernel of the principle of increase of entropy and, therefore, the preliminary condition for the existence of entropy. Without elementary disorder there is neither entropy nor irreversible process.[1] Therefore, a single atom can never possess an entropy; for we cannot speak of disorder in connection with it. But with a fairly large number of atoms, say 100 or 1000, the matter is quite different. Here, one can certainly speak of a disorder, in case that the values of the coordinates and the velocity components are distributed among the atoms in accordance with the laws of accident. Then it is possible to calculate the probability for a given state. But how is it with regard to the increase of entropy? May we assert that the motion of 100 atoms is irreversible? Certainly not; but this is only because the state of 100 atoms cannot be defined in a thermodynamic sense, since the process does not proceed in a unique manner from the standpoint of a macro-observer, and this requirement forms, as we have seen above, the foundation and preliminary condition for the definition of a thermodynamic state.

If one therefore asks: How many atoms are at least necessary in order that a process may be considered irreversible?, the answer is: so many atoms that one may form from them definite mean values which define the state in a macroscopic sense. One must reflect that to secure the validity of the principle of increase of entropy there must be added to the condition of elementary disorder still another, namely, that the number of the elements under consideration be sufficiently large to render possible the formation of definite mean values. The second law has a meaning for these mean values only; but for them, it is quite exact, just as exact as the law of the calculus of probability, that the mean value, so far as it may be defined, of a sufficiently large number of throws with a six-sided die, is 3½.

These considerations are, at the same time, capable of throwing light upon questions such as the following: Does the principle of increase of entropy possess a meaning for the so-called Brownian molecular movement of a suspended particle? Does the kinetic energy of this motion represent useful work or not? The entropy principle is just as little valid for a single suspended particle as for an atom, and therefore is not valid for a few of them, but only when there is so large a number that definite mean values can be formed. That one is able to see the particles and not the atoms makes no material difference; because the progress of a process does not depend upon the power of an observing instrument. The question with regard to useful work plays no role in this connection; strictly speaking, this possesses, in general, no objective physical meaning. For it does not admit of an answer without

---

[1] To those physicists who, in spite of all this, regard the hypothesis of elementary disorder as gratuitous or as incorrect, I wish to refer the simple fact that in every calculation of a coefficient of friction, of diffusion, or of heat conduction, from molecular considerations, the notion of elementary disorder is employed, whether tacitly or otherwise, and that it is therefore essentially more correct to stipulate this condition instead of ignoring or concealing it. But he who regards the hypothesis of elementary disorder as self-evident, should be reminded that, in accordance with a law of H. Poincaré, the precise investigation concerning the foundation of which would here lead us too far, the assumption of this hypothesis for all time is unwarranted for a closed space with absolutely smooth walls,—an important conclusion, against which can only be urged the fact that absolutely smooth walls do not exist in nature.

reference to the scheme of the physicist or technician who proposes to make use of the work in question. The second law, therefore, has fundamentally nothing to do with the idea of useful work (*cf.* first lecture, p. 191).

But, if the entropy principle is to hold, a further assumption is necessary, concerning the various disordered elements,—an assumption which tacitly is commonly made and which we have not previously definitely expressed. It is, however, not less important than those referred to above. The elements must actually be of the same kind, or they must at least form a number of groups of like kind, *e.g.*, constitute a mixture in which each kind of element occurs in large numbers. For only through the similarity of the elements does it come about that order and law can result in the larger from the smaller. If the molecules of a gas be all different from one another, the properties of a gas can never show so simple a law-abiding behavior as that which is indicated by thermodynamics. In fact, the calculation of the probability of a state presupposes that all complexions which correspond to the state are *a priori* equally likely. Without this condition one is just as little able to calculate the probability of a given state as, for instance, the probability of a given throw with dice whose sides are unequal in size. In summing up we may therefore say: the second law of thermodynamics in its objective physical conception, freed from anthropomorphism, relates to certain mean values which are formed from a large number of disordered elements of the same kind.

The validity of the principle of increase of entropy and of the irreversible progress of thermodynamic processes in nature is completely assured in this formularization. After the introduction of the hypothesis of elementary disorder, the microscopic observer can no longer confidently assert that each process considered by him in a collection of atoms is reversible; for the motion occurring in the reverse order will not always obey the requirements of that hypothesis. In fact, the motions of single atoms are always reversible, and thus far one may say, as before, that the irreversible processes appear reduced to a reversible process, but the phenomenon as a whole is nevertheless irreversible, because upon reversal the disorder of the numerous individual elementary processes would be eliminated. Irreversibility is inherent, not in the individual elementary processes themselves, but solely in their irregular constitution. It is this only which guarantees the unique change of the macroscopic mean values.

Thus, for example, the reverse progress of a frictional process is impossible, in that it would presuppose elementary arrangement of interacting neighboring molecules. For the collisions between any two molecules must thereby possess a certain distinguishing character, in that the velocities of two colliding molecules depend in a definite way upon the place at which they meet. In this way only can it happen that in collisions like directed velocities ensue and, therefore, visible motion.

Previously we have only referred to the principle of elementary disorder in its application to the atomic theory of matter. But it may also be assumed as valid, as I wish to indicate at this point, on quite the same grounds as those holding in the case of matter, for the theory of radiant heat. Let us consider, *e.g.*, two bodies at different temperatures between which exchange of heat occurs through radiation. We can in this case also imagine a microscopic observer, as opposed to the ordinary macroscopic observer, who possesses insight into all the particulars of electromagnetic processes which are connected with emission and absorption, and the propagation of heat rays. The microscopic observer would declare the whole process reversible because all electrodynamic processes can also take place in the reverse direction, and the contradiction may here be referred back to a difference in definition of the state of a heat ray. Thus, while the macroscopic observer completely defines a monochromatic ray through direction, state of polarization, color, and intensity, the microscopic observer, in order to possess a complete knowledge of an electromagnetic state, necessarily requires the specification of all the numerous irregular variations of amplitude and phase to which the most homogeneous heat ray is actually subject. That such irregular variations actually exist follows immediately from the well known fact that two rays of the same color never interfere, except when they originate in the same source of light. But until these fluctuations are given in all particulars, the micro-observer can say nothing with regard to the progress of the process. He is also unable to specify whether the exchange of heat radiation between the two bodies leads to a decrease or to an increase of their difference in temperature. The principle of elementary disorder first furnishes the adequate criterion of the tendency of the radiation process, *i.e.*, the warming of the colder body at the expense of the warmer, just as the same principle conditions the irreversibility of exchange of heat through conduction. However, in the two cases compared, there is indicated an essential difference in the kind of the disorder. While in heat conduction the disordered elements may be represented as associated with the various molecules, in heat radiation there are the numerous vibration periods, connected with a heat ray, among which the energy of radiation is irregularly distributed. In other words: the disorder among the molecules is a material one, while in heat radiation it is one of energy distribution. This is the most important difference between the two kinds of disorder; a common feature exists as regards the great number of uncoordinated elements required. Just as the entropy of a body is defined as a function of the macroscopic state, only when the body contains so many atoms that from them definite mean values may be formed, so the entropy principle only possesses a meaning with regard to a heat ray when the ray comprehends so many periodic vibrations, *i.e.*, persists for so long a time, that a definite mean value for the intensity of the ray may be obtained from the successive irregular fluctuating amplitudes.

Now, after the principle of elementary disorder has been introduced and accepted by us as valid throughout nature, the fundamental question arises as to the calculation of the probability of a given state, and the actual derivation of the entropy therefrom. From the entropy all the laws of thermodynamic states of equilibrium, for material substances, and also for energy radiation, may be uniquely derived. With regard to the connection between entropy and probability, this is inferred very simply from the law that the probability of two independent configurations is represented by the product of the individual probabilities:

$$W = W_1 \cdot W_2, \qquad (15.1)$$

while the entropy S is represented by the sum of the individual entropies:

$$S = S_1 + S_2 \tag{15.2}$$

Accordingly, the entropy is proportional to the logarithm of the probability:

$$S = k \log W. \tag{15.3}$$

k is a universal constant. In particular, it is the same for atomic as for radiation configurations, for there is nothing to prevent us assuming that the configuration designated by 1 is atomic, while that designated by 2 is a radiation configuration. If k has been calculated, say with the aid of radiation measurements, then k must have the same value for atomic processes. Later we shall follow this procedure, in order to utilize the laws of heat radiation in the kinetic theory of gases. Now, there remains, as the last and most difficult part of the problem, the calculation of the probability W of a given physical configuration in a given macroscopic state. We shall treat today, by way of preparation for the quite general problem to follow, the simple problem: to specify the probability of a given state for a single moving material point, subject to given conservative forces. Since the state depends upon six variables: the three generalized coordinates ϕ_1, ϕ_2, ϕ_3, and the three corresponding velocity components $\dot{\phi}_1$, $\dot{\phi}_2$, $\dot{\phi}_3$, and since all possible values of these six variables constitute a continuous manifold, the probability sought is, that these six quantities shall lie respectively within certain infinitely small intervals, or, if one thinks of these six quantities as the rectilinear orthogonal coordinates of a point in an ideal six-dimensional space, that this ideal "state point" shall fall within a given, infinitely small "state domain." Since the domain is infinitely small, the probability will be proportional to the magnitude of the domain and therefore proportional to

$$\int d\phi_1 \cdot d\phi_2 \cdot d\phi_3 \cdot d\dot{\phi}_1 \cdot d\dot{\phi}_2 \cdot d\dot{\phi}_3 \tag{15.4}$$

But this expression cannot serve as an absolute measure of the probability, because in general it changes in magnitude with the time, if each state point moves in accordance with the laws of motion of material points, while the probability of a state which follows of necessity from another must be the same for the one as the other. Now, as is well known, another integral quite similarly formed, may be specified in place of the one above, which possesses the special property of not changing in value with the time. It is only necessary to employ, in addition to the general coordinates ϕ_1, ϕ_2, ϕ_3, the three so-called momenta ψ_1, ψ_2, ψ_3, in place of the three velocities $\dot{\phi}_1$, $\dot{\phi}_2$, $\dot{\phi}_3$, as the determining coordinates of the state. These are defined in the following way:

$$\psi_1 = \left(\frac{\partial H}{\partial \dot{\phi}_1}\right), \quad \psi_2 = \left(\frac{\partial H}{\partial \dot{\phi}_2}\right), \quad \psi_3 = \left(\frac{\partial H}{\partial \dot{\phi}_3}\right), \tag{15.5}$$

wherein H denotes the kinetic potential (Helmholtz). Then, in Hamiltonian form, the equations of motion are:

$$\dot{\psi}_1 = \frac{d\psi_1}{dt} = -\left(\frac{\partial E}{\partial \phi_1}\right)_\psi, \dots, \quad \dot{\phi}_1 = \frac{d\phi_1}{dt} = \left(\frac{\partial E}{\partial \psi_1}\right)_\phi, \dots, \tag{15.6}$$

(E is the energy), and from these equations follows the "condition of incompressibility":

$$\frac{\partial \dot{\phi}_1}{\partial \phi_1} + \frac{\partial \dot{\psi}_1}{\partial \psi_1} + \ldots = 0 \tag{15.7}$$

Referring to the six-dimensional space represented by the coordinates ϕ_1, ϕ_2, ϕ_3, ψ_1, ψ_2, ψ_3, this equation states that the magnitude of an arbitrarily chosen state domain, *viz*.:

$$\int d\phi_1 \cdot d\phi_2 \cdot d\phi_3 \cdot d\psi_1 \cdot d\psi_2 \cdot d\psi_3 \tag{15.8}$$

does not change with the time, when each point of the domain changes its position in accordance with the laws of motion of material points. Accordingly, it is made possible to take the magnitude of this domain as a direct measure for the probability that the state point falls within the domain. From the last expression, which can be easily generalized for the case of an arbitrary number of variables, we shall calculate later the probability of a thermodynamic state, for the case of radiant energy as well as that for material substances.

15.3   Study Questions

QUES. 15.1. Is entropy an objective or a subjective quantity?

a) In what sense is the purely thermodynamic definition of entropy based on the limited mental capacities of man? Is entropy thus defined an objective quantity?
b) What are some idealizations employed by physicists and chemists? Is the definition of entropy dependent on these idealizations?
c) How is Boltzmann's definition of entropy different than the thermodynamic definition? In particular, how are thermodynamic states analogous to the possible outcomes of rolling a set of dice? Is entropy thus defined an objective quantity?
d) Is the atomistic hypothesis implied by Boltzmann's concept of entropy?

QUES. 15.2. Does electromagnetic radiation have entropy?

a) Can the atomistic hypothesis be applied to non-material bodies? How would this imply an atomistic understanding of energy itself?
b) Does Planck wish to overthrow Maxwell's differential equations governing electromagnetic fields? Where, then, does he situate the source of the "discrete elements of heat radiation"?

QUES. 15.3. Does the concept of atomism imply the validity of the second law? More specifically, is the probabilistic definition of entropy sufficient to explain the second law of thermodynamics?

QUES. 15.4. Is it possible for a thermodynamic process be both reversible and irreversible at the same time?

a) What is meant by the "state of a system"? How does a microscopic observer define the (micro)-state of the system? How does a macroscopic observer define the (macro)-state of the system?

b) Can a microscopic and a macroscopic observer come to different conclusions regarding the reversibility of a given change of state? Which must keep track of fewer variables? Which observes the change of state to be reversible?

c) Are the general laws of dynamics reversible? If so, then from where does irreversibility arise?

QUES. 15.5. What is meant by "disorder"?

a) What is the "principle of elementary disorder"? In particular, what does Planck mean when he says that micro-processes are governed by the "laws of accident"? Is this last term an oxymoron?

b) How does the principle of elementary disorder assure the uniqueness of a particular macro-state (given an enormously large number of micro-states)?

c) Are all combinations of the positions and velocities of the atoms comprising a macro-state possible—at least in principle? Are there any constraints on the micro-states which are typically realized for a given macro-state?

QUES. 15.6. Can entropy be assigned to a single atom? Can it be assigned to radiation?

a) How many atoms are necessary in order that a process may be considered irreversible? And what are the necessary conditions so as to ensure the principle of increase of entropy?

b) Can the thermodynamic laws which govern, say, a gas be computed, based on probability, if each molecule is unique?

c) Does the principle of elementary disorder apply to radiation as well as to matter? How does the concept of disorder differ for matter and for radiant heat?

QUES. 15.7. How are entropy and probability connected mathematically?

a) What separate conditions must be satisfied by the probability and by the entropy?

b) Is Eq. 15.3 the only equation that fulfills both these conditions? If not, what other equation might work?

c) Does the value of k depend on the type of process involved? What does this imply about k?

QUES. 15.8. What is the entropy of a system of particles?

a) How can one compute the probability of a given state for a single moving material point?

b) Upon upon how many (and what) variables does this probability depend? How are the allowed values of these variables summed over a small range of values?

c) How can this probability be re-written in terms of generalized coordinates and generalized momenta? What is the virtue of this alternative form?

d) How can this probability for a single particle be generalized so as to account for a system of particles?

4

15.4   Exercises

EX. 15.1 (DRUNKEN SAILOR GAME). Suppose a very drunk sailor stands at a lamp post. Every 10 s he takes one step either north, south, east, or west. How far will he be from the lamp post after 5 min? Let us model this situation by playing a game using a piece of quad-ruled graph paper and a four-sided (tetrahedral) die. This exercise is most easily and efficiently carried out by a group of perhaps ten or fifteen students, each of which can generate his or her own data set. First, each student should make a mark near the center of his or her graph paper and label it with a small zero. This denotes the lamp post. Now roll the die. Make a new mark, labelled with a "1", to the top, right, bottom, or left of the previous mark, depending on the outcome of your die throw. Using a ruler, measure and record the distance r between the lamp post and the sailor. To be realistic, you might assume that 1 in. on your graph paper corresponds to 1 yard for the sailor, for instance. The sailor has just completed his $N = 1$ step. Repeat this procedure until the sailor has taken 30 steps. At each step, measure the distance between the lamp post and the sailor. Now draw a graph of r^2 vs. N for $N = 1 \ldots 30$. Describe the relationship between r^2 and N for your drunken sailor game. Is there a pattern?

We have just considered the case of a single drunken sailor. Now what if we have a bunch of drunken sailors standing at the lamp post? What can we say about the average distance of all of the drunken sailors from the lamp post? Check with your classmates. Record each of their r^2 vs. N data in a table, and then compute the average value of r^2 at each step, N. Plot the mean square distance $\overline{r^2}$ vs. N data on the same graph as your own r^2 vs. N data. What do you notice about the motion of the random walkers as the number of walkers increases? Based on what you have learned, estimate the distance of an average drunken sailor from the lamp post after half an hour. Which is more predictable, the distance of a single drunken sailor or the average distance of a collection of drunken sailors?

EX. 15.2 (BOLTZMANN'S EQUATION—COINS IN A BOX). Clausius introduced his second law of thermodynamics, and the concept of entropy, in order to understand the types of processes which occur spontaneously in nature. According to Ludwig Boltzmann (1844–1906), the entropy of a system may be determined from the configuration of its individual components. In particular, the entropy associated with a particular state of a system may be determined by simply counting the number of ways, g, that the particular state can be realized:

$$S = k \ln g. \tag{15.9}$$

This is known as *Boltzmann's equation*, and Boltzmann's constant, $k = 1.38 \times 10^{-23}$ J/K, is a universal constant of nature which can be related to Avogadro's number and the gas constant.[2] As an exercise in working with entropy thus defined,

---

[2] See Eq. 11.13 in the discussion of diffusion and the kinetic theory of gases in Chap. 11 of the present volume.

suppose you place four coins heads-up in a box. After shaking the box vigorously, you open it and observe the state of the coins. Let us classify the possible final states according to how many heads are showing: (i) 4 heads, (ii) 3 heads, (iii) 2 heads, (iv) 1 heads, or (v) no heads.

a) Consider every possible outcome. How many ways can each of the states (i)–(v) be realized? Which state is most likely to occur? Is your answer consistent with common sense?

b) What is the entropy of each state? Which state has the highest entropy? Is the state with the highest entropy the most likely state to occur? Does shaking the box tend to increase the entropy, thus defined?

EX. 15.3 (ORDER, ENTROPY AND CARD SHUFFLING). The second law of thermodynamics can be understood as a statement that natural processes occur in such a way as to achieve the most likely final state. And the most likely final state is simply the one that can be realized in the largest number of ways. Put this way, the second law is hardly surprising. Nonetheless, there are subtleties involving how we define, or classify, the possible states of a system. For example, consider a fresh deck of playing cards, which is organized in ascending order according to suits (*i.e.* A♠, 2♠, 3♠,... Q♠, K♠,... A♡, 2♡, 3♡,... Q♡, K♡).

a) Does the deck of cards become more disordered by the process of shuffling? In what sense? In particular, what is an "ordered" state, as opposed to a "disordered" state, of the deck? Does the deck have more ordered states or more disordered states?

b) Does the entropy of the deck of cards increase by the process of shuffling? Explain your reasoning.

c) Now, suppose that the playing cards are originally unmarked (*i.e.* they are all identical and blank.) Does the entropy of this deck increase by the process of shuffling? Explain.

d) Based on your previous considerations, does the act of shuffling itself lead to an increase in the entropy of the deck? Is there a relationship between the entropy of a system and the distinguishability of its parts? More generally, is the entropy of a system an objective quantity, or does it depend subjectively on how an ordered configuration of its individual components is defined?[3]

EX. 15.4 (INDISTINGUISHABILITY AND STATISTICAL LAWS). Planck states that if each molecule is unique (distinguishable) then "the properties of a gas can never show so simple a law-abiding behavior as that which is indicated by thermodynamics." In other words, the laws of thermodynamics require *indistinguishability* on a microscopic level. If Planck is correct, then what does this imply about the use of statistical methods to derive or establish social laws for human populations?

---

[3] For a discussion of this concept, see Kenneth Denbigh's 1981 article entitled "How subjective is entropy" in Leff, H. S., and A. F. Rex (Eds.), *Maxwell's Demon: Entropy, Information, Computing*, Princeton Series in Physics, Princeton University Press, 1990.

15.5   Vocabulary

1. Atomism
2. *A priori*
3. Provisional
4. Suffice
5. Liberty
6. Dissociate
7. Ion
8. Emancipate
9. Auxiliary
10. Configuration
11. Complexion
12. Differentiate
13. Discrete
14. Homogeneous
15. Diathermanous
16. Micro-state
17. Macro-state
18. Manifold
19. Assimilate
20. Tacit
21. Ensue
22. Monochromatic
23. Constitute
24. Domain

Chapter 16
Corpuscles of Light

> *The energy of a light ray spreading out from a point source is not continuously distributed over an increasing space but consists of a finite number of energy quanta which are localized at points in space, which move without dividing, and which can only be produced and absorbed as complete units.*
>
> —A. Einstein

16.1 Introduction

By the end of the nineteenth century, the kinetic theory of gases had achieved great success in explaining a number of seemingly unrelated phenomena. For example, Maxwell and Boltzmann had shown that the empirical gas laws of Charles and Boyle could be seen as a consequence of the motion of countless microscopic gas molecules, each obeying Newton's laws of motion. After the discovery of the electron by J. J. Thomson in 1897, the kinetic theory was used by Drude to explain the electrical and thermal conductivities of metals; he treated the electrons within the metal as a gas of charged particles, each responding to the presence of electrical potential or thermal gradients.

Unfortunately, the kinetic theory had been less successful in explaining the color of light emitted by vibrating electrical charges. Consider: many heated objects (such as the coils in a toaster oven) glow brightly when heated. Why is this? According to the classical theory of electromagnetism, a charged particle which is vibrating (or in any way accelerating, for that matter) will emit electromagnetic radiation—light—whose color depends on the frequency of vibration. The vibration frequency, in turn, is related to the temperature of the body in which the charged particle is situated. Therefore the color of light emitted by a body depends essentially upon its temperature. For certain types of bodies, the so-called *blackbodies*, the emission spectrum depends only on the temperature of the body. Radiation from such bodies is called *blackbody radiation*.

© Springer International Publishing Switzerland 2016
K. Kuehn, *A Student's Guide Through the Great Physics Texts,*
Undergraduate Lecture Notes in Physics, DOI 10.1007/978-3-319-21828-1_16

Now here is the problem: the emission spectrum of a heated blackbody is very different than that predicted using this classical theory of electromagnetic radiation.[1] In fact, according to the classical theory, the radiant energy emitted from a heated blackbody grows without limit in the short-wavelength (ultra-violet) region of the electromagnetic spectrum. This problem, known as the *ultraviolet catastrophe*, was finally solved by Max Planck in 1900. Planck proposed that radiant energy can only be emitted or absorbed by matter in discrete amounts, or *quanta*.[2] This is often hailed as the birth of quantum theory.

In the following paper, translated from German by Arons and Peppard in 1965, Einstein extends Planck's quantum concept, claiming that not only the emission and absorption of radiant energy, but even light *itself* is quantized. Essentially, Einstein treats light not as a classical electromagnetic wave, but rather as a gas of corpuscles, each of which carries a discrete amount of energy that depends on its wavelength. At the outset of the reading selection, Einstein introduces his paradoxical new concept of light.[3] Then in the next several sections, Einstein motivates and explains this concept by applying Boltzmann's principle of entropy to a gas of light corpuscles. These sections are rather technical, so feel free to skim over them for now; you might come back and re-read these after having studied a bit of quantum theory, variational calculus and statistical mechanics. In the mean time, be sure to carefully study Sect. 16.2.8, wherein Einstein describes how his concept of the light quantum may be used to understand the *photoelectric effect*. This curious emission of cathode rays (electrons) from illuminated metal bodies had been observed and noted by Heinrich Hertz in the 1880's. Einstein's novel theory of the photoelectric effect was originally published in 1905; it earned him a Nobel Prize in physics 16 year later.

16.2 Reading: Einstein, *Concerning a Heuristic Point of View Toward the Emission and Transformation of Light*

Arons, A., and M. B. Peppard, Einstein's Proposal of the Photon Concept—a Translation of the Annalen der Physik Paper of 1905, *American Journal of Physics*, *33*(5), 367–374, 1965.

A profound formal distinction exists between the theoretical concepts which physicists have formed regarding gases and other ponderable bodies and the Maxwellian theory of electromagnetic processes in so-called empty space. While we consider the state of a body to be completely determined by the positions and velocities of a very large, yet finite, number of atoms and electrons, we make use of continuous

---

[1] The problem of blackbody radiation was mentioned briefly in Chap. 14 of the present volume.

[2] See Planck's lecture in Chap. 15 of the present volume.

[3] Recall, however, that the concept of a light corpuscle was introduced some 300 years earlier by Isaac Newton in Book III of his *Opticks*; see, for example, Chap. 18 of volume III.

spatial functions to describe the electromagnetic state of a given volume, and a finite number of parameters cannot be regarded as sufficient for the complete determination of such a state. According to the Maxwellian theory, energy is to be considered a continuous spatial function in the case of all purely electromagnetic phenomena including light, while the energy of a ponderable object should, according to the present conceptions of physicists, be represented as a sum carried over the atoms and electrons. The energy of a ponderable body cannot be subdivided into arbitrarily many or arbitrarily small parts, while the energy of a beam of light from a point source (according to the Maxwellian theory of light or, more generally, according to any wave theory) is continuously spread over an ever increasing volume.

The wave theory of light, which operates with continuous spatial functions, has worked well in the representation of purely optical phenomena and will probably never be replaced by another theory. It should be kept in mind, however, that the optical observations refer to time averages rather than instantaneous values. In spite of the complete experimental confirmation of the theory as applied to diffraction, reflection, refraction, dispersion, *etc.*, it is still conceivable that the theory of light which operates with continuous spatial functions may lead to contradictions with experience when it is applied to the phenomena of emission and transformation of light.

It seems to me that the observations associated with blackbody radiation, fluorescence, the production of cathode rays by ultraviolet light, and other related phenomena connected with the emission or transformation of light are more readily understood if one assumes that the energy of light is discontinuously distributed in space. In accordance with the assumption to be considered here, the energy of a light ray spreading out from a point source is not continuously distributed over an increasing space but consists of a finite number of energy quanta which are localized at points in space, which move without dividing, and which can only be produced and absorbed as complete units.

In the following I wish to present the line of thought and the facts which have led me to this point of view, hoping that this approach may be useful to some investigators in their research.

16.2.1   Concerning a Difficulty with Regard to the Theory of Blackbody Radiation

We start first with the point of view taken in the Maxwellian and the electron theories and consider the following case. In a space enclosed by completely reflecting walls, let there be a number of gas molecules and electrons which are free to move and which exert conservative forces on each other on close approach; *i.e.* they can collide

with each other like molecules in the kinetic theory of gases.[4] Furthermore, let there
be a number of electrons which are bound to widely separated points by forces
proportional to their distances from these points. The bound electrons are also to
participate in conservative interactions with the free molecules and electrons when
the latter come very close. We call the bound electrons "oscillators"; they emit and
absorb electromagnetic waves of definite periods.

According to the present view regarding the origin of light, the radiation in
the space we are considering (radiation which is found for the case of dynamic
equilibrium in accordance with the Maxwellian theory) must be identical with
the blackbody radiation—at least if oscillators of all the relevant frequencies are
considered to be present.

For the time being, we disregard the radiation emitted and absorbed by the oscillators and inquire into the condition of dynamical equilibrium associated with the
interaction (or collision) of molecules and electrons. The kinetic theory of gases
asserts that the average kinetic energy of an oscillator electron must be equal to the
average kinetic energy of a translating gas molecule. If we separate the motion of an
oscillator electron into three components at right angles to each other, we find for
the average energy $\overline{E}$ of one of these linear components the expression

$$\overline{E} = (R/N)T \tag{16.1}$$

where R denotes the universal gas constant, N denotes the number of "real
molecules" in a gram equivalent, and T the absolute temperature. The energy $\overline{E}$
is equal to two-thirds the kinetic energy of a free monatomic gas particle because
of the equality between the time average values of the kinetic and potential energies
of the oscillator. If through any cause—in our case through radiation processes—it
should occur that the energy of an oscillator takes on a time average value greater
or less than $\overline{E}$, then the collisions with the free electrons and molecules would lead
to a gain or loss of energy by the gas, different on the average from zero. Therefore,
in the case we are considering, dynamic equilibrium is possible only when each
oscillator has the average energy $\overline{E}$.

We shall now proceed to present a similar argument regarding the interaction between the oscillators and the radiation present in the cavity. Herr Planck
has derived[5] the condition for the dynamical equilibrium in this case under the

---

[4] This assumption is equivalent to the supposition that the average kinetic energies of gas molecules
and electrons are equal to each other at thermal equilibrium. It is well known that, with the help of
this assumption, Herr Drude derived a theoretical expression for the ratio of thermal and electrical
conductivities of metals.

[5] M. Planck, Ann. Physik **1**, 99 (1900).

supposition that the radiation can be considered a completely random process.[6] He found

$$(\overline{E}_\nu) = (L^3/8\pi \nu^2)\rho_\nu, \tag{16.5}$$

where $(\overline{E}_\nu)$ is the average energy (per degree of freedom) of an oscillator with eigen-frequency ν, L the velocity of light, ν the frequency, and $\rho_\nu d\nu$ the energy per unit volume of that portion of the radiation with frequency between ν and $\nu + d\nu$.

If the radiation energy of frequency ν is not continually increasing or decreasing, the following relations must obtain:

$$
\begin{aligned}
(R/N)T = \overline{E} = (\overline{E}_\nu) = (L^3/8\pi \nu^2)\rho_\nu, \\
\rho_\nu = (R/N)(8\pi \nu^2/L^3)T
\end{aligned}
\tag{16.6}
$$

These relations, found to be the conditions of dynamic equilibrium, not only fail to coincide with experiment, but also state that in our model there can be no talk of a definite energy distribution between ether and matter. The wider the range of wavenumbers of the oscillators, the greater will be the radiation energy of the space, and in the limit we obtain

$$\int_0^\infty \rho_\nu \, d\nu = \frac{R}{N}\frac{8\pi}{L^3}T \int_0^\infty \nu^2 \, d\nu = \infty \tag{16.7}$$

---

[6] This problem can be formulated in the following manner. We expand the Z component of the electrical force (Z) at an arbitrary point during the time interval between $t = 0$ and $t = T$ in a Fourier series in which $A_\nu \geq 0$ and $0 \leq \alpha_\nu \leq 2\pi$; the time T is taken to be very large relative to all the periods of oscillation that are present:

$$Z = \sum_{\nu=1}^{\nu=\infty} A_\nu \sin\left(2\pi\nu\frac{t}{T} + \alpha_\nu\right). \tag{16.2}$$

If one imagines making this expansion arbitrarily often at a given point in space at randomly chosen instants of time, one will obtain various sets of values of A_ν and α_ν. There then exist for the frequency of occurrence of different sets of values of A_ν and α_ν (statistical) probabilities dW of the form;

$$dW = f(A_1, A_2 \ldots \alpha_1, \alpha_2)dA_1 \, dA_2 \ldots d\alpha_1 \, d\alpha_2 \tag{16.3}$$

The radiation is then as disordered as conceivable if

$$f(A_1, A_2 \ldots \alpha_1, \alpha_2 \ldots) = F_1(A_1)F_2(A_2)\ldots f_1(\alpha_1)f_2(\alpha_2)\ldots, \tag{16.4}$$

i.e., if the probability of a particular value of A or α is independent of other values of A or α. The more closely this condition is fulfilled (namely, that the individual pairs of values of A_ν and α_ν are dependent upon the emission and absorption processes of *specific* groups of oscillators) the more closely will radiation in the case being considered approximate a perfectly random state.

16.2.2 Concerning Planck's Determination of the Fundamental Constants

We wish to show in the following that Herr Planck's determination of the fundamental constants is, to a certain extent, independent of his theory of blackbody radiation.

Planck's formula,[7] which has proved adequate up to this point, gives for ρ_v

$$\rho_v = \frac{\alpha v^3}{e^{\beta v/T} - 1},$$
$$\alpha = 6.10 \times 10^{-56},$$
$$\beta = 4866 \times 10^{-11}$$

(16.8)

For large values of T/v; i.e. for large wavelengths and radiation densities, this equation takes the form

$$\rho_v = (\alpha/\beta)v^2 T. \tag{16.9}$$

It is evident that this equation is identical with the one obtained in Sect. 1 from the Maxwellian and electron theories. By equating the coefficients of both formulas one obtains

$$(R/N)(8\pi/L^3) = (\alpha/\beta) \tag{16.10}$$

or

$$N = (\beta/\alpha)(8\pi R/L^3) = 6.17 \times 10^{23} \tag{16.11}$$

i.e., an atom of hydrogen weighs $1/N$ g $= 1.62 \times 10^{-24}$ g. This is exactly the value found by Herr Planck, which in turn agrees with values found by other methods.

We therefore arrive at the conclusion: the greater the energy density and the wavelength of a radiation, the more useful do the theoretical principles we have employed turn out to be; for small wavelengths and small radiation densities, however, these principles fail us completely.

In the following we shall consider the experimental facts concerning blackbody radiation without invoking a model for the emission and propagation of the radiation itself.

16.2.3 Concerning the Entropy of Radiation

The following treatment is to be found in a famous work by Herr W. Wien and is introduced here only for the sake of completeness.

---

[7] M. Planck, Ann. Physik **4**, 561 (1901).

Suppose we have radiation occupying a volume v. We assume that the observable properties of the radiation are completely determined when the radiation density $\rho(v)$ is given for all frequencies.[8] Since radiations of different frequencies are to be considered independent of each other when there is no transfer of heat or work, the entropy of the radiation can be represented by

$$S = v \int_0^\infty \phi(\rho, v)\, dv, \tag{16.12}$$

where ϕ is a function of the variables ρ and v.

ϕ can be reduced to a function of a single variable through formulation of the condition that the entropy of the radiation is unaltered during adiabatic compression between reflecting walls. We shall not enter into this problem, however, but shall directly investigate the derivation of the function ϕ from the blackbody radiation law.

In the case of blackbody radiation, ρ is such a function of v that the entropy is a maximum for a fixed value of energy; *i.e.*,

$$\delta \int_0^\infty \phi(\rho, v)\, dv = 0 \tag{16.13}$$

providing

$$\delta \int_0^\infty \rho\, dv = 0 \tag{16.14}$$

From this it follows that for every choice of $\delta\rho$ as a function of v

$$\delta \int_0^\infty \left(\frac{\partial \phi}{\partial \rho} - \lambda \right) \delta\rho\, dv = 0 \tag{16.15}$$

where λ is independent of v. In the case of blackbody radiation, therefore, $\partial\phi/\partial\rho$ is independent of v.

The following equation applies when the temperature of a unit volume of blackbody radiation increases by dT

$$dS = \int_{v=0}^{v=\infty} \left(\frac{\partial \phi}{\partial \rho} \right) d\rho\, dv, \tag{16.16}$$

or, since $\partial\phi/\partial\rho$ is independent of v,

$$dS = (\partial\phi/\partial\rho)dE. \tag{16.17}$$

Since dE is equal to the heat added and since the process is reversible, the following statement also applies

$$dS = (1/T)dE. \tag{16.18}$$

---

[8] This assumption is an arbitrary one. One will naturally cling to this simplest assumption as long as it is not controverted by experiment.

By comparison one obtains

$$\partial\phi/\partial\rho = 1/T \tag{16.19}$$

This is the law of blackbody radiation. Therefore one can derive the law of black-body radiation from the function ϕ, and, inversely, one can derive the function ϕ by integration, keeping in mind the fact that ϕ vanishes when $\rho = 0$.

16.2.4 Asymptotic Form for the Entropy of Monochromatic Radiation at Low Radiation Density

From existing observations of the blackbody radiation, it is clear that the law originally postulated by Herr W. Wien,

$$\rho = \alpha v^3 \epsilon^{-\beta v/T}, \tag{16.20}$$

is not exactly valid. It is, however, well confirmed experimentally for large values of v/T. We shall base our analysis on this formula, keeping in mind that our results are only valid within certain limits.

This formula gives immediately

$$(1/T) = -(1/\beta v)\ln(\rho/\alpha v^3) \tag{16.21}$$

and then, by using the relation obtained in the preceding section,

$$\phi(\rho, v) = \frac{\rho}{\beta v}\left[\ln\left(\frac{\rho}{\alpha v^3}\right) - 1\right]. \tag{16.22}$$

Suppose that we have radiation of energy E, with frequency between v and $v + dv$ enclosed in volume v. The entropy of this radiation is:

$$s = v\phi(\rho, v)dv = -\frac{E}{\beta v}\left[\ln\left(\frac{E}{v\alpha v^3 dv}\right) - 1\right]. \tag{16.23}$$

If we confine ourselves to investigating the dependence of the entropy on the volume occupied by the radiation, and if we denote by S_0 the entropy of the radiation at volume v_0, we obtain

$$S - S_0 = (E/\beta v)\ln(v/v_0). \tag{16.24}$$

This equation shows that the entropy of a monochromatic radiation of sufficiently low density varies with the volume in the same manner as the entropy of an ideal gas or a dilute solution. In the following, this equation will be interpreted in accordance with the principle introduced into physics by Herr Boltzmann, namely that the entropy of a system is a function of the probability its state.

16.2.5   Molecular-Theoretic Investigation of the Dependence of the Entropy of Gases and Dilute Solutions on the Volume

In the calculation of entropy by molecular-theoretic methods we frequently use the word "probability" in a sense differing from that employed in the calculus of probabilities. In particular, "cases of equal probability" have frequently been hypothetically established when the theoretical models being utilized are definite enough to permit a deduction rather than a conjecture. I will show in a separate paper that the so-called "statistical probability" is fully adequate for the treatment of thermal phenomena, and I hope that by doing so I will eliminate a logical difficulty that obstructs the application of Boltzmann's Principle. Here, however, only a general formulation and application to very special cases will be given.

If it is reasonable to speak of the probability of the state of a system, and furthermore if every entropy increase can be understood as a transition to a state of higher probability, then the entropy S_1 of a system is a function of W_1, the probability of its instantaneous state. If we have two noninteracting systems S_1 and S_2, we can write

$$S_1 = \phi_1(W_1)$$
$$S_2 = \phi_2(W_2) \tag{16.25}$$

If one considers these two systems as a single system of entropy S and probability W, it follows that

$$S = S_1 + S_2 = \phi(W) \tag{16.26}$$

and

$$W = W_1 \cdot W_2 \tag{16.27}$$

The last equation says that the states of the two systems are independent of each other.

From these equations it follows that

$$\phi(W_1 \cdot W_2) = \phi(W_1) + \phi(W_2), \tag{16.28}$$

and finally

$$\phi_1(W_1) = C \ln(W_1) + \text{const},$$
$$\phi_2(W_2) = C \ln(W_2) + \text{const}, \tag{16.29}$$
$$\phi_3(W) = C \ln(W) + \text{const}.$$

The quantity C is therefore a universal constant; the kinetic theory of gases shows its value to be R/N, where the constants R and N have been defined above. If S_0 denotes the entropy of a system in some initial state and W denotes the relative probability of a state of entropy S, we obtain in general

$$S - S_0 = (R/N) \ln W. \tag{16.30}$$

First we treat the following special case. We consider a number (n) of movable
points (*e.g.*, molecules) confined in a volume v_0. Besides these points, there can
be in the space any number of other movable points of any kind. We shall not
assume anything concerning the law in accordance with which the points move in
this space except that with regard to this motion, no part of the space (and no direction within it) can be distinguished from any other. Further, we take the number
of these movable points to be so small that we can disregard interactions between
them.

This system, which, for example, can be an ideal gas or a dilute solution, possesses an entropy S_0. Let us imagine transferring all n movable points into a volume
v (part of the volume v_0) without anything else being changed in the system. This
state obviously possesses a different entropy (S), and we now wish to evaluate the
entropy difference with the help of the Boltzmann Principle.

We inquire: How large is the probability of the latter state relative to the original
one? Or: How large is the probability that at a randomly chosen instant of time all n
movable points in the given volume v_0 will be found by chance in the volume v?

For this probability, which is a "statistical probability," one obviously obtains:

$$W = (v/v_0)^n. \tag{16.31}$$

By applying the Boltzmann Principle, one then obtains

$$S - S_0 = R(n/N)\ln(v/vo). \tag{16.32}$$

It is noteworthy that in the derivation of this equation, from which one can easily
obtain the law of Boyle and Gay-Lussac as well as the analogous law of osmotic
pressure thermodynamically,[9] no assumption had to be made as to a law of motion
of the molecules.

16.2.6   Interpretation of the Expression for the Volume Dependence of the Entropy of Monochromatic Radiation in Accordance with Boltzmann's Principle

In Sect. 16.2.4, we found the following expression for the dependence of the entropy
of monochromatic radiation on the volume

$$S - S_0 = (E/\beta v)\ln(v/v_0). \tag{16.35}$$

---

[9] If E is the energy of the system, one obtains:

$$-d(E - TS) = pdv = TdS = RT(n/N)(dv/v); \tag{16.33}$$

therefore

$$pv = R(n/N)T. \tag{16.34}$$

If one writes this in the form

$$S - S_0 = (R/N) \ln \left[(v/v_0)^{(N/R) \cdot (E/\beta v)} \right], \tag{16.36}$$

and if one compares this with the general formula for the Boltzmann principle

$$S - S_0 = (R/N) \log W, \tag{16.37}$$

one arrives at the following conclusion:

If monochromatic radiation of frequency v and energy E is enclosed by reflecting walls in a volume v_0, the probability that the total radiation energy will be found in a volume v (part of the volume v_0) at any randomly chosen instant is

$$W = (v/v_0)^{(N/R) \cdot (E/\beta v)} \tag{16.38}$$

From this we further conclude that: Monochromatic radiation of low density (within the range of validity of Wien's radiation formula) behaves thermodynamically as though it consisted of a number of independent energy quanta of magnitude $R\beta v/N$.

We still wish to compare the average magnitude of the energy quanta of the black-body radiation with the average translational kinetic energy of a molecule at the same temperature. The latter is $\frac{3}{2}(R/N)T$, while, according to the Wien formula, one obtains for the average magnitude of an energy quantum

$$\int_0^\infty \alpha v^3 \epsilon^{-\beta v/T} \, dv \Big/ \int_0^\infty \frac{N}{R\beta v} \alpha v^3 \epsilon^{-\beta v/T} \, dv = 3(RT/N) \tag{16.39}$$

If the entropy of monochromatic radiation depends on volume as though the radiation were a discontinuous medium consisting of energy quanta of magnitude $R\beta v/N$, the next obvious step is to investigate whether the laws of emission and transformation of light are also of such a nature that they can be interpreted or explained by considering light to consist of such energy quanta. We shall examine this question in the following.

16.2.7   Concerning Stoke's Rule

According to the result just obtained, let us assume that, when monochromatic light is transformed through photoluminescence into light of a different frequency, both the incident and emitted light consist of energy quanta of magnitude $R\beta v/N$, where v denotes the relevant frequency. The transformation process is to be interpreted in the following manner. Each incident energy quantum of frequency v_1 is absorbed and generates by itself—at least at sufficiently low densities of incident energy quanta—a light quantum of frequency v_2; it is possible that the absorption of the incident light quantum can give rise to the simultaneous emission of light quanta of frequencies v_3, v_4, *etc.*, as well as to energy of other kinds, *e.g.*, heat. It does not matter what intermediate processes give rise to this final result. If the fluorescent substance is not a perpetual source of energy, the principle of conservation of energy requires that the energy of an emitted energy quantum cannot be greater than that of the incident light quantum; it follows that

$$R\beta v_2/N \geq R\beta v_1/N \tag{16.40}$$

or

$$v_3 \leq v_1. \tag{16.41}$$

This is the well-known Stokes's Rule.

It should be strongly emphasized that according to our conception the quantity of light emitted under conditions of low illumination (other conditions remaining constant) must be proportional to the strength of the incident light, since each incident energy quantum will cause an elementary process of the postulated kind, independently of the action of other incident energy quanta. In particular, there will be no lower limit for the intensity of incident light necessary to excite the fluorescent effect.

According to the conception set forth above, deviations from Stokes's Rule are conceivable in the following cases:

1. when the number of simultaneously interacting energy quanta per unit volume is so large that an energy quantum of emitted light can receive its energy from several incident energy quanta;
2. when the incident (or emitted) light is not of such a composition that it corresponds to blackbody radiation within the range of validity of Wein's Law, that is to say, for example, when the incident light is produced by a body of such high temperature that for the wavelengths under consideration Wein's Law is no longer valid.

The last-mentioned possibility commands especial interest. According to the conception we have outlined, the possibility is not excluded that a "non-Wien radiation" of very low density can exhibit an energy behavior different from that of a blackbody radiation within the range of validity of Wein's Law.

16.2.8 Concerning the Emission of Cathode Rays Through the Illumination of Solid Bodies

The usual conception, that the energy of light is continuously distributed over the space through which it propagates, encounters very serious difficulties when one attempts to explain the photoelectric phenomena, as has been pointed out in Herr Lenard's pioneering paper.[10]

---

[10] P. Lenard, Ann. Physik **8**, 169, 170 (1902).

According to the concept that the incident light consists of energy quanta of magnitude $R\beta v/N$ however, one can conceive of the ejection of electrons by light in the following way. Energy quanta penetrate into the surface layer of the body, and their energy is transformed, at least in part, into kinetic energy of electrons. The simplest way to imagine this is that a light quantum delivers its entire energy to a single electron; we shall assume that this is what happens. The possibility should not be excluded, however, that electrons might receive their energy only in part from the light quantum.

An electron to which kinetic energy has been imparted in the interior of the body will have lost some of this energy by the time it reaches the surface. Furthermore, we shall assume that in leaving the body each electron must perform an amount of work P characteristic of the substance. The ejected electrons leaving the body with the largest normal velocity will be those that were directly at the surface. The kinetic energy of such electrons is given by

$$R\beta v/N - P \tag{16.42}$$

If the body is charged to a positive potential Π and is surrounded by conductors at zero potential, and if Π is just large enough to prevent loss of electricity by the body, it follows that:

$$\Pi\epsilon = R\beta v/N - P \tag{16.43}$$

where ϵ denotes the electronic charge, or

$$\Pi E = R\beta v - P' \tag{16.44}$$

where E is the charge of a gram equivalent of a monovalent ion and P' is the potential of this quantity of negative electricity relative to the body.[11]

If one takes $E = 9.6 \times 10^3$, then $\Pi \cdot 10^{-8}$ is the potential in volts which the body assumes when irradiated in a vacuum.

In order to see whether the derived relation yields an order of magnitude consistent with experience, we take $P' = 0$, $v = 1.03 \times 10^{15}$ (corresponding to the limit of the solar spectrum toward the ultraviolet) and $\beta = 4866 \times 10^{-11}$. We obtain $\Pi \cdot 10^7 = 4.3$ V , a result agreeing in order magnitude with those of Herr Lenard.[12]

If the derived formula is correct, then Π, when represented in Cartesian coordinates as a function of the frequency of the incident light, must be a straight line whose slope is independent of the nature of the emitting substance.

As far as I can see, there is no contradiction between these conceptions and the properties of the photoelectric effect observed by Herr Lenard. If each energy quantum of the incident light, independently of everything else, delivers its energy to

---

[11] If one assumes that the individual electron is detached from a neutral molecule by light with the performance of a certain amount of work, nothing in the relation derived above need be changed; one can simply consider P' as the sum of two terms.

[12] P. Lenard, Ann. Physik **8**, pp. 165, 184 and Table I, Fig. 2 (1902).

electrons, then the velocity distribution of the ejected electrons will be independent of the intensity of the incident light; on the other hand the number of electrons leaving the body will, if other conditions are kept constant, be proportional to the intensity of the incident light.[13]

Remarks similar to those made concerning hypothetical deviations from Stokes's Rule can be made with regard to hypothetical boundaries of validity of the law set forth above.

In the foregoing it has been assumed that the energy of at least some of the quanta of the incident light is delivered completely to individual electrons. If one does not make this obvious assumption, one obtains, in place of the last equation:

$$\Pi E + P' \leq R\beta v. \tag{16.45}$$

For fluorescence induced by cathode rays, which is the inverse process to the one discussed above, one obtains by analogous considerations:

$$\Pi E + P' \geq R\beta v. \tag{16.46}$$

In the case of the substances investigated by Herr Lenard, PE[14] is always significantly greater than $R\beta v$, since the potential difference, which the cathode rays must traverse in order to produce visible light, amounts in some cases to hundreds and in others to thousands of volts.[15] It is therefore to be assumed that the kinetic energy of an electron goes into the production of many light energy quanta.

16.2.9   Concerning the Ionization of Gases by Ultraviolet Light

We shall have to assume that, in the ionization of a gas by ultraviolet light, an individual light energy quantum is used for the ionization of an individual gas molecule. From this it follows immediately that the work of ionization (*i.e.*, the work theoretically needed for ionization) of a molecule cannot be greater than the energy of an absorbed light quantum capable of producing this effect. If one denotes by J the (theoretical) work of ionization per gram equivalent, then it follows that:

$$R\beta v \geq J. \tag{16.47}$$

According to Lenard's measurements, however, the largest effective wavelength for air is approximately 1.9×10^{-5} cm; therefore:

$$R\beta v = 6.4 \times 10^{12} \text{ erg} \geq J. \tag{16.48}$$

---

[13] P. Lenard, Ref. 9, p. 150 and p. 166–168.

[14] Should be ΠE (translator's note).

[15] P. Lenard, Ann. Physik **12**, 469 (1903).

An upper limit for the work of ionization can also be obtained from the ionization potentials of rarefied gases. According to J. Stark[16] the smallest observed ionization potentials for air (at platinum anodes) is about 10 V.[17] One therefore obtains 9.6×10^{12} as an upper limit for J, which is nearly equal to the value found above.

There is another consequence the experimental testing of which seems to me to be of great importance. If every absorbed light energy quantum ionizes a molecule, the following relation must obtain between the quantity of absorbed light L and the number of gram molecules of ionized gas j:

$$j = L/R\beta\nu. \tag{16.49}$$

If our conception is correct, this relationship must be valid for all gases which (at the relevant frequency) show no appreciable absorption without ionization.

Bern, 17 March 1905

RECEIVED 18 MARCH 1905.

16.3   Study Questions

QUES. 16.1. What is Einstein's conception of light?

a) What is the fundamental distinction between the theoretical concepts of matter and of light? Is matter infinitely divisible? Are electromagnetic fields infinitely divisible?

b) What phenomena are well described using the wave theory of light? What types of phenomena, according to Einstein, may *not* be well-described by the wave theory?

c) What alternative conception of light does Einstein suggest?

QUES. 16.2. Are there any problems with the classical theory of blackbody radiation?

a) What is meant by the classical theory of blackbody radiation? According to this theory, is there light inside a chamber completely enclosed by reflecting walls? If so, what is its cause, or origin?

b) According to the kinetic theory of gases, what is the relationship between the gas molecules and the electrons in the chamber? Are they in (thermal) equilibrium?

c) What, now, is the relationship between the vibrating matter and any radiation existing in the chamber? Are they in (thermal) equilibrium? How is this equilibrium condition expressed mathematically?

---

[16] J. Stark, *Die Elektrizität in Gasen* (Leipzig, 1902), p. 57.

[17] In the interior of gases the ionization potential for negative ions is, however, five times greater.

d) According to this (classical) theory of electromagnetic radiation in a chamber, what is the total (radiant) energy in such a chamber? Is this a problem? Why?
e) How did Max Planck solve this problem? In particular, what quantity did he introduce? And how did it solve the problem of the radiant blackbody energy?

QUES. 16.3. Consider a gas of particles confined in a chamber. What is the probability that all of the particles happen to be located in one corner of this chamber? According to Boltzmann's Principle, what is the relationship between the probability of this configuration and the entropy of this configuration? Does this relationship depend in any way on Newton's laws of motion?

QUES. 16.4. In what sense does monochromatic radiation, confined in a chamber, behave like a gas of particles? What is the energy of each of these particles? How does this compare to the average translational kinetic energy of a gas molecule at the same temperature? What does this suggest regarding the nature of light?

QUES. 16.5. How does Einstein explain the emission of cathode rays from solid substances when they are illuminated?

a) What is the photoelectric effect? Who discovered it? And what picture does Einstein propose to explain this effect?
b) How does increasing the frequency of radiation affect the emission of cathode rays? How does the intensity of the incident radiation affect the emission of cathode rays?
c) What difficulty does the wave theory encounter in describing the photoelectric effect? How does Einstein's theory address this problem? Does Einstein's conception of light raise any difficulties?

QUES. 16.6. How does Einstein's apply his theory of the photoelectric to understand the ionization of gases?

16.4   Exercises

EX. 16.1 (STOPPING VOLTAGE). Einstein mentions that if a substance from which electrons are being ejected by incident light is held at an electric potential, Π, above its surroundings, then it is able to slow down, or even stop the emitted photoelectrons. Suppose that a particular substance requires a minimum work of 2 electron-volts to unbind an electron. (This is called the *work function* of the substance.)

a) What is the minimum wavelength of light required to eject an electron from this substance? What happens if light of this minimum wavelength is used and its intensity is gradually increased?
b) If ultra-violet light, having a wavelength of 200 nm, is incident upon the substance, then what is the minimum value of the potential, Π, such that ejected electrons return to the substance, rather than escaping from the substance entirely?

Ex. 16.2 (MEASURING PLANCK'S CONSTANT USING THE PHOTOELECTRIC EFFECT). In this laboratory experiment, we will use the photoelectric effect to measure Planck's constant, h, and to explore the validity of Einstein's quantum theory of light. A photo-electric apparatus consists of a specially designed photo-tube, a sensitive galvanometer and several monochromatic light sources having different wavelengths.[18] The photo-tube is an evacuated glass tube containing two electrodes which are maintained at a slight voltage difference. The negative electrode (the cathode) has a photosensitive surface. When light having sufficient energy strikes the cathode, electrons are ejected. Some of these photo-electrons are collected by the positive electrode (the anode) producing a small photo-current which can be measured using a sensitive galvanometer. By reversing the electric potential between the cathode and the anode, this photo-current can be slowed down or even stopped entirely. By measuring the minimum *stopping voltage* with a digital voltmeter, one can estimate the kinetic energy of the most energetic photo-electrons. This is because the kinetic energy of an electron is equal to the work required to bring it to a complete stop.[19] Einstein claimed that each photo-electron is ejected by a light particle—a *photon*—whose energy, E, depends on the color (and *not* the intensity) of the incident light:

$$E = h\nu \tag{16.50}$$

Here, h is Planck's constant and ν is the frequency of the photon, as it were. By measuring photo-electron stopping voltages for several known light frequencies, one can infer the energy of the incident photons. Is the photon energy indeed proportional to its frequency? If so, what is the proportionality constant? Also, is there a minimum energy required to eject photo-electrons? If so, what is its value? Do you think it depends on the nature of the photo-sensitive surface?[20] Finally, do the results of your experiments support Einstein's photon theory of light?

16.5   Vocabulary

1. Corpuscles
2. Phosphorescence
3. Cathode
4. Maltese cross

---

[18] A complete Photoelectric Apparatus (Model EP-07)—containing a phototube, an amplifier and three color filters—is available form Daedalon, Downeast Maine. Additional light sources, such as a mercury arc and a He-Ne laser may be used to extend the measurements to additional wavelengths. A digital voltmeter is also required.

[19] See Ex. 16.1, above.

[20] For precise measurements of the Planck's constant using the photoelectric effect for several surfaces, see Millikan, R. A., A direct photoelectric determination of Planck's "h", *Physical Review*, 7(3), 355, 1916.

5. Luminosity
6. Alkali metals
7. Supersaturated
8. Mahomet's coffin
9. Electrolysis
10. Radium

Chapter 17
The Discovery of the Electron

*The atom is not the ultimate limit to the subdivision of matter;
we may go further and get to the corpuscle, and at this stage the
corpuscle is the same from whatever source it may be derived.*

—J. J. Thomson

17.1 Introduction

According to the ancient doctrines of Democritus and Epicurus, atoms are tiny indivisible masses rattling about through the vacuum of space. This atomic hypothesis was preserved—in one form or another—for over 2000 years. Yet by the close of the nineteenth century it was still not universally accepted. The most convincing evidence for atomism was rather indirect, having come out of (i) the study of chemical reactions, (ii) the kinetic theory of gases and (iii) Boltzmann's probabilistic interpretation of entropy.[1] Even among the atomists there was considerable disagreement regarding the nature and structure of the atom itself. For example, the followers of Roger Boscovich[2] believed atoms to be little more than point-like mathematical centers of force.[3] Boscovich's atoms possessed an unchangeable mass, they could move through the vacuum of space, and they were endowed with an irreducible power to attract or repel other atoms. A very different view of the atom was maintained by William Thomson (Lord Kelvin). Inspired by the earlier work of Hermann von Helmholtz, Kelvin imagined atoms to be like minuscule smoke-rings traveling through an all-pervasive frictionless fluid medium—the æther.[4] These so-called *vortex ring* atoms were indivisible local excitations which could emit light by vibrating like the rim of a tiny ringing bell. In this way, the vortex ring model of the atom

---

[1] See, for example, the discussions of atomism by Clausius, Maxwell, and Planck which are presented in Chaps. 6, 11 and 15 of the present volume.

[2] Roger Boscovich was an eighteenth century natural philosopher and Jesuit priest.

[3] See Part I, Sect. 7 of Boscovich, R. J., *A Theory of Natural Philosophy*, Open Court Publishing Co., Chicago and London, 1922. Boscovich originally published this work in 1763.

[4] See Thomson, W., On Vortex Atoms, *Proceedings of the Royal Society of Edinburgh*, 6, 1867.

© Springer International Publishing Switzerland 2016
K. Kuehn, *A Student's Guide Through the Great Physics Texts*,
Undergraduate Lecture Notes in Physics, DOI 10.1007/978-3-319-21828-1_17

offered the possibility of explaining the complicated emission spectra of atomic gasses.[5]

Kelvin's vortex theory of the atom aroused the interest of another Thomson, J.J., who published his *Treatise on the motion of vortex rings* in 1883 while studying at the University of Cambridge. A few years later, J.J. Thomson would propose the existence of subatomic particles which weighed considerably less than the smallest known atom. Joseph John Thomson (1856–1940) was born in Cheetham Hill, near Manchester, England. He attended Owens College and then Trinity College, where he became a Fellow in 1880. He was appointed Cavendish Professor of Experimental Physics at Cambridge in 1884. Over the course of his career, Thomson's work focused on the nature of electricity, magnetism, and the atom. His works, for example, include a textbook on the *Elements of the Mathematical Theory of Electricity and Magnetism* (1895), lectures on the *Discharge of Electricity through Gases* (1897) and on the *Conduction of Electricity through Gases* (1903), and books on *The Structure of Light* (1907), *The Corpuscular Theory of Matter* (1907), *Rays of Positive Electricity* (1913) and *The Electron in Chemistry* (1923). J.J. Thomson won the Nobel prize in physics in 1906 "in recognition of the great merits of his theoretical and experimental investigations on the conduction of electricity by gases." Among his famous students was Ernest Rutherford, who is today recognized as the father of nuclear physics.[6] Thomson's son, George, shared the Nobel prize in Physics in 1937 with Clinton Davisson "for their experimental discovery of the diffraction of electrons by crystals."[7] In the reading selection below, Thomson describes his famous 1897 discovery of subatomic electrically charged particles. It is taken from the first chapter of his 1907 book on *The Corpuscular Theory of Matter*.

17.2   Reading: Thomson, *The Corpuscular Theory of Matter*

Thomson, J. J., *The Corpuscular Theory of Matter*, Charles Scribner's Sons, New York, 1907. Chap. I.

The theory of the constitution of matter which I propose to discuss in these lectures, is one which supposes that the various properties of matter may be regarded as arising from electrical effects. The basis of the theory is electricity, and its object is to construct a model atom, made up of specified arrangements of positive and negative electricity, which shall imitate as far as possible the properties of the real atom. We shall postulate that the attractions and repulsions between the electrical charges

---

[5] Niels Bohr discusses atomic emission spectra in the context of his quantized model of the atom; see Chap. 28 of the present volume.

[6] Rutherford describes his work with α-particles, radioactivity and nuclear fission in the readings included in Chaps. 18–21 of the present volume.

[7] See, for example, Davisson's paper on electron diffraction which is included in Chap. 26 of the present volume.

in the atom follow the familiar law of the inverse square of the distance, though, of course, we have only direct experimental proof of this law when the magnitude of the charges and the distances between them are enormously greater than those which can occur in the atom. We shall not attempt to go behind these forces and discuss the mechanism by which they might be produced. The theory is not an ultimate one; its object is physical rather than metaphysical. From the point of view of the physicist, a theory of matter is a policy rather than a creed; its object is to connect or co-ordinate apparently diverse phenomena, and above all to suggest, stimulate and direct experiment. It ought to furnish a compass which, if followed, will lead the observer further and further into previously unexplored regions. Whether these regions will be barren or fertile experience alone will decide; but, at any rate, one who is guided in this way will travel onward in a definite direction, and will not wander aimlessly to and fro.

The corpuscular theory of matter with its assumptions of electrical charges and the forces between them is not nearly so fundamental as the vortex atom theory of matter, in which all that is postulated is an incompressible, frictionless liquid possessing inertia and capable of transmitting pressure. On this theory the difference between matter and non-matter and between one kind of matter and another is a difference between the kinds of motion in the incompressible liquid at various places, matter being those portions of the liquid in which there is vortex motion. The simplicity of the assumptions of the vortex atom theory are, however, somewhat dearly purchased at the cost of the mathematical difficulties which are met with in its development; and for many purposes a theory whose consequences are easily followed is preferable to one which is more fundamental but also more unwieldy. We shall, however, often have occasion to avail ourselves of the analogy which exists between the properties of lines of electric force in the electric field and lines of vortex motion in an incompressible fluid.

To return to the corpuscular theory. This theory, as I have said, supposes that the atom is made up of positive and negative electricity. A distinctive feature of this theory—the one from which it derives its name—is the peculiar way in which the negative electricity occurs both in the atom and when free from matter. We suppose that the negative electricity always occurs as exceedingly fine particles called corpuscles, and that all these corpuscles, whenever they occur, are always of the same size and always carry the same quantity of electricity. Whatever may prove to be the constitution of the atom, we have direct experimental proof of the existence of these corpuscles, and I will begin the discussion of the corpuscular theory with a description of the discovery and properties of corpuscles.

17.2.1 Corpuscles in Vacuum Tubes

The first place in which corpuscles were detected was a highly exhausted tube through which an electric discharge was passing. When I send an electric discharge through this highly exhausted tube you will notice that the sides of the tube glow

Fig. 17.1 Perrin's apparatus
for measuring charge carried
by the cathode rays.—[*K.K.*]

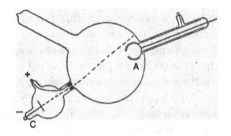

with a vivid green phosphorescence. That this is due to something proceeding in straight lines from the cathode—the electrode where the negative electricity enters the tube—can be shown in the following way: the experiment is one made many years ago by Sir William Crookes. A Maltese cross made of thin mica is placed between the cathode and the walls of the tube. You will notice that when I send the discharge through the tube, the green phosphorescence does not now extend all over the end of the tube as it did in the tube without the cross. There is a well-defined cross in which there is no phosphorescence at the end of the tube; the mica cross has thrown a shadow on the tube, and the shape of the shadow proves that the phosphorescence is due to something, travelling from the cathode in straight lines, which is stopped by a thin plate of mica. The green phosphorescence is caused by cathode rays, and at one time there was a keen controversy as to the nature of these rays. Two views were prevalent, one, which was chiefly supported by English physicists, was that the rays are negatively electrified bodies shot off from the cathode with great velocity; the other view, which was held by the great majority of German physicists, was that the rays are some kind of ethereal vibrations or waves.

The arguments in favour of the rays being negatively charged particles are (1) that they are deflected by a magnet in just the same way as moving negatively electrified particles. We know that such particles when a magnet is placed near them are acted upon by a force whose direction is at right angles to the magnetic force, and also at right angles to the direction in which the particles are moving. Thus, if the particles are moving horizontally from east to west, and the magnetic force is horizontal and from north to south, the force acting on the negatively electrified particles will be vertical and downwards.

When the magnet is placed so that the magnetic force is along the direction in which the particle is moving the latter will not be affected by the magnet. By placing the magnet in suitable positions I can show you that the cathode particles move in the way indicated by the theory. The observations that can be made in lecture are necessarily very rough and incomplete; but I may add that elaborate and accurate measurements of the movement of cathode rays under magnetic forces have shown that in this respect the rays behave exactly as if they were moving electrified particles (Fig. 17.1).

The next step made in the proof that the rays are negatively charged particles, was to show that when they are caught in a metal vessel they give up to it a charge

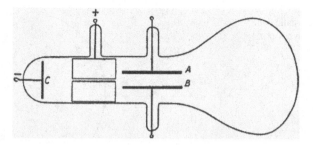

Fig. 17.2 Thomson's cathode ray tube.—[*K.K.*]

of negative electricity. This was first done by Perrin. I have here a modification of his experiment. *A* is a metal cylinder with a hole in it. It is placed so as to be out of the way of the rays coming from *C*, unless they are deflected by a magnet, and is connected with an electroscope. You see that when the rays do not pass through the hole in the cylinder the electroscope does not receive a charge. I now, by means of a magnet, deflect the rays so that they pass through the hole in the cylinder. You see by the divergence of the gold-leaves that the electroscope is charged, and on testing the sign of the charge we find that it is negative.

17.2.2   Deflection of the Rays by a Charged Body

If the rays are charged with negative electricity they ought to be deflected by an electrified body as well as by a magnet. In the earlier experiments made on this point no such deflection was observed. The reason of this has been shown to be that when the cathode rays pass through a gas they make it a conductor of electricity, so that if there is any appreciable quantity of gas in the vessel through which the rays are passing, this gas will become a conductor of electricity, and the rays will be surrounded by a conductor which will screen them from the effects of electric force just as the metal covering of an electroscope screens off all external electric effects. By exhausting the vacuum tube until there was only an exceedingly small quantity of air left in to be made a conductor, I was able to get rid of this effect and to obtain the electric deflection of the cathode rays. The arrangement I used for this purpose is shown in Fig. 17.2. The rays on their way through the tube pass between two parallel plates, *A*, *B*, which can be connected with the poles of a battery of storage cells. The pressure in the tube is very low. You will notice that the rays are very considerably deflected, when I connect the plates with the poles of the battery, and that the direction of the deflection shows that the rays are negatively charged.

We can also show the effect of magnetic and electric force on these rays if we avail ourselves of the discovery made by Wehnelt, that lime when raised to a red heat emits when negatively charged large quantities of cathode rays. I have here a tube whose cathode is a strip of platinum on which there is a speck of lime. When

Fig. 17.3 Electrostatic deflection of cathode rays.—[*K.K.*]

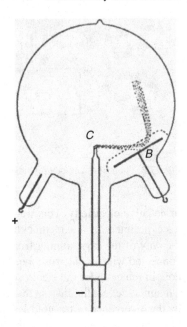

the piece of platinum is made very hot, a potential difference of 100 V or so is
sufficient to make a stream of cathode rays start from this speck; you will be able
to trace the course of the rays by the luminosity they produce as they pass through
the gas. You can see the rays as a thin line of bluish light coming from a point on
the cathode; on bringing a magnet near it the line becomes curved, and I can bend
it into a circle or a spiral, and make it turn round and go right behind the cathode
from which it started. This arrangement shows in a very striking way the magnetic
deflection of the rays. To show the electrostatic deflection I use the tube shown in
Fig. 17.3. I charge up the plate *B* negatively so that it repels the pencil of rays which
approach it from the spot of lime on the cathode, *C*. You see that the pencil of rays
is deflected from the plate and pursues a curved path whose distance from the plate
I can increase or diminish by increasing or diminishing the negative charge on the
plate.

We have seen that the cathode rays behave under every test that we have applied
as if they are negatively electrified particles; we have seen that they carry a negative charge of electricity and are deflected by electric and magnetic forces just as
negatively electrified particles would be.

Hertz showed, however, that the cathode particles possess another property which
seemed inconsistent with the idea that they are particles of matter, for he found that
they were able to penetrate very thin sheets of metal, for example, pieces of gold-
leaf placed between them and the glass, and produce appreciable luminosity on
the glass after doing so. The idea of particles as large as the molecules of a gas
passing through a solid plate was a somewhat startling one in an age which knew
not radium—which does project particles of this size through pieces of metal much

thicker than gold-leaf—and this led me to investigate more closely the nature of the particles which form the cathode rays.

The principle of the method used is as follows: When a particle carrying a charge e is moving with the velocity v across the lines of force in a magnetic field, placed so that the lines of magnetic force are at right angles to the motion of the particle, then if H is the magnetic force, the moving particle will be acted on by a force equal to Hev. This force acts in the direction which is at right angles to the magnetic force and to the direction of motion of the particle, so that if the particle is moving horizontally as in the figure and the magnetic force is at right angles to the plane of the paper and towards the reader, then the negatively electrified particle will be acted on by a vertical and upward force (See Fig. 17.4). The pencil of rays will therefore be deflected upwards and with it the patch of green phosphorescence where it strikes the walls of the tube. Let now the two parallel plates A and B (Fig. 17.2) between which the pencil of rays is moving be charged with electricity so that the upper plate is negatively and the lower plate positively electrified, the cathode rays will be repelled from the upper plate with a force Xe where X is the electric force between the plates. Thus, if the plates are charged, when the magnetic field is acting on the rays, the magnetic force will tend to send the rays upwards, while the charge on the plates will tend to send them downwards. We can adjust the electric and magnetic forces until they balance and the pencil of rays passes horizontally in a straight line between the plates, the green patch of phosphorescence being undisturbed. When this is the case, the force Hev due to the magnetic field is equal to Xe—the force due to the electric field—and we have

$$Hev = Xe \tag{17.1}$$

or

$$v = \frac{X}{H} \tag{17.2}$$

Thus, if we measure, as we can without difficulty, the values of X and H when the rays are not deflected, we can determine the value of v, the velocity of the particles. The velocity of the rays found in this way is very great; it varies largely with the pressure of the gas left in the tube. In a very highly exhausted tube it may be ⅓ the velocity of light or about 60,000 miles/s; in tubes not so highly exhausted it may not be more than 5000 miles/s, but in all cases when the cathode rays are produced in tubes their velocity is much greater than the velocity of any other moving body with which we are acquainted. It is, for example, many thousand times the average velocity with which the molecules of hydrogen are moving at ordinary temperatures, or indeed at any temperature yet realised.

17.2.3 Determination of e/m

Having found the velocity of the rays, let us in the preceding experiment take away the magnetic force and leave the rays to the action of tho electric force alone. Then

Fig. 17.4 A force acts on
a charge moving through a
magnetic field.—[K.K.]

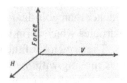

the particles forming the rays are acted upon by a constant vertical downward force
and the problem is practically that of a bullet projected horizontally with a velocity
v and falling under gravity. We know that in time t the body will fall a depth equal
to $\frac{1}{2} gt^2$ where g is the vertical acceleration; in our case the vertical acceleration
is equal to Xe/m where m is the mass of the particle, the time it is falling is l/v
where l is the length or path measured horizontally, and v the velocity of projection.
Thus, the depth the particle has fallen when it reaches the glass, *i.e.*, the downward
displacement of the patch of phosphorescence where the rays strike the glass, is
equal to

$$\frac{1}{2} \frac{Xe}{m} \frac{l^2}{v^2} \qquad (17.3)$$

We can easily measure d the distance the phosphorescent patch is lowered, and
as we have found v and X and l are easily measured, we can find e/m from the
equation:

$$\frac{e}{m} = \frac{2d}{X} \frac{v^2}{l^2} \qquad (17.4)$$

The results of the determinations of the values of e/m made by this method are
very interesting, for it is found that however the cathode rays produced we always
get the same value of e/m for all the particles in the rays. We may, for example,
by altering the shape of the discharge tube and the pressure of the gas in the tube,
produce great changes in the velocity of the particles, but unless the velocity of the
particles becomes so great that they are moving nearly as fast as light, when, as we
shall see, other considerations have to be taken into account, the value of e/m is
constant. The value of e/m is not merely independent of tho velocity. What is even
more remarkable is that it is independent of the kind of electrodes we use and also
of the kind of gas in the tube. The particles which form the cathode rays must come
either from the gas in the tube or from the electrodes; we may, however, use any
kind of substance we please for the electrodes and fill the tube with gas of any kind,
and yet the value of e/m will remain unaltered.

This constant value is, when we measure e/m in the C. G. S. system of magnetic
units, equal to about 1.7×10^7. If we compare this with the value of the ratio of
the mass to the charge of electricity carried by any system previously known, we
find that it is of quite a different order of magnitude. Before the cathode rays were
investigated the charged atom of hydrogen met with in the electrolysis of liquids
was the system which had the greatest known value for e/m, and in this case the
value is only 10^4; hence for the corpuscle in the cathode rays the value of e/m is
1700 times the value of the corresponding quantity for the charged hydrogen atom.
This discrepancy must arise in one or other of two ways, either the mass of the

corpuscle must be very small compared with that of the atom of hydrogen, which until quite recently was the smallest mass recognised in physics, or else the charge on the corpuscle must be very much greater than that on the hydrogen atom. Now it has been shown by a method which I shall shortly describe that the electric charge is practically the same in the two cases; hence we are driven to the conclusion that the mass of the corpuscle is only about 1/1700 of that of the hydrogen atom. Thus the atom is not the ultimate limit to the subdivision of matter; we may go further and get to the corpuscle, and at this stage the corpuscle is the same from whatever source it may be derived.

17.2.4   Corpuscles Very Widely Distributed

It is not only from what may be regarded as a somewhat artificial and sophisticated source, *viz.*, cathode rays, that we can obtain corpuscles. When once they had been discovered it was found that they were of very general occurrence. They are given out by metals when raised to a red heat : you have already seen what a copious supply is given out by hot lime. Any substance when heated gives out corpuscles to some extent; indeed, we can detect the emission of them from some substances, such as rubidium and the alloy of sodium and potassium, even when they are cold; and it is perhaps allowable to suppose that there is some emission by all substances, though our instruments are not at present sufficiently delicate to detect it unless it is unusually large.

Corpuscles are also given out by metals and other bodies, but especially by the alkali metals, when these are exposed to light. They are being continually given out in large quantities, and with very great velocities by radio-active substances such as uranium and radium; they are produced in large quantities when salts are put into flames, and there is good reason to suppose that corpuscles reach us from the sun.

The corpuscle is thus very widely distributed, but wherever it is found it preserves its individuality, e/m being always equal to a certain constant value.

The corpuscle appears to form a part of all kinds of matter under the most diverse conditions; it seems natural, therefore, to regard it as one of the bricks of which atoms are built up.

17.2.5   Magnitude of the Electric Charge Carried by the Corpuscle

I shall now return to the proof that the very large value of e/m for the corpuscle as compared with that for the atom of hydrogen is due to the smallness of m the mass, and not to the greatness of e the charge. We can do this by actually measuring the value of e, availing ourselves for this purpose of a discovery by C.T.R. Wilson, that a charged particle acts as a nucleus round which water vapour condenses, and forms

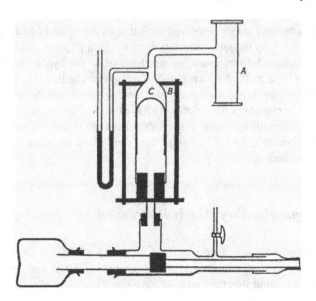

Fig. 17.5 Thomson's cloud chamber apparatus.—[K.K.]

drops of water. If we have air saturated with water vapour and cool it so that it would be supersaturated if there were no deposition of moisture, we know that if any dust is present, the particles of dust act as nuclei round which the water condenses and we get the too familiar phenomena of fog and rain. If the air is quite dust-free we can, however, cool it very considerably without any deposition of moisture taking place. If there is no dust, C.T.R. Wilson has shown that the cloud does not form until the temperature has been lowered to such a point that the supersaturation is about eightfold. When, however, this temperature is reached, a thick fog forms, even in dust-free air. When charged particles are present in the gas, Wilson showed that a much smaller amount of cooling is sufficient to produce the fog, a fourfold supersaturation being all that is required when the charged particles are those which occur in a gas when it is in the state in which it conducts electricity. Each of the charged particles becomes the centre round which a drop of water forms; the drops form a cloud, and thus the charged particles, however small to begin with, now become visible and can be observed. The effect of the charged particles on the formation of a cloud can be shown very distinctly by the following experiment (see Fig. 17.5). The vessel *A*, which is in contact with water, is saturated with moisture at the temperature of the room. This vessel is in communication with *B*, a cylinder in which a large piston, *C*, slides up and down; the piston, to begin with, is at the top of its travel; then by suddenly exhausting the air from below the piston, the pressure of the air above it will force it down with great rapidity, and the air in the vessel *A* will expand very quickly. When, however, air expands it gets cool; thus the air in *A* gets colder, and as it was saturated with moisture before cooling, it is now supersaturated. If there is no dust present, no deposition of moisture will take place unless the air in A is cooled to such a low temperature that the amount of moisture required to saturate it is only about ⅛ of that actually present. Now the amount of cooling, and therefore of supersaturation, depends upon the travel of the piston; the greater the travel the greater the cooling. I can regulate this travel so that the supersaturation is less than eightfold, and greater than fourfold. We now free the air from dust by forming cloud after cloud in the dusty air, as the clouds fall they carry the dust down with them, just as in nature the air is cleared by showers. We find at last that when we make the expansion no cloud is visible. We now put the gas in a conducting state by bringing a little radium near the vessel A; this fills the gas with large quantities of both positively and negatively electrified particles. On making the expansion now, an exceedingly dense cloud is formed. That this is due to the electrification in the gas can be shown by the following experiment: Along the inside walls of the vessel A we have two vertical insulated plates which can be electrified; if these plates are electrified they will drag the charged particles out of the gas as fast as they are formed, so that by electrifying the plates we can get rid of, or at any rate largely reduce, the number of electrified particles in the gas. I now repeat the experiment, electrifying the plates before bringing up the radium. You see that the presence of the radium hardly increases the small amount of cloud. I now discharge the plates, and on making the expansion the cloud is so dense as to be quite opaque.

We can use the drops to find the charge on the particles, for when we know the travel of the piston we can deduce the amount of supersaturation, and hence the amount of water deposited when the cloud forms. The water is deposited in the form of a number of small drops all of the same size; thus the number of drops will be the volume of the water deposited divided by the volume of one of the drops. Hence, if we find the volume of one of the drops we can find the number of drops which are formed round the charged particles. If the particles are not too numerous, each will have a drop round it, and we can thus find the number of electrified particles.

If we observe the rate at which the drops slowly fall down we can determine the size of the drops. In consequence of the viscosity or friction of the air small bodies do not fall with a constantly accelerated velocity, but soon reach a speed which remains uniform for the rest of the fall; the smaller the body the slower this speed, and Sir George Stokes has shown that v, the speed at which a drop of rain falls, is given by the formula—

$$v = \frac{2}{9}\frac{ga^2}{\mu} \qquad (17.5)$$

where a is the radius of the drop, g the acceleration due to gravity, and μ the coefficient of viscosity of the air. If we substitute the values of g and μ we get

$$v = 1.28 \times 10^6 a^2 \qquad (17.6)$$

Hence, if we measure v we can determine a, the radius of the drop. We can, in this way, find the volume of a drop, and may therefore, as explained above, calculate the number of drops, and therefore the number of electrified particles. It is a simple matter to find, by electrical methods, the total quantity of electricity on these particles; and hence, as we know the number of particles, we can deduce at once the charge on each particle.

This was the method by which I first determined the charge on the particle. H.A. Wilson has since used a simpler method founded on the following principles. C.T.R. Wilson has shown that the drops of water condense more easily on negatively electrified particles than on positively electrified ones. Thus, by adjusting the expansion, it is possible to get drops of water round the negative particles and not round the positive; with this expansion, therefore, all the drops are negatively electrified. The size of these drops, and therefore their weight, can, as before, be determined by measuring the speed at which they fall under gravity. Suppose now, that we hold above the drops a positively electrified body, then since the drops are negatively electrified they will be attracted towards the positive electricity and thus the downward force on the drops will be diminished, and they will not fall so rapidly as they did when free from electrical attraction. If we adjust the electrical attraction so that the upward force on each drop is equal to the weight of the drop, the drops will not fall at all, but will, like Mahomet's coffin, remain suspended between heaven and earth. If, then, we adjust the electrical force until the drops are in equilibrium and neither fall nor rise, we know that the upward force on the drop is equal to the weight of the drop, which we have already determined by measuring the rate of fall when the drop was not exposed to any electrical force. If X is the electrical force, e the charge on the drop, and w its weight, we have, when there is equilibrium—

$$Xe = w. \tag{17.7}$$

Since X can easily be measured, and w is known, we can use this relation to determine e, the charge on the drop. The value of e found by these methods is $3.1 \times 10^{-10} = 10$ electrostatic units, or 10^{-20} electromagnetic units. This value is the same as that of the charge carried by a hydrogen atom in the electrolysis of dilute solutions, an approximate value of which has long been known.

It might be objected that the charge measured in the preceding experiments is the charge on a molecule or collection of molecules of the gas, and not the charge on a corpuscle. This objection does not, however, apply to another form in which I tried the experiment, where the charges on the particles were got, not by exposing the gas to the effects of radium, but by allowing ultra-violet light to fall on a metal plate in contact with the gas. In this case, as experiments made in a very high vacuum show, the electrification which is entirely negative escapes from the metal in the form of corpuscles. When a gas is present, the corpuscles strike against the molecules of the gas and stick to them. Thus, though it is the molecules which are charged, the charge on a molecule is equal to the charge on a corpuscle, and when we determine the charge on the molecules by the methods I have just described, we determine the charge carried by the corpuscle. The value of the charge when the electrification is produced by ultra-violet light is the same as when the electrification is produced by radium.

We have just seen that e, the charge on the corpuscle, is in electromagnetic units equal to 10^{-20}, and we have previously found that e/m, m being the mass of a corpuscle, is equal to 1.7×10^7, hence $m = 6 \times 10^{-28}$ g.

We can realise more easily what this means if we express the mass of the corpuscle in terms of the mass of the atom of hydrogen. We have seen that for the corpuscle $e/m = 1.7 \times 10^7$; while if E is the charge carried by an atom of hydrogen in the electrolysis of dilute solutions, and M the mass of the hydrogen atom, $E/M = 10^4$; hence $e/m = 1700\,E/M$. We have already stated that the value of e found by the preceding methods agrees well with the value of E, which has long been approximately known. Townsend has used a method in which the value of e/E is directly measured and has showed in this way also that e is equal to E; hence, since $e/m = 1700\,E/M$, we have $M = 1700\,m$, *i.e.* the mass of a corpuscle is only about $1/1700$ part of the mass of the hydrogen atom.

In all known cases in which negative electricity occurs in gases at very low pressures it occurs in the form of corpuscles, small bodies with an invariable charge and mass. The case is entirely different with positive electricity.

17.3   Study Questions

QUES. 17.1. Do you agree with Thomson's general approach to science?

QUES. 17.2. What is the vortex theory of the atom? Who developed this theory? What are its features and merits?

QUES. 17.3. What did Thomson discover, and how is this discovery related to the structure of the atom?

a) What are cathode rays? And what were the two competing views on the nature of cathode rays at the time of Thomson's experiments?
b) Which did Thomson support, and for what reason? In particular, what was determined, regarding the nature of the rays, by the experiments of Perrin?
c) What initial difficulty did Thomson overcome when studying whether cathode rays are affected by an electric field? And what did his results imply?
d) What puzzle was raised by Hertz? And what did this lead Thomson to do?
e) How, precisely, can a charged particle be affected by a magnetic field? And how did this allow Thomson to measure the speed of the cathode rays? What was most notable about his measurements?
f) How, then, did Thomson measure the value of e/m of the cathode rays? In particular, what was his experimental apparatus, what did he measure, and how did he arrive at his conclusions?

QUES. 17.4. Where, besides in cathode rays, can Thomson's corpuscles be found? And what does this imply about the nature of matter?

QUES. 17.5. What is the magnitude of the electric charge carried by electric corpuscles?

a) How can one produce a supersaturated vapor in a controlled fashion? What effect does the presence of dust or other particles have on such a vapor?
b) How, by measuring the speed of a falling droplet was Thomson able to measure the size, and then the charge of the corpuscles?

Fig. 17.6 Schematic diagram
of a mass spectrometer used to
separate uranium isotopes.—
[*K.K.*]

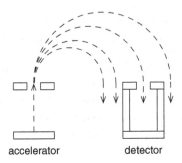

accelerator   detector

c) How, alternatively, was H.A. Wilson able to make a similar measurement? What
    was remarkable about both of their results?

17.4   Exercises

EX. 17.1 (ELECTRON ESSAY).Do electrons exist? Be sure to give a clear justification
for your beliefs.

EX. 17.2 (DROPLET SUSPENSION). What potential difference must be applied
between horizontally aligned parallel plates separated by 15 mm in order to suspend
an electron around which a $1.0\,\mu$m diameter droplet of water has condensed? If this
electric potential were suddenly reduced to zero, what would be the terminal velocity of the droplet when falling through air at standard temperature and pressure?
(ANSWER: 380 V)

EX. 17.3 (ELECTRON TRAJECTORY). Consider a cathode ray tube, as shown in
Fig. 17.2, which uses an accelerating voltage of 1000 V to generate a beam of high-
speed electrons. The electrons then pass between the parallel plates AB, which are
5 cm long, are separated by 15 mm, and are maintained at a potential difference
of 100 V. (a) What (uniform) magnetic field must be applied so that the electrons
continue a straight-line trajectory between the parallel plates? (b) In which direction
must the magnetic field be applied? (c) If the magnetic field were turned off, would
the cathode rays strike one of the plates?

EX. 17.4 (URANIUM PRODUCTION). Suppose that a mass spectrometer is designed to
separate doubly-ionized uranium-235 isotopes from a sample of uranium. To do so,
a beam of mixed uranium isotopes is first accelerated through a potential difference
of 100 kV. A magnetic field is then used to deflect the beam so that the desired
isotopes pass along a semi-circular path with a radius of curvature of 50 cm (see
Fig. 17.6). They then pass through a narrow slit and are collected in a detector.
The slit is 1 mm wide and 1 cm tall. What is the magnitude and orientation of the
required magnetic field? If a separation rate of 1 micro-gram per hour is required,
what cooling power (in watts) is required to keep the collection cup at a constant
temperature? (ANSWER: 1 T)

Fig. 17.7 Nakamura e/m apparatus and power supply with patch cords attached

Ex. 17.5 (THOMSON e/m LABORATORY). In this laboratory experiment, we will attempt to measure the ratio of the electron's charge to its mass. The apparatus we will be using consists of a vacuum discharge tube situated between a pair of Helmholtz coils (see Fig. 17.7).[8] The vacuum discharge tube generates a thin beam of high-speed electrons that travel through the low-pressure gas inside the tube. The Helmholtz coils generate a magnetic field which exerts a Lorentz force on the electron beam. By measuring the curvature of the beam, the ratio of e/m can be determined.

Figure 17.8 illustrates how the electron beam is generated. A potential difference is maintained between a cathode and an anode by a high-voltage power supply. A low-voltage power supply drives an electrical current passing through a small heater wire adjacent to the cathode, causing electrons to be emitted from the cathode. These electrons are then accelerated towards the anode. The kinetic energy of the electron beam passing through the anode can be determined from the accelerating voltage using the principle of conservation of energy.

The Helmholtz coils consist of two circular coils of wire, mounted in parallel on a common axis. When an electrical current is passed through the coils, a uniform magnetic field is formed around the common axis between the coils. When the electrons in the beam pass through the magnetic field, they experience a Lorentz

---

[8] I have had success with two different pieces of equipment: the Nakamura Model B10-7350, and the Sargent Welch Model P63412. Detailed instructions on the operation of these pieces of equipment are included with the equipment. The most expensive component of the apparatus is the bulb, which costs about 800 USD. In addition, you will need a low-voltage power supply to heat the lamp filament, a high-voltage (500 V) power supply to accelerate the electrons, current source (3 amps) to drive the helmholtz coils, and a hand-held digital multimeter.

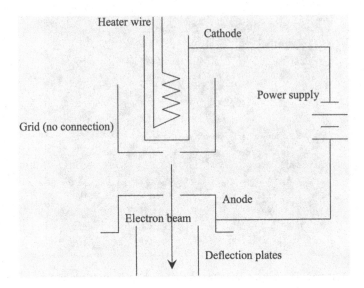

Fig. 17.8 Schematic diagram of the apparatus used to produce and accelerate electrons

force perpendicular to both their velocity and the magnetic field. Since the electron speed is constant, and the magnetic field is constant, the Lorentz force will cause the electrons to move in a circular orbit. The charge to mass ratio of the electron may thereby be determined in terms of the orbital radius, the magnetic field strength, and the accelerating voltage. Does your measured value fall within your uncertainty estimate of the accepted value?

17.5   Vocabulary

1. Corpuscle
2. Phosphorescence
3. Cathode
4. Maltese cross
5. Luminosity
6. Supersaturated
7. Electrolysis
8. Radium

Chapter 18
The Birth of Nuclear Physics

The α radiation from radium is very complex, and consists of four groups of α particles, each of which is made up of α particles which escape at widely different velocities.

—Ernest Rutherford

18.1 Introduction

Ernest Rutherford (1871–1937) was born in Spring Grove, a rural community near Nelson on the South Island of New Zealand.[1] He was the fourth of twelve children born to a flax-miller and a schoolteacher who had emigrated from the United Kingdom. He attended Canterbury College in Christchurch, from which he earned his Bachelor of Arts (1892), Master of Arts (1893), and Bachelor of Science (1894) degrees. While at Christchurch, he carried out a set of experiments in which he demonstrated that when a steel needle is subjected to high frequency magnetic fields, only the surface of the needle is magnetized; this he proved by slowly dissolving the outer layers of the magnetized needle with acid.

He left New Zealand in 1895 and went to work for Professor J.J. Thomson at the Cavendish Laboratory in London—the famous laboratory's first research student who had not graduated from the University of Cambridge. While there, Rutherford's initial interest was in the transmission and detection of low-frequency electromagnetic "Hertzian" waves. But with the recent (1896) discoveries of X-rays by Wilhelm Röntgen and radioactive uranium by Henri Becquerel, Rutherford quickly turned his attention to the cause and nature of radioactivity. In 1902, at the age of 27, he accepted a professorship at McGill University in Montreal. During his time at McGill, he discovered the radioactive gas radon, characterized radioactive decay rates of various elements, and published a famous book on these topics in 1904 entitled *Radioactivity*. For this work, he was later awarded the Nobel Prize in Chemistry in 1908. In 1907 he returned to Britain—to the University of Manchester—where he succeeded in identifying the nature of the α-particles which had been previously

---

[1] The biographical information contained in this introduction is based on Campbell, J., *Rutherford: Scientist Supreme*, AAS Publications, Christchurch, New Zealand, 1999.

© Springer International Publishing Switzerland 2016
K. Kuehn, *A Student's Guide Through the Great Physics Texts*,
Undergraduate Lecture Notes in Physics, DOI 10.1007/978-3-319-21828-1_18

observed in certain radioactive decay processes. His subsequent studies of the scattering of these same α-particles from gold-foil, carried out with the help of Hans Geiger and Ernest Marsden in 1911, led him to the so-called "planetary model" of the atom. This model would later serve as the basis for the early quantum theory developed by the Danish physicist Niels Bohr. Today, Rutherford is widely recognized as the father of nuclear physics. He was knighted in 1914, and in 1931 he was awarded the title of Baron Rutherford of Nelson. He is buried in Westminster Abbey.

The next reading selection, which is divided between the next two chapters of this volume, originally appeared in 1906 in the *Philosophical Journal*. Herein, Rutherford addresses the question: what is the identity of the newly-discovered α-particle? In order to answer this question, Rutherford first needed to determine the properties of the α-particle—specifically its speed and charge-to-mass ratio. After all, this is how his mentor, J.J. Thomson, characterized the unique properties of the electron a few years earlier. Rutherford was not the first to attempt such measurements on α-particles. What difficulties did previous researchers encounter? And how were Rutherford's experiments different?

18.2   Reading: Rutherford, *The Mass and Velocity of the α Particles Expelled from Radium and Actinium*

Rutherford, E., XLI. The Mass and Velocity of the α particles expelled from Radium and Actinium., *Philosophical Magazine Series 6, 12*(70), 348–371, 1906.

The present paper contains an account of investigations that have been made to determine, as accurately as possible, the mass and velocity of the α particles expelled from some of the products of radium and actinium. At the present stage of our knowledge of radioactivity, such measurements have an important theoretical value in throwing light on the following questions:—

(1) Has the α particle expelled from all radioactive products the same mass?
(2) Does the value of $\frac{e}{m}$ of the α particle vary in its passage through matter?
(3) What is the connexion between the velocity of the a particle and its range of ionization in air?
(4) What is the connexion, if any, between the α particle and the helium atom?
(5) Is the heating effect of radium or other radioactive substance due to the bombardment of the radioactive matter by the α particles expelled throughout its own mass?

In the course of these investigations, sufficient data have been accumulated, if not to answer completely all of the above questions, at least to indicate with some certainty the relations that exist between the various quantities.

The experiments outlined in this paper have been in progress for more than a year,[2] but publication has been delayed in order to determine the mass of the α particle from thorium and actinium as well as from radium.

The investigations on the mass of the α particle from thorium have been made in conjunction with Dr. Hahn, and are described in a following paper.

Determinations of the mass and velocity of the α particles from radium have been made by several observers. In 1902, using the electroscopic method and radium of activity 19000, I showed that the α particles from radium consisted of positively charged particles which were appreciably deflected in intense magnetic and electric fields.[3] I deduced that the value of $\frac{e}{m}$—the ratio of the charge on the α particle to its mass—was about 6×10^3, and that the swiftest α particles emitted from radium had a velocity of about 2.5×10^9 cm/s. Shortly afterwards, these experiments were repeated by Des Coudres,[4] using the photographic method and with pure radium bromide as a source of rays. He found the value of $\frac{e}{m}$ to be 6.3×10^3, and the average velocity to be 1.65×10^9 cm/s.

On account of the difficulty of obtaining a sufficiently large deflexion of the α rays in passing through an electric field, the values of $\frac{e}{m}$ and of the velocity of the α particles obtained by Rutherford and Des Coudres could only be considered as a first approximation to the true values.

Recently the question has again been attacked by Mackenzie,[5] using the photographic method and pure radium bromide as a source of rays. Fairly large deflexions of the pencil of rays were obtained by using strong magnetic and electric fields. He showed that the α particles emitted by a thick layer of radium bromide were unequally deflected in a magnetic and electric field, and presumably consisted of α particles moving with different velocities. By assuming that the value of $\frac{e}{m}$ was the same for all the α particles, he deduced that the value of $\frac{e}{m}$ for the average ray was 4.6×10^3, and that the average velocity was 1.37×10^9 cm/s.

It will be seen that all of these investigators have used a thick layer of radium in radioactive equilibrium as a source of rays. We know that the α particles from radium in equilibrium come from four distinct α ray products. The α particles from each of these products have different ranges of ionization in air and different velocities of projection. In addition, the α particles from each single product reach the surface from different depths of radioactive matter, and consequently have different velocities. It is thus seen that the α radiation from radium is very complex, and consists of four groups of α particles, each of which is made up of α particles which escape at widely different velocities.

---

[2] A preliminary account of the measurements of the value of $\frac{e}{m}$ for the particle from radium C was given before the American Physical Society; December 1905. An abstract of the results appeared in the Physical Review, Feb. 1906.

[3] Rutherford, *Phys. Zeit.* iv. p. 235 (1902); Phil. Mag. Feb. 1903.

[4] Des Coudres, *Phys. Zeit.* iv. p. 483 (1903).

[5] Mackenzie, Phil. Mag. Nov. 1905.

On account of the dispersion of the pencil of rays in passing through an electric
and magnetic field, it is difficult to interpret with certainty the deflexions observed.
The difficulties which arise are clearly pointed out by Mackenzie in his paper (*loc.
cit.*)

In a previous paper (Phil. Mag. July 1905) I pointed out that these difficulties
would disappear if a homogeneous pencil of α rays was employed. I showed that
such a homogeneous pencil could be obtained by using as a source of rays a small
wire which had been made very active by exposure to the radium emanation. Fifteen
minutes after removal from the emanation, radium A has been transformed, and the
α particles are then emitted only from radium C.

An examination of the deflexion of the rays in a magnetic field showed that such
an active wire fulfilled the conditions necessary for a homogeneous source of rays.
The α particles all escaped from the thin film of radioactive matter at the same speed,
and all suffered the same reduction of velocity in passing through an absorbing
screen. On account of the rapid decay of the activity of the active deposit, it is
necessary to employ an intensely active wire to obtain a strong photographic effect.
In most of the experiments described later, the active deposit was concentrated on
the wire by making it the only negatively charged surface in a vessel containing a
large quantity of the radium emanation. In this way, very active wires were obtained
which served as suitable sources of homogeneous α rays.

18.2.1 Electric Deflexion of the α Rays

The determination of $\frac{e}{m}$, and of the velocity of the α particle was made in the usual
way by measuring the deflexion of a pencil of rays in passing through both a magnetic and electric field of known strength. The method employed for measuring
the magnetic deflexion has already been described in a previous paper. After some
preliminary experiments, the following arrangement was adopted to determine the
deflexion of the α rays in passing through an electric field. The rays from the active
wire W (Fig. 18.1), after traversing a thin mica plate in the base of the brass vessel M, passed between two parallel insulated plates A and B about 4 cm high and
0.21 mm apart. The distance between the plates was fixed by thin strips of mica
placed at the four corners, and the plates were rigidly held together by rubber bands.
The terminals of a storage-battery were connected with A and B so that a strong
electric field could be produced between the two plates. The pencil of rays, after
emerging from the plates, fell on a photographic plate P. The latter was rigidly
fixed to a ground-brass plate which fitted accurately on the top surface of the vessel.
The ground surfaces were air-tight, and the photographic plate could thus easily be
placed in position or removed without disturbing the rest of the apparatus. The vessel was connected to a mercury pump and exhausted to a low vacuum. If necessary,
the exhaustion was completed by means of a side tube filled with cocoanut charcoal
and immersed in liquid air.

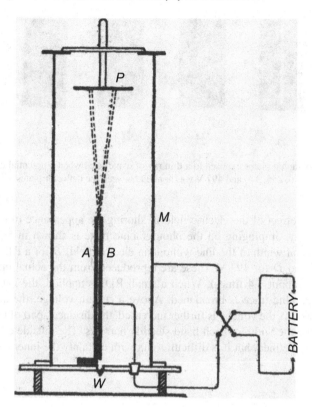

Fig. 18.1 Rutherford's apparatus for measuring the deflection of α-particles passing through an electric field.—[*K.K.*]

The plates *A* and *B* were placed close together for several reasons. In the first place, a strong electric field could be produced between the plates for a comparatively small voltage. The greatest P.D. necessary in the experiments was about 500 V. Since the plates were about one-fifth of a millimetre apart, this voltage produced an electric field between the plates corresponding to 25000 V/cm. One advantage of the arrangement lies in the fact that, provided the P.D. is below about 350 V, there is no danger of a discharge between the plates, even if there is not a good vacuum. This is particularly convenient where it is found necessary to expose the photographic plate to a weak source of radiation for several days, for there is no necessity to continually watch the state of the vacuum.

On account of the small distance between the plates, there is no necessity to correct for the disturbance of the electric field near the ends of the plates. In addition, the parallel plates acted as a slit in order to obtain a narrow pencil of rays. In its passage through the electric field each α particle describes a parabolic path, and after emergence travels in a straight line to the photographic plate. By reversing the electric field at intervals, the direction of deflexion of the pencil of rays is reversed.

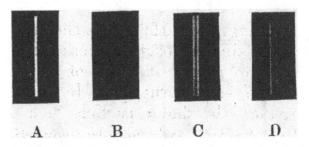

A   B   C   D

Fig. 18.2 Photographic plates exposed to a thin ray of α-particles when a potential difference of (from *left* to *right*) 0, 255, 340 and 497 V was applied between the deflector plates.—[*K.K.*]

The general effect of the electric field in altering the appearance of the trace of the pencil of rays impinging on the photographic plate is shown in Fig. 18.2. *A* shows the natural width of the line without an electric field, *B* for a P.D. of 255 V, *C* for 340 V, and *D* for 497 V. These are reproduced from the actual photographs (magnification about 1.4 times). When a small P.D. is applied, the natural width of the photographic trace is broadened. Above a certain voltage, the single band breaks into two. As the voltage is further increased, the distance apart of these bands increases while the width of each band steadily narrows. The outside edge of each band is sharply defined, but it is difficult to fix with certainty the inner boundary of the bands.

18.2.2   Theory of the Experiment

The theory of the experimental arrangement where the parallel plates act both as a slit and a means of applying the electric field, is more complicated than the ordinary case where a narrow pencil of α rays is made to pass between the two parallel plates of the condenser without impinging on the sides.

A diagram of the experimental arrangement is shown in Fig. 18.3a, b. *AB* and *CD* are the two charged parallel plates, and *CE* the radiant source which was of greater width than the distance between the plates. It is required to find the width of the trace on the photographic plate when a P.D. V is applied between the plates.

Let m = mass of α particle,

   e = charge on α particle,

   u = velocity of α particle in passing between plates,

   $AB = l_1$, $CD = l_2$, $Bb = l_3$,

   d = distance between plates,

   D = distance between extreme edges of the photographic trace for reversal

     of the electric field.

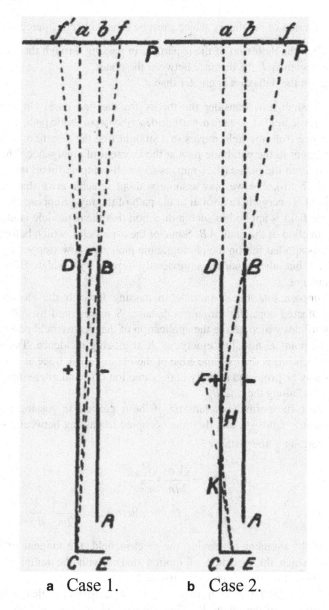

a Case 1.   **b** Case 2.

Fig. 18.3 Diagrams depicting the change in trajectory of an α-particle passing between charged condenser plates when the particle's deflection is less than (case 1) and greater than (case 2) the plate separation.—[K.K.]

There are two cases of the theory which must be separately considered:—

Case 1.   when the deflexion of the α particle in passing through the electric field is less than d, the distance between the plates,

Case 2.   when the deflexion is greater than d.

Case 1.   We shall now consider the theory for the first case. On entering the electric field at A, each α particle describes a parabolic path, and on emergence from the field moves in a straight line the direction of which is a tangent to the parabolic path at the moment of emergence. The distance between the plates (0.21 mm) is so small compared with the length AB (3.77 cm), that we may assume without sensible error that the electric field is everywhere normal to the path of the rays. Suppose that the electric field is applied in such a direction that the α particle is urged in the direction of the plate AB. Some of the α particles, which before the field was applied fell on the photographic plate, are now stopped by the plate AB, but other α particles previously stopped by the plate CD are able to emerge.

Suppose that the α particle in passing through the electric field is deflected normally through a distance S represented by FB. All the α particles which before the application of the electric field passed through the point F now just emerge at B at grazing incidence. The α particle which forms the extreme edge of the photographic trace at f must obviously be projected initially in the direction CF, and after emergence will travel along the line Bf.

Since the normal acceleration of the α particle in passing through the electric field $\frac{Ve}{dm}$, and the time occupied in passing between the charged plates is $\frac{l_1}{u}$ the distance

$$FB = s = \frac{Ve}{2dm} \cdot \frac{l_1^2}{u^2}$$

or $s = \lambda l_1^2$,   where $\lambda = \frac{e}{2m} \cdot \frac{V}{du^2}$.

At the moment of leaving the electric field, the tangent of the angle θ, which the direction of motion makes with the initial direction of projection CF, is given by $\tan \theta = 2\lambda l_1$.

If the angle $DCF = \theta_1$, the emerging ray makes an angle $\theta + \theta_1$ with the direction of the plates Bb. The distance $bf = l_3 \tan(\theta + \theta_1)$. Since the angles θ and θ_1 are small,

$$bf = l_3(\theta + \theta_1)$$

$$= l_3 \left(2\lambda l_1 + \frac{d - s}{l_2} \right).$$

In a similar way when the electric field is reversed, the corresponding distance

$$af' = l_3 \left(2\lambda l_1 + \frac{d-s}{l_1} \right).$$

In this case, the α particle which is most deflected enters the electric field at grazing incidence at the point A.

The distance D between the extreme edges ff' of the photographic impression is consequently given by

$$D = bf + ab + af'$$

$$= 4\lambda l_1 l_3 + l_3(d-s) \left(\frac{1}{l_1} + \frac{1}{l_2} \right) + d.$$

Substituting the value $S = \lambda l_1^2$,

$$D = \lambda l_1 l_3 \left(3 - \frac{l_1}{l_2} \right) + \left(\frac{l_3}{l_2} + \frac{l_3}{l_2} + 1 \right) d.$$

But it is easily seen that the natural width of the photographic band without the electric field is given by

$$\left(\frac{l_3}{l_2} + \frac{l_3}{l_1} + 1 \right) d.$$

Therefore the *increase* D_1 of the breadth of the band by the reversal of the electric field is given by

$$D_1 = \lambda l_1 l_3 \left(3 - \frac{l_1}{l_2} \right).$$

Substituting the value of λ,

$$\frac{mu^2}{e} = \frac{V l_1 l_3}{2 d D_1} \left(3 - \frac{l_1}{l_2} \right). \tag{18.1}$$

This gives the formula required for determining the value of $\frac{mu^2}{e}$ for case 1.

Case 2.   In this case the electric field is supposed to be sufficiently strong to deflect the α particle in passing between the charged plates through a distance greater than d.

Suppose the electric field urges the α particle towards the plate AB (Fig. 18.3b). A little consideration shows that the α particle which forms the extreme edge of the photographic impression at f must touch at grazing incidence the plate CD. Let LKF be the direction of projection of such an α particle, intersecting the plate CD at K. The path of the α particle under the action of the electric field is shown by the dotted line in the figure. The path touches the plate CD at H and emerges at B at grazing incidence.

Let $DH = y$.

Then with the same notation as before, $d = \lambda y^2$.

The angle θ which the tangent to the parabola at B makes with the direction of the plate AB is given by

$$\tan \theta = 2\lambda y.$$

The distance $bf = l_3 \tan \theta = 2\lambda y l_3$.

The total distance D between the extreme edges of the photographic impression on the plate P by reversal of the field is consequently given by

$$D = 4\lambda y l_3 + d.$$

Then

$$(D - d)^2 = 16\lambda d l_3^2.$$

Substituting the value of λ as before,

$$\frac{mu^2}{e} = \frac{8Vl_3^2}{(D - d)^2} \tag{18.2}$$

It is interesting to note that this formula for the determination of $\frac{mu^2}{e}$ does not involve l_1 or l_2, and only involves the distance d to a subordinate extent. For example, in one experiment where $l_3 = 10$ cm, the value of d was only $\frac{1}{15}D$. This is a great advantage, as the distance d is difficult to measure with accuracy.

While the distance l_1 is not involved in the final formula, it must be remembered that the formula (18.2) only applies when the α particle is deflected through a distance greater than d in passing between the plates. If l_1 is made smaller, the value of V must be made correspondingly greater before the formula can be applied.

It has been mentioned that the width of two deflected traces obtained by reversal of the electric field decreases in width with increase in strength of the electric field. The reason of this can readily be shown from theoretical considerations. For example, it is seen that the inside edge of the deflected pencil (Fig. 18.3b) is produced by the α particles whose paths touch at grazing incidence the plate AB at A and also touch the plate CD. These conditions determine the direction of the α particle in entering the electric field at A. All α particles passing through A which make a greater angle with the plate AB than the above α particle are stopped by the plate CD. As the width of the trace is not required in the experiment, it has not been thought necessary to include here the connexion between the width of the deflected pencil and the strength of the electric field. The calculations, though a little long, are not difficult.

It is now necessary to consider how we are able to know from the photographs obtained whether the formula (18.1) or (18.2) is to be applied. The formula (18.1) holds provided the distance of deflexion of the α particle is not greater than d. Suppose that the α particle is deflected through a distance d in passing between the charged plates. The outside edge of the photographic impression, for example, on the right of the plate P (Fig. 18.3a, b) is due to the α particles which start from the point C parallel to the plate CD. With the same notation as before,

$$d = \lambda l_1^2.$$

Following the same method of calculation as for case (2), it is seen that the value of D is given by

$$D = 4\lambda l_1 l_3 + d.$$

Substituting the value $\lambda = \frac{d}{l_1^2}$,

$$D = d \left(\frac{4l_3}{l_1} + 1 \right)$$

In most of the experiments to be described later,

$$l_3 = 3.94 \text{ cm}, \qquad\qquad l_1 = 3.77 \text{ cm}, \qquad\qquad d = 0.21 \text{ mm}.$$

Consequently,

$$D = 1.09 \text{ mm}.$$

From the data given below it will be seen that the formula (18.2) applies for all voltages greater than about 300, while the formula (18.1) applies for all values smaller than this.

18.2.3   Results of Experiments

The electrostatic deflexion of the α rays from radium C was first determined for different voltages between the plates. The rays from the active wire passed through a mica plate in the base of the vessel, equivalent in stopping power to about 3.5 cm of air. The extreme distance D between the outside edges of the photographic impressions obtained on the plate by reversal of the electric field was measured by the lantern method described in a previous paper.[6]

In most of the experiments, the value l_3 of the distance of the photographic plate above the parallel plates was 3.94 cm. In one experiment this distance was 10.00 cm.

$$l_1 = 3.77 \text{ cm}, \qquad\qquad l_2 = 4.165 \text{ cm}, \qquad\qquad d = 0.210 \text{ mm}.$$

Volts between plates	l_3 (cm)	D (mm)	$\frac{mu^2}{e}$
171	3.94	0.857	5.1×10^{14}
255	"	0.995	4.9×10^{14}
340	"	1.136	4.93×10^{14}
497	"	1.346	4.79×10^{14}
508.6	10.00	3.10	4.87×10^{14}

Table 18.1 Measured values of $\frac{mu^2}{e}$ for α-particles deflected by different voltages.—[K.K.]

The values of $\frac{mu^2}{e}$ obtained for different voltages and distances of the photographic plate are tabulated below (Table 18.1).

Each of the values of D given above is the mean of a large number of separate measurements which agreed closely among themselves. The values for 171 and 255 V are calculated from formula (18.1), the natural width of the photographic trace being 0.61 mm, and for the higher voltages from formula (18.2). Some of the photographs from which the measurements were made are reproduced in Fig. 18.2, magnification about 1.4.

Two good photographs were obtained with a P.D. of 340 V. In each case two active wires were used successively to give a strong photographic impression. On account of the greater distance, the photographic impression was not so strongly marked for the distance 10 cm.

Giving a weight 1 to the measurement of $\frac{mu^2}{e}$ for 497 volts, and a weight 2 for both 340 and 508.6 V, the mean value is given by

$$\frac{mu^2}{e} = 4.87 \times 10^{14} \quad \text{electromagnetic units} \qquad (18.3)$$

By measurement of the magnetic deflexion, the maximum value of $\frac{mu}{e}$ for the α rays from radium C was found to be 4.06×10^5. The mica screen cut down the velocity of the rays to 0.763 of the initial velocity, so that the value of $\frac{mu}{e}$ for the rays which passed through the electric field is given by

$$\frac{mu}{e} = 3.10 \times 10^5 \dots \qquad (18.4)$$

By combining equations (18.3) and (18.4)

$$u = 1.57 \times 10^9 \text{ cm/s},$$

$$\frac{e}{m} = 5.07 \times 10^3 \text{ electromagnetic units}.$$

I think that the values of u and $\frac{e}{m}$ are certainly correct within 2 %.

The initial velocity of the α particles expelled from radium C is consequently 2.06×10^9 cm/s.

---

[6] Rutherford, Phil. Mag. August 1906.

Table 18.2 Computed values of $\frac{e}{m}$ for α-particles.—[K.K.]

Interposed absorbing screen in terms of air	Volts	D	$\frac{mu}{e}$	$\frac{mu^2}{e}$	$\frac{e}{m}$
0	340	0.88 mm	4.06×10^5	9.4×10^{14}	5.7×10^3
3.5 cm	...	...	3.10×10^5	4.87×10^{14}	5.07×10^3
6.5 cm	340	1.62 mm	2.11×10^5	2.11×10^{14}	4.8×10^3

18.2.4   Does the Value of $\frac{e}{m}$ for the α Particle Vary in Its Passage Through Matter?

In order to test this point, the value of $\frac{e}{m}$ for the α particle was determined under the following conditions:—

(1) The active wire was placed on top instead of under the mica screen, so that the electrostatic deflexion was determined for the α particle from the unscreened wire.

(2) The α particles passed through a mica screen equivalent in stopping power to 3.5 cm of air. The value of $\frac{e}{m}$ under these conditions has been determined in the previous section.

(3) The α particles passed through a screen of mica and aluminium equivalent to about 6.5 cm of air.

The magnetic and electrostatic deflexion were separately determined. The former gives the value of $\frac{mu}{e}$ and the latter $\frac{mu^2}{e}$. The results of the measurements are collected in the following table, where D has the same meaning as before (see Table 18.2). The value of l_3 in all cases was 3.94 cm.

The deflexion for the rays from the bare wire was small, but could be measured with fair certainty. I think the value of $\frac{e}{m}$ obtained in this case, *viz.* 5.7×10^3, is undoubtedly too high. On removing the active wire after completion of the experiment, it was noticed that its position was displaced somewhat to the side of the opening of the parallel plates. This would tend to make the observed width of the photographic trace too small, and consequently to give too great a value of $\frac{e}{m}$, when calculated from formula (18.1), which is based on the assumption that the active source completely covers the opening between the parallel plates.

The impression on the photographic plate due to the α particles which have passed through an absorbing screen equivalent to 6.5 cm of air was weak but clearly defined, and admitted of fairly accurate measurement. Allowing for an error in the estimation of $\frac{e}{m}$ for the α particles from the bare wire, I think the agreement of the values of $\frac{e}{m}$ obtained under the different conditions is sufficiently close to prove definitely that the value of $\frac{e}{m}$ for the α particle is unaltered in its passage through matter.

18.3 Study Questions

QUES. 18.1. What was Rutherford's goal? And what questions was he able to address as a result of his studies?

QUES. 18.2. How was Rutherford able to avoid the difficulties that plagued previous measurements of the speed and charge-to-mass ratio of α-particles?

a) Do all α-particles travel at the same speed? What kind of α-source did Rutherford employ?
b) What are the components of the apparatus shown in Fig. 18.1? In particular, where is the source of the α-particles? Where is the electric-field region? And how were the α-particles detected? How (and why) did Rutherford evacuate the brass vessel?
c) What happens to the trajectory of the α-particles as the potential difference between the condenser plates is gradually increased? What is the maximum electric field strength Rutherford could establish between the plates?

QUES. 18.3. What is the value of $\frac{mu^2}{e}$ for an α-particle?

a) When an α-particle is traveling through a transverse electric field, what is its acceleration? How is the acceleration computed?
b) In which direction is the α-particle deflected? By how much? What happens when the electric field is reversed?
c) How is the increase in width of the distribution of α-particles on the photographic plate used to compute the ratio $\frac{mu^2}{e}$?
d) What two cases does Rutherford consider when computing $\frac{mu^2}{e}$? Under what condition(s) are each of the two cases realized?
e) What were the measured values of $\frac{mu^2}{e}$ from Rutherford's experiments? In what units were these values measured?

QUES. 18.4. What is the value of $\frac{mu}{e}$ for an α-particle?

QUES. 18.5. How can the speed and the charge-to-mass ratio of Rutherford's α-particle be derived from his measurements? How do his results compare to those of previous researchers? And does the charge-to-mass ratio vary when α-particles travel through matter?

18.4 Exercises

EX. 18.1 (HELIUM DEFLECTION BY A MAGNETIC FIELD). Suppose a beam of singly-ionized $^4_2$He atoms travels at a speed u through a magnetic field. Demonstrate that the deflection of the beam is proportional to $\frac{e}{mu}$, where e and m are the charge and mass of the helium-4 atoms.

Fig. 18.4 The uranium series

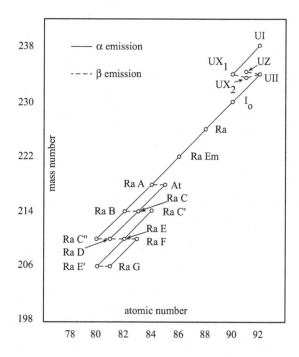

EX. 18.2 (HELIUM DEFLECTION BY AN ELECTRIC FIELD). Suppose a beam of singly-ionized $_2^4$He atoms is emitted vertically at 1 % the speed of light from a source at the base of the apparatus depicted in Fig. 18.1. How large a potential difference can be applied between the condenser plates before the beam strikes one of the plates?

EX. 18.3 (α-PARTICLES AND ELECTRONS). How do the speed and the charge-to-mass ratio of Rutherford's α-particles compare to those of an electron? What does this imply?

EX. 18.4 (URANIUM SERIES). In the reading selection of this (and the following) chapter, Rutherford employs isotope names with which you are probably unfamiliar. Identify the modern name and the half-life for each isotope in the *uranium series* depicted in Fig. 18.4.

18.5   Vocabulary

1. Radium
2. Actinium
3. Radioactivity
4. Thorium
5. Conjunction
6. Ionization
7. Consequent
8. Impinge

Chapter 19
Radioactivity

> *The α particle constitutes one of the fundamental units of matter of which the atoms of these elements are built up.*
>
> —Ernest Rutherford

19.1 Introduction

In the previous chapter we studied the first half of Rutherford's 1906 *Philosophical Magazine* article in which he described in detail how he was able to accurately measure the properties of α-particles expelled from radium. By allowing them to pass between a set of electrified plates, he was able to first deduce the ratio $\frac{mu^2}{e}$. Here m, u, and e are the mass, velocity and charge of the particles, respectively. Next, by measuring the deflection of α-particles in a magnetic field, he was able to deduce the ratio $\frac{mu}{e}$. Finally, by combining these results he was able to obtain the velocity and the charge-to-mass ratio of the α-particles. Now we turn to the second half of Rutherford's paper, in which he broadens his study by considering α-particles ejected from a host of other radioactive isotopes. Rutherford continues to use the original names which were adopted for these isotopes (*e.g.* radium A, radium C, radium F and radium emanation) so you may find it helpful to spend a few minutes looking up the modern naming conventions for each of these isotopes before proceeding.[1] What do the α-particles ejected from all these isotopes have in common? Is Rutherford able to finally determine the identity of these mysterious particles?

19.2 Reading: Rutherford, *The Mass and Velocity of the α Particles Expelled from Radium and Actinium*

Rutherford, E., XLI. The Mass and Velocity of the α particles expelled from Radium and Actinium., *Philosophical Magazine Series 6*, *12*(70), 348–371, 1906.

---

[1] See Ex. 18.4 at the end of the previous chapter.

© Springer International Publishing Switzerland 2016
K. Kuehn, *A Student's Guide Through the Great Physics Texts*,
Undergraduate Lecture Notes in Physics, DOI 10.1007/978-3-319-21828-1_19

19.2.1 Value of e/m for the α Particles from Radium A

In a previous paper (*loc. cit.*) I gave the results of the measurements by the photographic method of $\frac{mu}{e}$ for the α particle emitted from radium A. For the unscreened wire the value of $\frac{mu}{e} = 3.67 \times 10^5$, and for the wire covered with a mica plate of the same thickness as that over the opening in the base of the electrostatic apparatus, I found the value of $\frac{mu}{e} = 2.19 \times 10^5$.

I pointed out in that paper the difficulty of accurately measuring, the magnetic deflexion of the α particles from such a rapidly changing product as radium A, which is half transformed in 3 min. The difficulty of obtaining a sufficiently marked photographic impression, in order to measure the electric deflexions of the rays, was still greater. It was found necessary to place 20 active wires successively in position under the base of the apparatus, in order to obtain a measurable darkening of the photographic plate. Each wire was exposed for 2 min as negative electrode in a vessel containing a large quantity of the radium emanation. For such a short exposure, the initial radiation from the wire mainly comes from radium A. The wire was rapidly removed from the emanation vessel and placed in position and left for 6 min. In that time, the α ray activity of radium A is reduced to one quarter of its initial value. The α rays from the active wire passed through the standard mica plate before entering the electric field. The electric deflexion of the α rays from radium A is considerably greater than that for the swifter α particles from radium C; so that there is no danger of confusion between the two types of rays, even though the photographic impression of the rays from radium C present on the active wire is comparable with that due to the rays from radium A.

In the experiment the voltage was 255; $l_3 = 3.94$ cm; and $D = 1.30$ mm. This gave

$$\frac{mu^2}{e} = 2.67 \times 10^{14}.$$

The value $\frac{mu}{e}$ for the α rays from radium A after traversing the standard mica was 2.19×10^5. This gives the values

$$u = 1.22 \times 10^9 \quad \text{cm/s}$$
$$e/m = 5.6 \times 10^3.$$

It was found by experiment that the measured value of D—the distance between the extreme edges of the trace—was always underestimated for a very feeble photographic trace. An underestimate of the value of D gives too large a value for e/m. Taking this factor into consideration, the values of e/m obtained for the α particles from radium A and from radium C agree within the limit of experimental error. This shows that the α particles expelled from radium A and C have the same mass and differ only in their initial velocities of projection.

19.2.2 Mass of the α Particle from Radium F

A bismuth rod coated with radiotellurium was used as a source of α rays. It is now definitely established that the active constituent in both radiotellurium and polonium is the same and consists of the transformation product of radium, radium *F*. The active matter is deposited in the form of a thin film on the bismuth rod, and the α particles all escape from the surface at practically the same velocity. A piece of the rod was placed in position inside the electrostatic apparatus, and the photographic plate exposed for 4 days to the action of the α rays from the bare rod. The value $l_3 = 10.00$ cm, the voltage $= 443$, and the observed value $D = 2.72$ mm.

This gives a value $\frac{mu^2}{e} = 5.63 \times 10^{14}$ for the α rays from the unscreened source. The value of $\frac{mu}{e} = 3.2 \times 10^5$ deduced by me in a previous paper from measurement of the range of the α particles from radium *F* in air. The experimental value found directly by Mackenzie (*loc. cit.*) was 3.3×10^5. Taking the mean value $\frac{mu}{e} = 3.25 \times 10^5$, we find that $e/m = 5.3 \times 10^3$, and $u = 1.73 \times 10^9$ cm/s.

The actual photograph obtained was very weak in intensity, and, for the reasons previously mentioned, there is no doubt that the value e/m obtained is too large. We may consequently conclude that the α particle from radium *F* has the same mass as that expelled from radium *C*. Using a more active rod or a longer time of exposure, the value of e/m should be obtained with much greater precision; but there can be no doubt that it would be found identical with that observed for the α particle from radium *C*.

19.2.3 Mass of the α Particles from Actinium

In order to obtain a homogeneous source of α rays, the active deposit of actinium was used. The active deposit was concentrated on a small copper plate by making it the negative electrode in a small vessel containing the emanating actinium compound. This active deposit consists of two products, actinium *A* and *B*, the former of which is rayless. The activity imparted to a plate, 10 min after removal from the emanation, decays exponentially with a period of 36 min. The rays emitted are all of one kind and have a range in air, found by Dr. Hahn in this laboratory, of 5.5 cm.

The preparation of actinium[2] employed was not very active, and the activity imparted to the copper plate was too weak to produce appreciable photographic action at the distances required. With the experience gained in the previous experiments with weak radioactive sources, it was recognized that at least 20 active wires, placed successively in position, would have been required to produce a measurable photographic effect in the magnetic deflexion apparatus. Three or four times this

---

[2] I am indebted to Mr. H. Lieber of New York for his kindness in lending me the sample of actinium used in this experiment.

number would have been necessary in the electrostatic experiment. In order to avoid
the necessity of such a procedure, the apparatus was constructed so that the copper plate could be kept active in the position required for any length of time. The
actinium compound, wrapped in thin paper, was placed round the sides of a small
brass vessel, which was attached to the base of the magnetic or electric deflexion
apparatus. A small insulated copper plate, with its plane slightly inclined to the
vertical, was placed below a narrow slit covered with mica in the base of the apparatus and was kept negatively charged. The activity on the plate reached a maximum
after 3 h and then remained constant. The radiation passing through the slit into the
magnetic deflexion apparatus was mainly due to the α rays from actinium B. The
photographic effect of the radiations from the emanation close to the active plate
was too weak to be observed. The magnetic deflexion of the pencil of the α rays
was determined under identically the same conditions as in the experiments using
radium C as a source of rays. The photographic plate was exposed for 10 h in a
constant magnetic field which was reversed at intervals. Two fine well-defined lines
were obtained on the plate. The amount of the magnetic deflexion was then directly
compared with that due to the rays from radium C under the same conditions. For
this purpose, the copper plate was removed and made active by exposure to the
radium emanation. It was then placed back in its original position, and another photograph taken. The mica plate, covering the opening in the base of the magnetic
apparatus was of the same thickness as that used in the base of the electrostatic
apparatus. The following numbers illustrate the results obtained:—

Distance between centres of deflected bands due to rays from actinium $B = 1.85$ mm.

Distance between centres of deflected bands due to rays from radium $C = 1.53$ mm.

We have previously shown that the value of $\frac{mu}{e}$ for the α particles of radium C after
passing through the standard mica screen, absorbing-power equal to 3.5 cm of air,
is 3.10×10^5. Consequently the value of $\frac{mu}{e}$ for the α particles from actinium B is

$$\frac{mu}{e} = \frac{1.53}{1.85} \times 3.10 \times 10^5 = 2.56 \times 10^5.$$

It is interesting to note that the comparative magnetic deflexions observed for the
rays of actinium B and of radium C agree with those to be expected from their
known ranges in air, assuming the value of e/m for the α particle to be the same in
both cases.

I have shown in a previous paper that the velocity V of an α particle of range r
cm in air is given by

$$\frac{V}{V_0} = 0.348\sqrt{r + 1.25},$$

where V_0 is the maximum velocity of the rays from radium C which have a range
of 7.06 cm. Now, after passing through the mica screen, the rays from actinium
B have a range $5.5 - 3.5 = 2.0$ cm, while those from radium C have a range
$7.06 - 3.5 = 3.56$ cm.

Consequently,

$$\frac{\text{velocity of rays from radium } C \text{ after passage through mica}}{\text{velocity of rays from actinium } B \text{ after passage through mica}} = \sqrt{\frac{3.56 + 1.25}{2.0 + 1.25}} = 1.22$$

The experimental ratio is 1.21. From the agreement between the experimental and the theoretical ratios, it could be concluded with confidence that the mass of the α particle from actinium is the same as that from radium C. This, however has been experimentally verified by measuring the electric deflexion of the a rays from actinium B.

The apparatus for determining the electric deflexion of the α rays was the same as that used in the radium experiments. The copper plate which served as a source of a rays was kept active by an arrangement similar to that used for the magnetic deflexion. The photographic plate was exposed for 6 days. Through an accident in the connexions, the electric field was not acting between the plates for the first 3 days. The photographic plate consequently showed the undeflected trace of the rays and the deflected trace on one side of it. The distance between the centre of the undeflected trace and the outside edge of the deflected trace gives the value of $\frac{D}{2}$. The value of l_3 was 10.00 cm; the voltage 340, and D was 3.174 cm. Consequently the value $\frac{mu^2}{e}$ for the rays after passing through the standard mica was

$$\frac{mu^2}{e} = 3.10 \times 10^{14}$$

We have previously shown that

$$\frac{mu}{e} = 2.56 \times 10^5,$$

Therefore

$$\frac{e}{m} = 4.7 \times 10^3,$$

and

$$u = 1.21 \times 10^9 \text{ cm/s.}$$

We may thus conclude that the α particle from actinium has the same mass as that from radium.

19.2.4   Connexion of the α Particle with the Helium Atom

We have seen that, within the limit of experimental error, the mass of the α particle expelled from radium A, radium C, radium F, or actinium B is the same. In a later paper, in conjunction with Dr. Hahn, it will be shown that the mass of the α particle expelled from thorium C is also identical with that expelled from the radium products. We have also shown in a previous paper that the amount of the magnetic deflexion of the α particle from radium itself is in agreement with that to be deduced from its range in air—a result which is only to be expected if the α particle from radium has the same mass as that from radium C.

There then remains only one α ray product of radium, *viz*. the emanation, whose radiation has not been closely examined. There is, however, no reason to suppose that the α particles from the emanation differ in mass from those of the other products. An examination of the complex pencil of rays from a layer of radium in equilibrium shows no evidence of the presence of α rays which suffer an abnormal amount of deflexion. I think there can be no doubt that the α particles emitted from the various products of radium have an identical mass, but differ only in the initial velocities of projection. Although the mass of the α particles has been determined for only a single product of thorium and of actinium, the analogy with radium would lead us to expect that the α particle has the same mass for all the products of these substances.

We may thus reasonably conclude that the α particles expelled from the different radio-elements have the same mass in all cases. This is an important conclusion; for it shows that uranium, thorium, radium, and actinium, which behave chemically as distinct elements, have a common product of transformation. The α particle constitutes one of the fundamental units of matter of which the atoms of these elements are built up. When it is remembered that in the process of their transformation radium and thorium each expel five α particles, actinium four, and uranium one, and that radium is in all probability a transformation product of uranium, it is seen that the α particle is an important fundamental constituent of the atoms of the radio-elements proper. I have often pointed out what an important part the α particles play in radioactive transformation. In comparison, the β and γ rays play quite a secondary role.

It is now necessary to consider what deductions can be drawn from the observed value of e/m found for the α particle. The value of e/m for the hydrogen ion in the electrolysis of water is known to be very nearly 10^4. The hydrogen ion is supposed to be the hydrogen atom with a positive charge, so that the value of e/m for the hydrogen atom is 10^4. The observed value of e/m for the α particle is 5.1×10^3, or, in round numbers, one half of that of the hydrogen atom. The density of helium has been found to be 1.98 times that of hydrogen, and from observations of the velocity of sound in helium, it has been deduced that helium is monatomic gas. From this it is concluded that the helium atom has an atomic weight of 3.96. If a helium atom carries the same charge as the hydrogen ion, the value of e/m for the helium atom should consequently be about 2.5×10^3. If we assume that the α particle carries the same charge as the hydrogen ion, the mass of the α particle is twice that of the hydrogen atom. We are here unfortunately confronted with several possibilities between which it is difficult to make a definite decision.

The value of e/m for the α particle may be explained on the assumptions that the α particle is (1) a *molecule* of hydrogen carrying the ionic charge of hydrogen, (2) a helium atom carrying *twice* the ionic charge of hydrogen, or (3) *one half* of the helium atom carrying a single ionic charge.

The hypothesis that the α particle is a molecule of hydrogen seems for many reasons improbable. If hydrogen is a constituent of radioactive matter, it is to be expected that it would be expelled in the atomic, and not in the molecular state. In addition, it seems improbable that, even if the hydrogen were initially projected in the molecular state, it would escape decomposition into its component atoms in passing through matter, for the α particle is projected at an enormous velocity, and the shock of the collisions of the α particle with the molecules of matter must be very intense, and tend to disrupt the bonds that hold the hydrogen atoms together. If the α particle is hydrogen, we should expect to find a large quantity of hydrogen present in the old radioactive minerals, which are sufficiently compact to prevent its escape. This does not appear to be the case, but, on the other hand, the comparatively large amount of helium present supports the view that the α particle is a helium atom. A strong argument in support of the view of a connexion between helium and the α particle rests on the observed facts[3] that helium is produced by actinium as well as by radium. The only point of identity between these two substances lies in the expulsion of α particles of the same mass. The production of helium by both substances is at once obvious if the helium is derived from accumulated α particles, but is difficult to explain on any other hypothesis. We are thus reduced to the view that either the α particle is a helium atom carrying twice the ionic charge of hydrogen, or is half of a helium atom carrying a single ionic charge.

The latter assumption involves the conception that helium, while consisting of a monovalent atom under ordinary chemical and physical conditions, may exist in a still more elementary state as a component of the atoms of radioactive matter, and that, after expulsion, the parts of the atom lose their charge and recombine to form atoms of helium; while such a view cannot be dismissed as inherently improbable, there is as yet no direct evidence in its favour. On the other hand, the second hypothesis has the merit of greater simplicity and probability.

On this view, the α particle is in reality a helium atom which is either expelled with a double ionic charge or acquires this charge in its passage through matter. Even if the α particle were initially projected without charge, it would certainly acquire one after the first few collisions with the molecules in its path. We know that the α particle is a very efficient ionizer, and there is every reason to suppose that it would itself be ionized by its collisions with the molecules in its path, *i.e.* it would lose one or more electrons and retain a positive charge. If the α particle can remain stable with the loss of two electrons, these electrons would almost certainly be removed as a result of the intense disturbance set up by the collision of the α particle with the molecules of matter. The α particle would then have twice the normal ionic charge, and the value of e/m, as found by measurement, would be quite consistent with the view that the α particle is an atom of helium.

In a previous paper[4] I showed, from measurement of the charge carried by the α rays, that 6.2×10^{10} α particles were expelled per second from 1 g of radium at its

---

[3] Debierne, *C. R. exli.* p. 383 (1905).

[4] Phil. Mag, August (1905).

minimum activity. This was based on the assumption that each α particle carried a positive charge equal to the ionic charge of hydrogen, *viz.* 3.4×10^{-10} electrostatic units. Assuming that the α particle carries two ionic charges, the corresponding number is reduced to one half of the above, *viz.* 3.1×10^{10}. This would make the calculated period of radium 2600 years instead of 1300 years (see 'Radioactivity,' second edition, 1905, p. 457). In a similar way, the calculated volume of the emanation released from 1 g of radium would be 0.4 cubic mm instead of 0.8 cubic mm. The calculated volume of helium produced per year per gram of radium would be 0.11 cubic cm ('Radioactivity,' p. 481).

On the hypothesis that the α particle is a helium atom, the atomic weight of each product is diminished by four units, in consequence of the expulsion of an α particle. On the hypothesis that the α particle is half a helium atom carrying a single ionic charge, the atomic weight is diminished by two units instead of four. Taking the latter hypothesis, the number of α particles expelled per second from 1 g of radium at its minimum activity is 6.2×10^{10}. The calculated volume of the emanation is 0.8 cubic mm, while the production of helium per year is 0.11 cubic cm per gram. The two hypotheses thus lead to the same rate of production of helium by radium.

19.2.5 Age of Radioactive Minerals

I have previously pointed out that the age of the radioactive minerals can be calculated from the amount of helium contained in them. The method is based on the assumption that, in a compact mineral, the greater part of the helium is mechanically imprisoned in the mineral, and is unable to escape. Let us consider, for example, the mineral fergusonite which was found by Ramsay and Travers to contain 1.81 c.c. of helium per gram of the mineral. The fergusonite contains about 7 % of uranium. The amount of helium per gram of uranium is consequently 26 c.c. Now we have seen that 1 g of radium produces 0.11 c.c. of helium per year. The content of radium per gram of uranium is 3.8×10^{-7} g.[5] Supposing that uranium emits only one α particle corresponding to the five emitted by radium in equilibrium where the product radium F is present, the production of helium per year per gram of uranium is $\frac{6}{4} \times .11 \times 3.8 \times 10^{-7}$ or 6.3×10^{-8} c.c. per year. Assuming as a first approximation that the rate of production of helium has been constant since the formation of the mineral, the time required for a production of 26 c.c. of helium is about 400 million years. This is a minimum estimate, for some of the helium has probably escaped from the mineral.

As another example, consider the mineral thorianite, which contains about 72 % of thorium and 10 % of uranium. The evolution of helium per gram of the mineral was found by Ramsay to be 9.5 c.c. Bragg (Phil. Mag. June 1906) has shown that thorium breaks up at 0.26 of the rate of uranium. This was based on measurements

---

[5] Rutherford and Boltwood, Amer. Journ. Sci. July 1906.

Table 19.1 The velocity
of α-particles emitted by
various radioactive materials
as deduced from their range
in air.—[K.K.]

Product	Range of α particle in cm	Velocity in cm/s	$\frac{mu^2}{e}$
Radium	3.50	1.56×10^9	4.78×10^{14}
Emanation	4.36	1.70×10^9	5.65×10^{14}
Radium A	4.83	1.77×10^9	6.12×10^{14}
Radium C	7.06	2.06×10^9	8.37×10^{14}
Radium F*	3.86	1.61×10^9	5.15×10^{14}

made with ordinary commercial thorium. Boltwood (Amer. Journ. Sci. June 1906) has, however, drawn attention to the fact that ordinary commercial thorium has in many cases only about one half of the activity obtained by direct preparation of the thorium from the radioactive minerals. This would double the rate of breaking up of thorium observed by Bragg. Remembering that a thorium atom during its transformations emits five α particles, and assuming that thorium breaks up at half the rate of uranium, it is seen that 72 % of thorium in a mineral corresponds as a producer of helium to about $\frac{5}{6} \times \frac{72}{2} = 30\%$ of uranium. The amount of helium corresponding to 1 g of uranium or its equivalent in the mineral is consequently 24 c.c. As before, the age of the mineral works out to be about 400 million years.

Numerous other examples may be given, but these serve to illustrate the method of calculation from radioactive data of the age of some radioactive minerals, and indirectly, in some cases, of the geological strata in which they are found.

19.2.6   Velocity and Energy of the α Particles Expelled from Radium Products

If the value of e/m is the same for the α particle expelled from the various radium products, the maximum velocity of each set of a particles can he deduced from their range in air, knowing the velocity of the α particles expelled from radium C. The velocities so determined are probably more accurate than those obtained by direct measurement under difficult conditions. In the following table, the second column gives the range in air of the α particles from the radium products, found by Bragg and Kleeman; in the fourth column is given the value of $\frac{mu^2}{e}$ where u is the initial velocity of projection of the α particles (Table 19.1).

Disregarding radium F, the average energy of the α particle expelled from radium in equilibrium is $3.11 \times 10^{14}e$, where e is the charge carried by the α particle. Assuming that the heating effect of radium is a measure of the kinetic energy of the expelled α particles, we can at once deduce the total number of α particles expelled per second per gram of radium in equilibrium. One gram of radium in equilibrium emits 100 gram-calories of heat per hour. This rate of emission of energy is mechanically equivalent to 1.16×10^8 ergs per second. Since the average energy of

Table 19.2 A comparison of the calculated and observed heating effects of various radioactive materials.—[K.K.]

Product	Calculated heating effect	Observed heating effect
Radium	19.2	23
Emanation	22.7	45
Radium A	24.5	45
Radium C	33.6	32

the expelled α particle is $3.11 \times 10^{14}e$, the number of α particles expelled per second from 1 g of radium in equilibrium is $\frac{3.65}{10^9 e}$. The number previously found by the writer by measuring the total charge carried by the α particles was $\frac{2.82}{10^9 e}$, *i.e.* 77 % of the theoretical number. The agreement between theory and experiment is thus fairly good. In the above estimate, it is assumed that the heating effect is due entirely to the kinetic energy of the expelled α particles. It is known experimentally that the heating effect of the β and γ rays is only a small percentage of that due to the α rays. The expulsion of an α particle from an atom should lead to the recoil of the residue of the atom. Assuming that the momentum of the atom is equal and opposite to that of the α particle, the velocity of recoil of the atom can be simply calculated. Taking the mass of the α particle as 4 and of the radium atom as 225, the velocity of recoil of the disintegrated radium atom, for example, is $\frac{1}{55} \times 1.56 \times 10^9$ or 2.8×10^7 cm/s. The heating effect resulting from this recoil is thus only about 2 % of that due to the α particle.

Assuming that each α particle carries a single ionic charge of 1.13×10^{-20} electromagnetic units, the number of α particles which must be expelled per second from 1 g of radium in order to account entirely for the heating effect is 3.2×10^{11}. The experimental number is 2.5×10^{11}. If it is assumed that the α particle carries twice the usual ionic charge, each of these numbers is reduced by one half.

It is of interest to calculate the distribution of the heating effect of radium in equilibrium amongst the various α ray products. The theoretical percentages of the total heating effect are given in column 1. These are calculated from the known energy of the α particles expelled from each product. The observed percentages are deduced from the experimental numbers and curves given by Rutherford and Barnes (*Phil. Mag.* Feb. 1904) (Table 19.2).

The observed heating effects of the emanation and radium A are given together, as it is very difficult experimentally to determine their separate effects. It will be seen that there is a substantial agreement between the calculated and observed values.

19.2.7   Connexion between the Velocity and Amount of Ionization Produced by the α Particle

Bragg (*Phil. Mag.* Nov. 1905) has shown that the ionization produced by a single α particle increases with the distance from the source to nearly the end of its range,

when the ionization falls off very abruptly. He has shown that the ionization produced by the α particle at a distance r cm from the end of the path is inversely
proportional to $\sqrt{r+c}$, where c is a constant equal to 1.33. In a previous paper
(*Phil. Mag.* Aug. 1906) I have shown that the velocity of an α particle at a distance
r cm from the end of its range is proportional to $\sqrt{r+d}$, where d is a constant
equal to 1.25. The close agreement between these two expressions shows that the
ionization produced per unit path by the α particle is inversely proportional to its
velocity. This is in agreement with the theoretical views of Bragg, who supposed
that the rate of expenditure of energy of the α particle in ionization at any point is
inversely proportional to the energy of motion which it possesses.

A comparison of the velocities of the α particles expelled from the various products of the radio-elements, and a discussion of the connexion that exists between the
velocity of expulsion of the α particle and the character of the transformation will
be given in a later paper.

I desire to express my thanks to Dr. Hahn and Dr. Levin for their assistance in
the measurement of the numerous photographs obtained in this investigation.

Berkley, California, July 20, 1906.

19.3   Study Questions

QUES. 19.1. What is the identity of the α-particle?

a) What renders measurement of the deflection of α-particles from radium A so
difficult? How long does it take for the α ray activity to drop to ¼ of its initial
value?
b) How can the α rays emitted by radium A be distinguished from those emitted by
radium C? Why is it important to analyze these rays separately? Do they have
the same values of e/m?
c) How do the speeds and charge-to-mass ratios of the α-particles from radium A,
radium C, radium F, actinium B and thorium C compare? How do their masses
compare, and what important conclusion does he draw from this?
d) What unique role do α-particles play in the constitution of atoms? Where does
Rutherford suspect radium came from?
e) How do the charge-to-mass ratios of hydrogen and α-particles compare? What
logical possibilities does this suggest regarding the identity of the α-particle?
f) Which of these does Rutherford find most reasonable? Why? Do you agree with
his rationale and his conclusions? Consider: does a helium ion have twice, or
four times, the atomic weight of a hydrogen ion?

QUES. 19.2. How can the age of a radioactive mineral be computed?

a) How many α-particles per second are expelled from 1 g of radium? What does this imply about the rate of helium production in radium?
b) Are Rutherford's calculation of helium production dependent on the atomic weight assigned to the α-particle?
c) Upon what assumption(s) is Rutherford's method of radioactive dating based? How old are the minerals dated using Rutherford's method? Does this sound reasonable?

QUES. 19.3. Why does radium feel warm? And how can the rate of production of α-particles be deduced from the self-heating of radium?

QUES. 19.4. What is the velocity of an α-particle expelled from a radium product? How can this be inferred from its range? And how is an α-particle's velocity related to the ionization it produces?

19.4   Exercises

EX. 19.1 (HALF-LIFE AND RADIOACTIVE DECAY RATE). The emission of an α-particle by a radioactive isotope is an unpredictable process. This is true of all radioactive decays. Nonetheless, a *half-life* can be assigned to each radioactive decay process which indicates the time during which one-half of the isotopes present will have undergone such a decay. The measured half-lives of many of the isotopes comprising the uranium series are shown in Table 19.3.[6]

The rate of decay of a pure radioactive substance is (inversely) related to the half-life of its constituent isotopes: the longer the half-life, the smaller the decay rate. More precisely, if there are N isotopes present in a sample at a particular time, t, then one can define the *decay rate*, R, as

$$R = \frac{-dN/dt}{N}.$$

a) Demonstrate that the solution to this differential equation is

$$N(t) = N_0 \, e^{-Rt}. \tag{19.1}$$

Make a sketch of Eq. 19.1. What does N_0 represent?
b) Demonstrate that the half-life, $\tau_{1/2}$, can be expressed in terms of the decay rate as

$$\tau_{1/2} = \frac{\ln 2}{R}. \tag{19.2}$$

c) How many half-lives must elapse before only $1/100$ of the number of initial radioactive atoms in a sample are remaining?

---

[6] For more information on the uranium series, see Ex. 18.4 in the previous chapter.

Table 19.3 Half-lives for isotopes appearing in the uranium series.—[K.K.]

Isotope	$\tau_{1/2}$		Particle emitted
^{238}U	4.468×10^9	Years	α
^{234}Th	24.10	Days	β
^{234}Pa	1.17	Minutes	β
^{234}U	2.455×10^5	Years	α
^{230}Th	7.338×10^4	Years	α
^{226}Ra	1.600×10^3	Years	α
^{222}Rn	3.825	Days	α
^{218}Po	3.10	Minutes	α
^{214}Pb	26.8	Minutes	β
^{214}Bi	19.9	Minutes	β
^{214}Po	164.3	μs	α
^{210}Pb	22.3	Years	β
^{210}Bi	5.013	Days	β
^{210}Po	138.376	Days	α
^{206}Pb	Stable		

EX. 19.2 (ISOTOPE ACTIVITY). The *activity* of a particular radioactive sample is measured in *becquerels* (decays per second) using a radiation detector such as a Geiger-Muller tube or a scintillation counter. Suppose that the initial activity of a laboratory sample containing radon-222 isotopes is ten times the initial activity of another laboratory sample containing bismuth-210 isotopes. How many days must elapse before the activities of these two samples are equal?

EX. 19.3 (ATOMIC RECOIL AND HEATING). Based on Rutherford's data, what is the recoil speed of the residual (daughter) nucleus when an atom of radium F emits an α-particle? Calculate the kinetic energies of each of the decay products and the (initial) heating rate of a 1-g bulk sample of pure radium F. How would your answer change if a very thin layer of radon was employed so that nearly all of the α-particles escape from the surface of the sample, rather than being absorbed into its bulk?

EX. 19.4 (PLUTONIUM SELF HEATING). Uranium-235 (atomic mass $235.044\,u$) is produced when plutonium-239 (atomic mass $239.05216\,u$) undergoes α-decay:

$$^{239}_{94}\text{Pu} \longrightarrow \,^{235}_{92}\text{U} + \,^{4}_{2}\text{He}$$

The half-life of this decay process is about 24,100 years. Calculate (a) the speed of the emitted α particle and (b) the (initial) heating rate of a 1-g bulk sample of pure plutonium-239.

EX. 19.5 (RANGE OF α-PARTICLES LABORATORY). In this laboratory experiment, we will measure the range of α-particles in air and thereby deduce their speed and kinetic energy.[7] Place an α-source, such as polonium-210 near a geiger-mueller tube or an ionization chamber and plot the measured activity (counts per second) as a function of distance.[8] From a plot of your data, determine the average range of α-particles from polonium-210. The energy may be approximated using the formula

$$E \simeq r + 1.5$$

where r is the range in air (in centimeters) and E is the energy (in MeV). Compare the speed of your α-particles to Rutherford's α-particles from polonium-210. What effect do various materials have on the range of the α-particles?[9]

EX. 19.6 (BARIUM-137 HALF-LIFE LABORATORY). In this laboratory exercise, we will use a geiger-muller tube to measure the activity and half-life of a sample of radioactive barium-137m (a metastable state of barium-137, as described below). A safe and easy-to-use isotope generator kit, from which barium-137m may be extracted, is available commercially.[10] In this kit you should find (i) an isotope exchange column containing a small quantity of radioactive Cesium-137, (ii) a small bottle containing an eluting solution consisting of 0.9 % NaCl in 0.04 M HCl, (iii) a syringe for injecting about seven drops of eluting solution into the top of the exchange column, and (iv) a small steel planchet in which to capture the eluting solution forced through the exchange column.

The Cs-137 which resides in the exchange column undergoes β-decay, with a half life of 30.17 years, producing the meta-stable isotope Ba-137m. The eluting solution which is forced through the exchange column selectively extracts the Ba-137m. The large number of Cs-137 isotopes remaining in the column quickly regenerate the Ba-137m, so the exchange column can be "milked" repeatedly. The Ba-137m which is extracted by the eluting solution undergoes γ-decay to the stable isotope Ba-137 with a much shorter half-life; this may be measured by counting the emitted γ-particles using a geiger-muller tube held near the eluting solution in the steel planchet.[11] A data-collection time of up to 60 min may be necessary to ensure that all of the Ba-137m has decayed. This will allow for an accurate background

---

[7] A valuable resource for laboratory experiments involving radioactivity is Chase, G., S. Rituper, and J. Sulcoski, *Experiments in Nuclear Science*, Burgess Pub. Co., 1971. Also see Gastineau, J. E., *Nuclear Radiation with Computers and Calculators*, 3rd ed., Vernier Software & Technology, 2003.

[8] I have used a Geiger-Mueller tube (Vernier model RMD-BTD) together with a data-logger (Vernier LabPro), which are both available from Vernier Software & Technology, Beaverton, OR. ^{210}Po sources are available from Spectrum Techniques, Oak Ridge, TN. Alternatively, a small silver disk (such as a pre-1965 United States dime or quarter) can be immersed in RaDEF solution—consisting of RaD (^{210}Pb) in equilibrium with RaE (^{210}Bi) and RaF (^{210}Po)—for a few minutes and then washed with water and air-dried.

[9] A Calibrated absorber set (Model RAS 20) consisting of lead, aluminum and plastic plates of various thicknesses is available from Spectrum Techniques, Oak Ridge, TN.

[10] Cs/Ba-137m Isotope Generator, Spectrum Techniques, Oak Ridge, TN.

[11] Radiation monitor (Model RM-BTD), Vernier Software & Technology, Beaverton, OR.

count due to non-barium radioactive sources. The residual eluting solution containing stable Ba-137 may be simply washed from the planchets using water and discarded.

19.5 Vocabulary

1. Radium
2. Emanation
3. Radiotellurium
4. Polonium
5. Monovalent
6. Ionize
7. Fergusonite
8. Thorium
9. Actinium
10. Thorianite
11. Uranium
12. Disintegrate

Chapter 20
Atomic Fission

> *While it has long been known that helium is a product of the*
> *spontaneous transformation of some of the radio-active*
> *elements, the possibility of disintegrating the structure of stable*
> *atoms by artificial methods has been a matter of uncertainty.*
> —Ernest Rutherford

20.1   Introduction

In the previous two chapters we looked carefully at how Rutherford characterized the α-particles which were emitted by radioactive substances such as radium and actinium. These particles were all found to have nearly the same charge-to-mass ratios. They were also found to be traveling at enormous speeds—comparable to that of light itself. Based on these observations, Rutherford was able to identify the α-particle as the nucleus of a helium atom. This suggested that the radioactive elements, and perhaps even all the elements, were built up of helium nuclei. So a few years later he turned this experiment around. Rutherford, Geiger and Marsden employed a beam of high-speed α-particles from radium as a probe to study the structure of gold atoms. In particular, they measured the deflection of a beam of α-particles fired at a thin gold foil placed near a radioactive source in a vacuum chamber. The deflected α-particles were meticulously observed as they struck a zinc-sulphide screen placed behind the gold foil target, each time emitting a faint flash of light—a *scintillation*—which was visible when viewed with a microscope. These gold-foil experiments inspired Rutherford to construct a model of the atomic nucleus in which he likened it to a tiny solar system composed of negatively charged electrons orbiting a positively charged nucleus. This "planetary model" of the atom was very different than the "plum pudding" model suggested in 1904 by Rutherford's teacher, J. J. Thomson, who had himself discovered the electron in 1897. According to Thomson's model, tiny electrons were situated inside the atomic nucleus like negatively charged plums in a positively charged pudding.

In the reading selection below, taken from a lecture delivered in 1920, Rutherford begins by recounting the results of these gold-foil experiments. He seems particularly interested in determining the size of the atomic nucleus, and the strength of the forces which act in its vicinity. He clearly has an eye toward determining whether

© Springer International Publishing Switzerland 2016

K. Kuehn, *A Student's Guide Through the Great Physics Texts*,

Undergraduate Lecture Notes in Physics, DOI 10.1007/978-3-319-21828-1_20

atomic nuclei might possibly be broken apart by hitting them hard enough. He then proceeds to describe a new set of experiments in which he attempted to do just that. What was his experimental technique? Was he successful in splitting the atom?

20.2   Reading: Rutherford, *Nuclear Constitution of Atoms*

Rutherford, E., Bakerian Lecture: Nuclear Constitution of Atoms, *Proceedings of the Royal Society of London. Series A, Containing Papers of a Mathematical and Physical Character*, 97(686), 374–400, 1920.

20.2.1   Introduction

The conception of the nuclear constitution of atoms arose initially from attempts to account for the scattering of α-particles through large angles in traversing thin sheets of matter.[1] Taking into account the large mass and velocity of the α-particles, these large deflexions were very remarkable, and indicated that very intense electric or magnetic fields exist within the atom. To account for these results, it was found necessary to assume[2] that the atom consists of a charged massive nucleus of dimensions very small compared with the ordinarily accepted magnitude of the diameter of the atom. This positively charged nucleus contains most of the mass of the atom, and is surrounded at a distance by a distribution of negative electrons equal in number to the resultant positive charge on the nucleus. Under these conditions, a very intense electric field exists close to the nucleus, and the large deflexion of the α-particle in an encounter with a single atom happens when the particle passes close to the nucleus. Assuming that the electric forces between the α-particle and the nucleus varied according to an inverse square law in the region close to the nucleus, the writer worked out the relations connecting the number of α-particles scattered through any angle with the charge on the nucleus and the energy of the α-particle. Under the central field of force, the α-particle describes a hyperbolic orbit round the nucleus, and the magnitude of the deflection depends on the closeness of approach to the nucleus. From the data of scattering of α-particles then available, it was deduced that the resultant charge on the nucleus was about $\frac{1}{2}Ae$, where A is the atomic weight and e the fundamental unit of charge. Geiger and Marsden[3] made an elaborate series of experiments to test the correctness of the theory, and confirmed the main conclusions. They found the nucleus charge was about $\frac{1}{2}Ae$, but, from the nature of the experiments, it was difficult to fix the actual value within

---

[1] Geiger and Marsden, 'Roy. Soc. Proc.,' **A**, vol. 82, p. 495 (1909).

[2] Rutherford, 'Phil. Mag.,' vol. 21, p.669 (1911); vol. 27, p. 488 (1914).

[3] Geiger and Marsden, 'Phil. Mag.,' vol. 25, p. 604 (1913).

about 20 %. C. G. Darwin[4] worked out completely the deflexion of the α-particle and of the nucleus, taking into account the mass of the latter, and showed that the scattering experiments of Geiger and Marsden could not be reconciled with any law of central force, except the inverse square. The nuclear constitution of the atom was thus very strongly supported by the experiments on scattering of α-rays.

Since the atom is electrically neutral, the number of external electrons surrounding the nucleus must be equal to the number of units of resultant charge on the nucleus. It should be noted that, from the consideration of the scattering of X-rays by light elements, Barkla[5] had shown, in 1911, that the number of electrons was equal to about half the atomic weight. This was deduced from the theory of scattering of Sir J. J. Thomson, in which it was assumed that each of the external electrons in an atom acted as an independent scattering unit.

Two entirely different methods had thus given similar results with regard to the number of external electrons in the atom, but the scattering of α-rays had shown in addition that the positive charge must be concentrated on a massive nucleus of small dimensions. It was suggested by Van den Broek[6] that the scattering of α-particles by the atoms was not inconsistent with the possibility that the charge on the nucleus was equal to the atomic number of the atom, *i.e.*, to the number of the atom when arranged in order of increasing atomic weight. The importance of the atomic number in fixing the properties of an atom was shown by the remarkable work of Moseley[7] on the X-ray spectra of the elements. He showed that the frequency of vibration of corresponding lines in the X-ray spectra of the elements depended on the square of a number which varied by unity in successive elements. This relation received an interpretation by supposing that the nuclear charge varied by unity in passing from atom to atom, and was given numerically by the atomic number. I can only emphasise in passing the great importance of Moseley's work, not only in fixing the number of possible elements, and the position of undetermined elements, but in showing that the properties of an atom were defined by a number which varied by unity in successive atoms. This gives a new method of regarding the periodic classification of the elements, for the atomic number, or its equivalent the nuclear charge, is of more fundamental importance than its atomic weight. In Moseley's work, the frequency of vibration of the atom was not exactly proportional to N, where N is the atomic number, but to $(N - a)^2$, where a was a constant which had different values, depending on whether the K or L series of characteristic radiations were measured. It was supposed that this constant depended on the number and position of the electrons close to the nucleus.

---

[4] Darwin, 'Phil. Mag.,' vol. 27, p. 499 (1914).

[5] Barkla, 'Phil. Mag.,' vol. 21, p. 648 (1911).

[6] Van den Broek, 'Phys. Zeit.,' vol. 14, p. 32 (1913).

[7] Moseley, 'Phil. Mag.,' vol. 26, p. 1024 (1913); vol. 27, p. 703 (1914).

20.2.2 *Charge on the Nucleus*

The question whether the atomic number of an element is the actual measure of its nuclear charge is a matter of such fundamental importance that all methods of attack should be followed up. Several researches are in progress in the Cavendish Laboratory to test the accuracy of this relation. The two most direct methods depend on the scattering of swift α- and β-rays. The former is under investigation, using new methods, by Mr. Chadwick, and the latter by Dr. Crowther. The results so far obtained by Mr. Chadwick strongly support the identity of the atomic number with the nuclear charge within the possible accuracy of experiment, *viz.*, about 1 %.

It thus seems clear that we are on firm ground in supposing that the nuclear charge is numerically given by the atomic number of the element. Incidentally, these results, combined with the work of Moseley, indicate that the law of the inverse square holds with considerable accuracy in the region surrounding the nucleus. It will be of great interest to determine the extent of this region, for it will give us definite information as to the distance of the inner electrons from the nucleus. A comparison of the scattering of slow and swift β-rays should yield important information on this point. The agreement of experiment with theory for the scattering of α-rays between 5° and 150° shows that the law of inverse square holds accurately in the case of a heavy element like gold for distances between about 36×10^{-12} cm and 3×10^{-12} cm from the centre of the nucleus. We may consequently conclude that few, if any, electrons are present in this region.

An α-particle in a direct collision with a gold atom of nuclear charge 79 will be turned back in its path at a distance of 3×10^{-12} cm, indicating that the nucleus may be regarded as a point charge even for such a short distance. Until swifter α-particles are available for experiment, we are unable in the case of heavy elements to push further the question of dimensions of heavy atoms. We shall see later, however, that the outlook is more promising in the case of lighter atoms, where the α-particle can approach closer to the nucleus.

It is hardly necessary to emphasise the great importance of the nuclear charge in fixing the physical and chemical properties of an element, for obviously the number and the arrangements of the external electrons on which the great majority of the physical and chemical properties depend, is conditioned by the resultant charge on the nucleus. It is to be anticipated theoretically, and is confirmed by experiment, that the actual mass of the nucleus exercises only a second order effect on the arrangement of the external electrons and their rates of vibration.

It is thus quite possible to imagine the existence of elements of almost identical physical and chemical properties, but which differ from one another in mass, for, provided the resultant nuclear charge is the same, a number of possible stable modes of combination of the different units which make up a complex nucleus may be possible. The dependence of the properties of an atom on its nuclear charge and not on its mass thus offers a rational explanation of the existence of isotopes in which the chemical and physical properties may be almost indistinguishable, but the mass of the isotopes may vary within certain limits. This important question will be considered in more detail later in the paper in the light of evidence as to the nature of the units which make up the nucleus.

The general problem of the structure of the atom thus naturally divides itself into two parts:—

(1) Constitution of the nucleus itself.
(2) The arrangement and modes of vibration of the external electrons.

I do not propose to-day to enter into (2), for it is a very large subject, in which there is room for much difference of opinion. This side of the problem was first attacked by Bohr and Nicholson, and substantial advances have been made. Recently, Sommerfeld and others have applied Bohr's general method with great success in explaining the fine structure of the spectral lines and the complex modes of vibration of simple atoms involved in the Stark effect. Recently, Langmuir and others have attacked the problem of the arrangement of the external electrons from the chemical standpoint, and have emphasised the importance of assuming a more or less cubical arrangement of the electrons in the atom. No doubt each of these theories has a definite sphere of usefulness, but our knowledge is as yet too scanty to bridge over the apparent differences between them.

I propose to-day to discuss in some detail experiments that have been made with a view of throwing light on the constitution and stability of the nuclei of some of the simpler atoms. From a study of radio-activity we know that the nuclei of the radio-active elements consist in part of helium nuclei of charge $2e$. We also have strong reason for believing that the nuclei of atoms contain electrons as well as positively charged bodies, and that the positive charge on the nucleus represents the excess positive charge. It is of interest to note the very different rôle played by the electrons in the outer and inner atom. In the former case, the electrons arrange themselves at a distance from the nucleus, controlled no doubt mainly by the charge on the nucleus and the interaction of their own fields. In the case of the nucleus, the electron forms a very close and powerful combination with the positively charged units and, as far as we know, there is a region just outside the nucleus where no electron is in stable equilibrium. While no doubt each of the external electrons acts as a point charge in considering the forces between it and the nucleus, this cannot be the case for the electron in the nucleus itself. It is to be anticipated that under the intense forces in the latter, the electrons are much deformed and the forces may be of a very different character from those to be expected from an undeformed electron, as in the outer atom. It may be for this reason that the electron can play such a different part in the two cases and yet form stable systems.

It has been generally assumed, on the nucleus theory, that electric forces and charges play a predominant part in determining the structure of the inner and outer atom. The considerable success of this theory in explaining fundamental phenomena is an indication of the general correctness of this point of view. At the same time if the electrons and parts composing the nucleus are in motion, magnetic fields must arise which will have to be taken into account in any complete theory of the atom. In this sense the magnetic fields are to be regarded as a secondary rather than a primary factor, even though such fields may be shown to have an important bearing on the conditions of equilibrium of the atom.

20.2.3   *Dimension of Nuclei*

We have seen that in the case of atoms of large nuclear charge the swiftest α-particle is unable to penetrate to the actual structure of the nucleus so that it is possible to give only a maximum estimate of its dimensions. In the case of light atoms, however, when the nucleus charge is small, there is so close an approach during a direct collision with an α-particle that we are able to estimate its dimensions and form some idea of the forces in operation. This is best shown in the case of a direct collision between an α-particle and an atom of hydrogen. In such a case, the H atom is set in such swift motion that it travels four times as far as the colliding α-particle and can be detected by the scintillation produced on a zinc sulphide screen.[8] The writer[9] has shown that these scintillations are due to hydrogen atoms carrying unit positive charge recoiling with the velocity to be expected from the simple collision theory, *viz.*, 1.6 times the velocity of the α-particle. The relation between the number and velocity of these H atoms is entirely different from that to be expected if the α-particle and H atom are regarded as point charges for the distances under consideration. The result of the collision with swift α-particles is to produce H atoms which have a narrow range of velocity, and which travel nearly in the direction of the impinging particles. It was deduced that the law of inverse squares no longer holds when the nuclei approach to within a distance of 3×10^{-13} cm of each other. This is an indication that the nuclei have dimensions of this order of magnitude and that the forces between the nuclei vary very rapidly in magnitude and in direction for a distance of approach comparable with the diameter of the electron as ordinarily calculated. It was pointed out that in such close encounters there were enormous forces between the nuclei, and probably the structure of the nuclei was much deformed during the collision. The fact that the helium nucleus, which may be supposed to consist of four H atoms and two electrons, appeared to survive the collision is an indication that it must be a highly stable structure. Similar results[10] were observed in the collision between α–particles and atoms of nitrogen and oxygen for the recoil atoms appeared to be shot forward mainly in the direction of the α–particles and the region where special forces come into play is of the same order of magnitude as in the case of the collision of an α–particle with hydrogen.

No doubt the space occupied by a nucleus and the distance at which the forces become abnormal increase with the complexity of the nucleus structure. We should expect the H nucleus to be the simplest of all and, if it be the positive electron, it may have exceedingly small dimensions compared with the negative electron. In the collisions between α–particles and H atoms, the α–particle is to be regarded as the more complex structure.

---

[8] Marsden, 'Phil. Mag.,' vol. 27, p. 824 (1914).

[9] Rutherford, 'Phil. Mag.,' vol. 37, I and II, pp. 538–571 (1919).

[10] Rutherford, 'Phil. Mag.,' vol. 37, III, p. 571 (1919).

Table 20.1 Calculated ranges
of singly charged particles of
various masses set in motion
by close collisions with an
α-particle

Mass	1	Range	3.91	R
"	2	"	4.6	R
"	3	"	5.06	R
"	4	"	4.0	R

The diameter of the nuclei of the light atoms except hydrogen are probably of the order of magnitude 5×10^{-13} cm and in a close collision the nuclei come nearly in contact and may possibly penetrate each other's structure. Under such conditions, only very stable nuclei would be expected to survive the collision and it is thus of great interest to examine whether evidence can be obtained of their disintegration.

20.2.4 Long Range Particles from Nitrogen

In previous papers, *loc. cit.*, I have given an account of the effects produced by close collisions of swift α–particles with light atoms of matter with the view of determining whether the nuclear structure of some of the lighter atoms, could be disintegrated by the intense forces brought into play in such close collisions. Evidence was given that the passage of α–particles through dry nitrogen gives rise to swift particles which closely resembled in brilliancy of the scintillations and distance of penetration hydrogen atoms set in motion by close collision with α–particles. It was shown that these swift atoms which appeared only in dry nitrogen and not in oxygen or carbon dioxide could not be ascribed to the presence of water vapour or other hydrogen material, but must arise from the collision of α–particles with nitrogen atoms. The number of such scintillations due to nitrogen was small, *viz.*, about 1 in 12 of the corresponding number in hydrogen, but was two to three times the number of natural scintillations from the source. The number observed in nitrogen was on an average equal to the number of scintillations when hydrogen at about 6 cm pressure was added to oxygen or carbon dioxide at normal pressure.

While the general evidence indicated that these long range atoms from nitrogen were charged atoms of hydrogen, the preliminary experiments to test the mass of the particles by bending them in a strong magnetic field yielded no definite results.

From the data given in my previous paper (*loc. cit.*) several theories could be advanced to account for these particles. The calculated range of a singly charged atom set in motion by a close collision with an α–particle of range R cm in air was shown to be for

On account of the small number and weakness of the scintillations under the experimental conditions, the range of the swift atoms from nitrogen could not be determined with sufficient certainty to decide definitely between any of these possibilities. The likelihood that the particles were the original α-particles which had lost one of their two charges, *i.e.*, atoms of charge 1 and mass 4, was suggested by me to several correspondents, but there appeared to be no obvious reason why nitrogen,

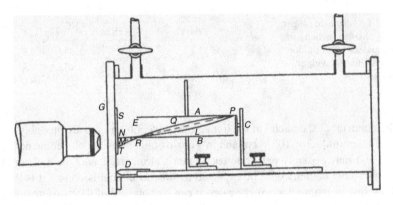

Fig. 20.1 Rutherford's arrangement for detecting any particles arising from gas struck by α-particles emitted from a radioactive source.—[K.K.]

of all the elements examined, should be the only one in which the passage of a swift α-particle led to the capture of a single electron.

If, however, a sufficient number of scintillations could be obtained under the experimental conditions, there should be no inherent difficulty in deciding between the various possibilities by examining the deflexion of the swift atoms by a magnetic field. The amount of deflexion of charged atoms in a magnetic field perpendicular to the direction of flight is proportional to e/mu. Assuming that the particles were liberated by a direct collision with an α-particle, the relative values of this quantity for different recoiling masses are easily calculated. Taking values MV/E for the α–particle as unity, the corresponding values of mu/e for atoms of charge 1 and mass 1, 2, 3, and 4 are 1.25, 0.75, 0.58, and 0.50 respectively. Consequently the H atoms should be more bent than the α–particles which produced them while the atoms of mass 2 or 3, or 4 would be more difficult to deflect than the parent α-particle.

On my arrival in Cambridge, this problem was attacked in several ways. By the choice of objectives of wide aperture, the scintillations were increased in brilliancy and counting thus made easier. A number of experiments were also made to obtain more powerful sources of radiation with the radium at my command, but finally it was found best, for reasons which need not be discussed here, to obtain the active source of radiation of radium C in the manner described in my previous paper. After a number of observations with solid nitrogen compounds, described later; a simple method was finally devised to estimate the mass of the particle by the use of nitrogen in the gaseous state. The use of the gas itself for this purpose had several advantages over the use of solid nitrogen compounds, for not only was the number of scintillations greater, but the absence of hydrogen or other hydrogen compounds could be ensured.

The arrangement finally adopted is shown in Fig. 20.1. The essential point lay in the use of wide slits, between which the α–particles passed. Experiment showed that the ratio of the number of scintillations on the screen arising from the gas to the number of natural scintillations from the source, increased rapidly with increased depth of the slits. For plates 1 mm, apart this ratio was less than unity, but for slits 8 mm apart the ratio had a value 2–3. Such a variation is to be anticipated on theoretical grounds if the majority of the particles are liberated at an angle with the direction of the incident α-particles.

The horizontal slits A, B were 6.0 cm long, 1.5 cm wide, and 8 mm deep, with the source, C of the active deposit of radium placed at one end and the zinc sulphide screen near the other. The carrier of the source and slits were placed in a rectangular brass box; through which a current of dry air or other gas was continuously passed to avoid the danger of radio-active contamination. The box was placed between the poles of a large electromagnet, so that the uniform field was parallel to the plane of the plates and perpendicular to their length. A distance piece, D, of length 1.2 cm, was added between the source and end of the slits, in order to increase the amount of deflexion of the radiation issuing from the slits. The zinc sulphide screen, S, was placed on a glass plate covering the end of the box. The distance between the source and the screen was 7.4 cm The recoil atoms from oxygen or nitrogen of range 9 cm could be stopped by inserting an aluminium screen of stopping power about 2 cm of air placed at the end of the slits.

With such deep slits it was impossible to bend the wide beam of radiation to the sides, but the amount of deflexion of the radiation issuing near the bottom of the slit was measured. For this purpose it was essential to observe the scintillations at a fixed point of the screen near M. The method of fixing the position of the counting microscope was as follows: The source, C, was placed in position, and the air exhausted to a pressure of a few centimetres. Without the field, the bottom edge of the beam was fixed by the straight line PM cutting the screen at M. The microscope was adjusted so that the boundary line of scintillations appeared above the horizontal cross wire in the microscope, marking the centre of the field.

On exciting the magnet to bend the rays upward (called the + field), the path of the limiting α-particles is marked by the curve $PLRN$ cutting the screen at N, so that the boundary of the scintillations appears to be displaced downwards in the field of view. On reversing the field (called the − field), the path of the limiting α-particle $PQRT$ cuts the screen at T, and the band of scintillations appears to be bent upwards. The strength of the magnetic field was adjusted so that, with a negative field, the scintillations were observed all over the screen, while, with a positive field, they were mainly confined below the cross wire. The appearance in the field of view of the microscope for the two fields is illustrated in Fig. 20.2, where the dots represent approximately the density of distribution of the scintillations. The horizontal boundaries of the field of view were given by a rectangular opening in a plate fixed in the position of the cross wires. A horizontal wire, which bisected the field of view, was visible under the conditions of counting, and allowed the relative numbers of scintillations in the two halves of the field to be counted if required. Since the number of scintillations in the actual experiments with nitrogen was much too small to mark directly the boundary of the scintillations, in order to estimate the bending of the rays, it was necessary to determine the ratio of the number of scintillations with the + and − field.

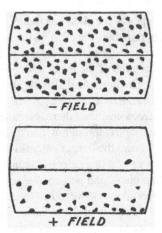

The position of the microscope and the strength of the magnetic field were in most experiments so adjusted that this ratio was about one-third. Preliminary observations showed that this ratio was sensitive to changes of the field and it thus afforded a suitable method for estimating the relative bending of any radiations under examination.

After the position of the microscope was fixed, air was let in, and a continuous flow of dry air maintained through the apparatus. The absorbing screen was introduced at E to stop the atoms from N and O of range 9 cm. The number of scintillations was then systematically counted for the two directions of the field, and a correction, if required, made for any slight radioactive contamination of the screen. The deflexion due to the unknown radiation was directly compared with that produced by a known radiation of α-rays. For this purpose, after removal of the source and absorbing screen, a similar plate, coated with a weak distribution of the active deposit of thorium, was substituted for the radium source. The α-particles from thorium C of range 8.6 cm produced bright scintillations in the screen after traversing the 7.4 cm of air in their path. The ratio of the number of scintillations with + and − fields was determined as before.

An example of such comparison is given below. For a current of 4.0 A through the electromagnet, the ratio for particles from nitrogen was found to be 0.33. The corresponding ratio for α-particles from thorium C was 0.44 for a current of 4 A and 0.31 for a current of 5 A It is thus seen that on the average, the particles from nitrogen are more bent in a given field than the α-particles from thorium C. In order, however, to make a quantitative comparison, it is necessary to take into account the reduction in velocity of the radiations in passing through the air. The value $\frac{mu}{e}$ for the α-ray of range 8.6 cm from thorium C is known to be 4.28×10^5. Since the rays pass through 7.4 cm of air in a uniform field before striking the screen, it can be calculated that the actual deflection corresponds to α-rays in a vacuum for which $\frac{mu}{e} = 3.7 \times 10^5$, about. Taking the deflection of the α-particles for a current of 4.8 A to be the same as for the nitrogen particles for a field of 4 A—ratio of fields 1.17—it is seen that the average deflexion of the nitrogen particles under the experimental conditions corresponds to a radiation in a vacuum for which the value of $\frac{mu}{e} = 3.1 \times 10^5$.

Bearing in mind that the particles under examination are produced throughout the volume of the gas between the slits, and that their distribution is unknown, and also that the particles are shot forward on an average at an angle with the incident α-particles, the experimental data are quite insufficient to calculate the average value of $\frac{mu}{e}$ to be expected under the experimental conditions for any assumed mass of projected particles. It seems probable that the majority of the particles which produce scintillations are generated in the first few centimetres of the air next the source. The actual deflection of a given particle by the magnetic field will depend on the distance of its point of origin from the source. These factors will obviously tend to make the average deflection of the particles to appear less than if they were all expelled with constant velocity from the source itself. Assuming that the correction for reduction of velocity of the long range particles in traversing the gas is 10 %, the average value of $\frac{mu}{e}$ is about 3.4×10^5. Since the value of $\frac{MV}{E}$ for the α-particle from radium C is 3.98×10^5, it is seen that under the experimental conditions the average value of $\frac{mu}{e}$ for the nitrogen particles is less than that of the α-particles which produce them.

From the data given earlier in the paper, this should only be true if the particles are comparable in mass with an atom of hydrogen, for singly charged particles of mass 2, 3, or 4 should suffer less deflexion than the α-particles. For example, if we assume that the particles were helium atoms carrying one charge, we should expect them to be deflected to about one-half of the extent of the α-particle. The experimental results thus afford strong presumptive evidence that the particles liberated from nitrogen are atoms of hydrogen.

A far more decisive test, however, can be made by comparing the deflexion of the nitrogen particles with that of H atoms under similar conditions. For this purpose, a mixture of about one volume of hydrogen to two of carbon dioxide was stored in a gas-holder and circulated in place of air through the testing apparatus. The proportions of the two gases were so adjusted that the stopping power of the mixture for α-rays was equal to that of air. Under these conditions, the H atoms, like the nitrogen particles, are produced throughout the volume of the gas, and probably the relative distribution of H atoms along the path of the α-rays is not very different from that of the nitrogen particles under examination. If the nitrogen particles are H atoms, we should expect the average deflexion to be nearly the same as for the H atoms liberated from the hydrogen mixture. A number of careful experiments showed that the ratio of the number of scintillations in + and − fields of equal value was so nearly identical in the two cases that the experiments were unable to distinguish between them. Since the two experiments were carried out under as nearly as possible identical conditions, the equality of the ratio shows that the long range particles liberated from nitrogen are atoms of hydrogen. The possibility that the particles may be of mass 2, 3, or 4 is definitely excluded.

In a previous paper I have given evidence that the long range particles observed in dry air and pure nitrogen must arise from the nitrogen atoms themselves. It is thus clear that some of the nitrogen atoms are disintegrated by their collision with swift α-particles and that swift atoms of positively charged hydrogen are expelled. It is to be inferred that the charged atom of hydrogen is one of the components of which the nucleus of nitrogen is built up.

While it has long been known that helium is a product of the spontaneous trans-formation of some of the radio-active elements, the possibility of disintegrating the structure of stable atoms by artificial methods has been a matter of uncertainty. This is the first time that evidence has been obtained that hydrogen is one of the components of the nitrogen nucleus.

It should be borne in mind that the amount of disintegration effected in nitrogen by the particles is excessively small, for probably on an average only one α-particle in about 300,000 is able to get near enough to the nitrogen nucleus to liberate the atom of hydrogen with sufficient energy to be detected by the scintillation method. Even if the whole α-radiation from 1 g of radium were absorbed in nitrogen gas, the volume of hydrogen set free would be only about 1/300,000 of the volume of helium due to the collected α-particles, *viz.*, about 5×10^{-4} cubic mm per year. It may be possible that the collision of an α-particle is effective in liberating the hydrogen from the nucleus without necessarily giving it sufficient velocity to be detected by scintillations. If this should prove to be the case, the amount of disintegration may be much greater than the value given above.

20.2.5   *Experiments with Solid Nitrogen Compounds*

A brief account will now be given of experiments with solid nitrogen compounds. Since the liberation of the particle from nitrogen is a purely atomic phenomenon, it was to be expected that similar particles would be liberated from nitrogen com-pounds in number proportional to the amount of nitrogen. To test this point, and also the nature of the particles, a number of compounds rich in nitrogen were examined. For this purpose I have employed the following substances, which were prepared as carefully as possible to exclude the presence of hydrogen in any form:—

(1) Boron nitride, kindly prepared for me by W. J. Shutt, in Manchester University.
(2) Sodium nitride, titanium nitride and para-cyanogen, kindly prepared for me by Sir William Pope and his assistants.

The apparatus used was similar in form to that given in Fig. 20.1, except that the plates were 4 cm long. By means of a fine gauze, the powdered material was sifted as uniformly as possible on a thin aluminium plate about 2 cm^2 in area. The weight of the aluminium plate was about 6 mg/cm^2, and usually about 4–5 mg of the material per square centimetre was used. The stopping power of the aluminium plate for α-particles corresponded to about 3.4 cm of air, and it was usually arranged that the average stopping power of the material was about the same as for the aluminium.

In order to make the material adhere tightly to the plate, a layer of alcohol was first brushed on and the material rapidly sifted into position, and the plate then dried. Experiment showed that no detectable hydrogen contamination was introduced by the use of alcohol in this way. The zinc sulphide screen was placed outside the box close to an aluminium plate of stopping power equal to 5.2 cm of air which covered an opening in the end of the brass box. The aluminium carrier was then placed in position to cover the end of the slits near the source, care being taken not to shake off any material. The air was exhausted and the number of scintillations on the screen counted.

(1) With material facing the source.
(2) With plate reversed.

In the former case, the α-particles were fired directly into the material under examination; in the latter case the α-particles only fell on the material when their range was reduced to about half, when their power of liberating swift atoms is much reduced. This method of reversal had the great advantage that no correction was necessary for unequal absorption of the α-particles from the source indifferent experiments.

In this way it was found that all the nitrogen compounds examined gave a larger number of scintillations in position (1). The nature of these particles was examined by a method similar to that employed in the case of nitrogen and a direct comparison was made of the deflexion of the particles with that of H atoms liberated from a film of paraffin put in place of the nitrogen compound. In all experiments, the particles were found to be deflected to the same degree as H atoms from the paraffin and no trace of particles of mass 2, 3 or 4 was detected.

For films of equal average stopping power for α-rays, it can readily be calculated from Bragg's rule that the relative stopping power of the nitrogen in the compounds is 0.67 for BN, 0.74 for C_2N_2, 0.40 for titanium nitride, taking the stopping power of sodium nitride as unity. Since the expulsion of long range nitrogen particles must be an atomic phenomenon, it was to be expected that the number of scintillations under the experimental conditions, after correction for the natural effect from the source, should be proportional to the relative values of stopping power given above. The observations with sodium nitride and titanium nitride were very consistent and the number of long range nitrogen particles was in the right proportion and about the same as that to be expected from the experiments with nitrogen gas. On the other hand, boron nitride and para-cyanogen gave between 1.5 and 2 times the number to be expected theoretically. In these experiments every precaution was taken to get rid of hydrogen and water vapour. Before use, the aluminium plates were heated in an exhausted quartz tube in an electric furnace nearly to its melting point to get rid of hydrogen and other gases. The films under examination were kept in a dessicator and heated in the electric furnace just before use and transferred at once to the testing vessel. Several control experiments were made, using preparations not containing nitrogen, *viz.*, pure graphite and silica which had been kindly prepared for me by Sir William Pope. In both of these cases, the number of scintillations observed with the material facing the α-rays was actually less than when the plate was reversed. This showed that some H atoms were liberated by the α-rays from the heated aluminium. The control experiments were thus very satisfactory in showing that H atoms were not present in materials not containing nitrogen. Incidentally, they show that H atoms do not arise in appreciable numbers from carbon, silicon, or oxygen.

The increased effect in boron nitride and para-cyanogen naturally led to the suspicion that these preparations contained some hydrogen although every precaution was taken to avoid such a possibility. In the case of boron nitride there is also the uncertainty whether boron itself emits H atoms. This point has not yet been properly examined. On account of these uncertainties, experiments on solid nitrogen compounds were abandoned for the time, and experiments already described made directly on gaseous nitrogen.

It is of interest to note that a considerable contamination with hydrogen is required to produce the number of H atoms observed in these compounds. In the case of sodium nitride at least 50 cc of hydrogen must be present per gram of material. I am inclined to think that the H atoms liberated by the α-rays from sodium nitride is due mainly, if not entirely to the nitrogen, and in the case of para-cyanogen, part of the effect is probably due to the presence of hydrogen or other hydrogen compound. It is hoped to examine this question in more detail later.

20.3 Study Questions

QUES. 20.1. What is the charge and the size of the atomic nucleus?

a) What law(s) governs the scattering of α-particles from an atomic nucleus? Where is the nuclear charge concentrated? How do you know?
b) Do all atoms with the same atomic number have the same atomic weight? What, in particular, is the relationship between nuclear charge and atomic number? Do Moseley's measurements of atomic x-ray spectra confirm or refute Van den Broek's assertions?
c) How can an atom maintain its overall neutrality? Where does Rutherford believe the atomic electrons reside?
d) How can the size of the atomic nucleus be measured? What limits the precision of such measurements? And upon what assumptions or theories do such a measurement rely?
e) What governs the chemical properties of the atom? What governs the stability and structure of the atomic nucleus? And under what conditions might nuclear disintegration occur?

QUES. 20.2. What happens when nitrogen and oxygen are bombarded with α-particles?

a) Describe Rutherford's experimental apparatus. How did it work? In particular, how was the magnetic field applied, and in which direction? And how were the recoil atoms detected?

Fig. 20.3 Before (*left*) and after (*right*) a nuclear fission reaction

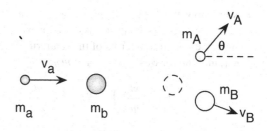

b) When the chamber was evacuated, what was the trajectory of the α particles with and without the magnetic field turned on?

c) On what basis does Rutherford determine that the long-range recoil particles, obtained after introducing dry nitrogen into the chamber, were *not* helium atoms? What were they? And where did they come from?

d) What was the nature of the short-range recoil atoms which he detected after introducing air into the chamber? In particular, how do they differ from the helium atoms which commonly arise from radioactive decay?

20.4 Exercises

Ex. 20.1 (ELASTIC HEAD-ON COLLISION). Suppose an α-particle moving at speed v collides head-on with an initially stationary hydrogen nucleus. Using the laws of conservation of energy and momentum, find the speed of the hydrogen particle after the collision. Is the value given by Rutherford on page 284 correct? What is the speed of the α-particle after the collision?

Ex. 20.2 (NUCLEAR REACTION Q-VALUE). Rutherford found that a high-speed α-particle can induce the fission of a target nitrogen nucleus. Consider such a nuclear fission reaction, depicted schematically in Fig. 20.3. In this problem, we will determine the mass of one of the fission products, m_B, based on the known masses of the incoming particle, m_a, the target particle, m_b, and the other fission product, m_A. But first we must find the Q-value of this nuclear reaction. The Q-value is defined as the difference between the kinetic energies of the products and the reactants:

$$Q = (K_A + K_B) - (K_a + K_b). \tag{20.1}$$

a) First, using the conservation of energy, demonstrate that the Q-value of this reaction may be expressed in terms of the difference between the rest energies of the reactants and the products:

$$Q = (m_a + m_b)c^2 - (m_A + m_B)c^2. \tag{20.2}$$

The Q-value is thus a measure of the amount of rest energy which is transformed into kinetic energy. A positive Q-value denotes an *exothermic* reaction; a negative Q-value denotes an *endothermic* reaction.

b) Now show that when all the particles' kinetic energies are small compared to their rest energies (semi-classical approximation), the Q-value of the reaction may be expressed in terms of the scattering angle, θ, the kinetic energies K_a and K_A, and the masses m_a, m_A and m_B:

$$Q = K_A \left(1 + \frac{m_A}{m_B}\right) - K_a \left(1 - \frac{m_a}{m_B}\right) - \frac{2\cos\theta}{m_B}\sqrt{K_a K_A m_a m_A}. \quad (20.3)$$

(HINT: Use the non-relativistic formula $K = p^2/2m$ to eliminate K_B in Eq. 20.1. Then use momentum conservation to write p_B in terms of p_a, p_A and θ.)

c) Now reconsider Rutherford's induced fission reaction:

$$^4_2\text{He} + ^{14}_7\text{N} \longrightarrow ^1_1\text{H} + ^{17}_8\text{O}$$

Suppose the incident α-particle has an energy of 7.69 MeV, and the produced hydrogen atom scatters at an angle of 90°. If the hydrogen atom is found to have an energy of 4.44 MeV, then what is the Q-value of this nuclear fission reaction? Is it endothermic or exothermic?

20.5   Vocabulary

1. Constitution
2. Hyperbolic
3. X-ray
4. Unity
5. Distinguishable
6. Isotope
7. Mode
8. Predominant
9. Impinge
10. Scintillation
11. Ascribe
12. Inherent
13. Liberate
14. Aperture
15. Presumptive
16. Infer
17. Spontaneous
18. Disintegrate
19. Adhere
20. Paraffin
21. Dessicator
22. Graphite

Chapter 21
Nuclear Structure

> *In considering the possible constitution of the elements, it is natural to suppose that they are built up ultimately of hydrogen nuclei and electrons.*
>
> —Ernest Rutherford

21.1 Introduction

In the first half of his 1920 Bakerian Lecture on the *Nuclear Constitution of Atoms*, Rutherford described in detail his experiments in which he fired high-speed α-particles across a chamber filled with dry nitrogen gas. On the opposite side of the chamber occasional scintillations from a zinc-sulphide detector screen signalled the arrival of tiny particles. Surprisingly, these particles did *not* have the known characteristics of α-particles. Rather, they had the charge and mass of hydrogen ions, despite the fact that every effort had been made to rid the gas of any hydrogen contamination. From this Rutherford concluded that "some of the nitrogen atoms are disintegrated by their collision with swift α-particles and that swift atoms of positively charged hydrogen are expelled." He also went on to say "this is the first time that evidence has been obtained that hydrogen is one of the components of the nitrogen nucleus." Now in the following reading selection—the second half of the same Bakerian Lecture—Rutherford continues to explore the artificial disintegration of atomic nuclei with high-speed α-particles. What new isotope does he discover? How does this discovery help to shape his theory of the internal structure of the atomic nucleus? And what new elementary particle does Rutherford postulate in the course of his analysis?

21.2 Reading: Rutherford, *Nuclear Constitution of Atoms*

Rutherford, E., Bakerian Lecture: Nuclear Constitution of Atoms, *Proceedings of the Royal Society of London. Series A, Containing Papers of a Mathematical and Physical Character*, 97(686), 374–400, 1920.

© Springer International Publishing Switzerland 2016
K. Kuehn, *A Student's Guide Through the Great Physics Texts*,
Undergraduate Lecture Notes in Physics, DOI 10.1007/978-3-319-21828-1_21

21.2.1 *Short Range Atoms from Oxygen and Nitrogen*

In addition to the long range H atoms liberated from nitrogen, the passage of α-particles through oxygen as well as through nitrogen gives rise to much more numerous swift atoms, which have a range in air of about 9.0 cm compared with that of 7.0 cm for the colliding α-particles. The method of determining the range and number of these atoms has been explained in a previous paper.[1] It is there shown that these projected atoms arise from the passage of the α-particles through the gas. Just beyond the range of the α-particles from radium C, the scintillations are much brighter than those due to H atoms, and more resemble α-particles.

In the absence of definite information as to the nature of these atoms, it was provisionally assumed that they were atoms of oxygen or nitrogen carrying a single charge set in swift motion by close collisions with α-particles, for the observed range of these particles was in approximate accord with that calculated on these assumptions. At the same time it was pointed out that the agreement of the ranges of the atoms set free in N and O was rather surprising, for it was to be anticipated that the range of the swifter N atoms should be about 19 % greater than for the slower O atoms. The possibility that these swift atoms might prove to be fragments of disintegrated atoms was always present, but up till quite recently, I did not see any method of settling the question.[2]

As soon as the use of wide slits had proved successful in deciding the nature of the long range particles from nitrogen, experiments were made with the same apparatus and method to test the nature of the short range particles in O and N.

First consider the relative deflexion to be expected for an O atom which is set in motion by a direct impact with an α-particle. The velocity of the O atom after the collision is $\frac{2}{5}V$, where V is the velocity of the incident α-particle. The value of $\frac{mu}{e}$ for the O atom carrying a single charge is easily seen to be 3.1 times that of the α-particle before impact. Consequently the O atom with a single charge should be much more difficult to deflect than the α-particle, and this is the case even if the former carries two charges.

To test these points, the apparatus was the same as that shown in Fig. 20.1. The source was 7.4 cm distant from the zinc sulphide screen, the end pieces, 1.2 cm long, being used as before to increase the deflexion of the rays. During an experiment, dried air or oxygen was circulated slowly through the apparatus to avoid radio-active contamination of the screen. In the case of oxygen, the scintillations observed on the screen were due to the O atoms with a small proportion of H atoms from the source. In the case of air, the scintillations on the screen were due partly to N atoms, some O atoms, and H atoms from the source and nitrogen. The actual number of short range N atoms appeared to be less than the number of O atoms under similar conditions.

---

[1] Rutherford, 'Phil. Mag.,' vol. 37, III, p. 571 (1919).

[2] Mr. G.S. Fulcher, of the National Research Council, U.S:A., sent me, in November, 1919, a suggestion that these atoms might prove to be α-particles.

The position of the microscope was fixed as before to give a convenient ratio for the number of scintillations on reversing the magnetic field. This ratio varied with the position of the microscope, and in the actual experiments had values between 0.2 and 0.4.

It was at once obvious that the atoms from O instead of being *less* deflected than the α-particles, as they should be if they were O atoms, were *more* deflected. This at once excluded the possibility that the atoms from oxygen were actual atoms of oxygen carrying either one or two charges. Since helium is expelled in so many radio-active changes, it might be expected to be one of the components of light atoms, liberated by the intense collision. The deflexion of the atoms from O was, however, much too large to be accounted for in this way. To test this point, at the conclusion of the experiments with oxygen, a plate which had been exposed to thorium emanation was substituted for the radium source, and the bending of the rays of range 8.6 cm from thorium C was examined in a similar way. If an α-particle were ejected from an O atom near the source, it would be bent like an α-particle of range 9.0 cm; if produced near the end of the range of α-rays, the amount of bending could not be more than for an α-particle of range 7.0 cm, *i.e.*, about 9 % more than in the first case. Even supposing the particles were liberated uniformly along the path of the α-rays and moved in the same line as the colliding particle, the average bending would not differ by 5 % from that of the α-particle from thorium C. If, as seems probable, some of the atoms are liberated at an angle with the incident particles, the average amount of bending of the beam would be less than the above, and in all probability less than for the α-particles from thorium C. Actually the bending observed was about 20 % greater, showing that the hypothesis that the atoms from O are charged atoms of helium is quite untenable.

If the atoms from O were H atoms, they would be more bent than the α-particles, but would have a maximum range of 28 cm instead of the 9.0 cm observed. It thus seemed clear from this evidence that the atom must be of mass intermediate between 1 and 4, while from consideration of the range of the particles and their amount of deflexion it was clear that the atom carried 2 units of charge.

In order to make a more decisive test, the deflexion of O atoms in a positive and negative field of given value was directly compared with the deflexion of H atoms from a mixture of hydrogen and carbonic acid, in the ratio of about 1–2 in volume. In order to absorb completely the O atoms from CO_2, aluminium foil was placed over the zinc sulphide screen, so that the total absorption between the source and screen corresponded to slightly more than 9 cm of air. In both experiments, the atoms under examination are produced in the gas between the slits, and probably the relative distribution along the path of the α-rays is not markedly different in the two cases.

The ratios for reversing the field in the two experiments were found to be nearly equal; but, as an average of a number of experiments, the H atoms were slightly more bent than the atoms from O. From a number of experiments it was concluded that the difference in deflexion did not on the average amount to more than 5 %, although from the nature of the observations it was difficult to fix the difference with any certainty.

From these data and the range of the atoms from O in air, we can deduce the mass of the particle liberated from oxygen.

Let $m = $ mass of the atom from O,

                                        $u = $ its maximum velocity near the source,

                                        $E = $ charge

Let M, V, E be the corresponding values for the incident α-particles and m', u', e the values for the H atoms liberated close to the source.

Taking into account that the particle from O of range 9 cm is steadily reduced in velocity in passing through the 7.4 cm of oxygen between the source and screen, it can easily be calculated that its average deflexion by the magnetic field is proportional to $1.14 \frac{E}{mu}$ in place of $\frac{E}{mu}$ in a vacuum.

In a similar way, the deflexion of the H atom is proportional to $1.05 \frac{e}{m'u'}$, the correction in this case for change of velocity being smaller, and estimated to be about 5 %. Now we have seen that the experimental results showed that the atoms from O were bent about 5 %, less than the H atoms. Consequently

$$1.14 \frac{E}{mu} = \frac{1.05}{1.05} \frac{e}{m'u'} = 1.25 \frac{E}{MV},$$

or

$$1.14 \, MV = 1.25 \, mu, \tag{21.1}$$

since it has been calculated and verified by experiment that the deflexion of the H atom in a magnetic field is 1.25 times that of the α-particle which sets it in motion (see Paper II, *loc. cit.*). Also in a previous paper, III, I have given reasons for believing that the range x of mass m and initial velocity u, carrying a double charge, is given by

$$\frac{x}{R} = \frac{m}{M} \left(\frac{u}{V}\right)^3, \tag{21.2}$$

where R is the range of the α-particle of mass M and velocity V. Since $x = 9.0$ cm for the atoms from O set in motion by collision with α-particles from radium C of range 7 cm,

$$\frac{x}{R} = 1.29, \tag{21.3}$$

and taking $M = 4$

$$mu^3 = 5.16 \, V^3,$$

A formula of this type has been shown to account for the range of the H atom, and there is every reason to believe it is fairly accurate over such a short difference of range. From (21.1) and (21.3)

$$u = 1.19 \, V,$$
$$m = 3.1.$$

Considering the difficulty of obtaining accurate data, the value $m = 3.1$ indicates that the atom has a mass about 3 and this value will be taken as the probable value in later discussions.

When air was substituted for oxygen it was not possible to distinguish any difference between the bending of the short range atoms in the two cases. Since the short range atoms from air arise mainly from the nitrogen, we may consequently conclude that the short range atoms liberated by the passage of particles through oxygen or nitrogen consist of atoms of mass 3, carrying a double charge, and initially projected with a velocity $1.19 \, V$, where V is the velocity of the colliding α-particle.

There seems to be no escape from the conclusion that these atoms of mass 3 are liberated from the atoms of oxygen or nitrogen as a result of an intense collision with an α-particle. It is thus reasonable to suppose that atoms of mass 3 are constituents of the structure of the nuclei of the atoms of both oxygen and nitrogen. We have shown earlier in the paper that hydrogen is also one of the constituents of the nitrogen nucleus. It is thus clear that the nitrogen nucleus can be disintegrated in two ways, one by the expulsion of an H atom and the other by the expulsion of an atom of mass 3 carrying two charges. Since now these atoms of mass 3 are 5–10 times as numerous as the H atoms, it appears that these two forms of disintegration are independent and not simultaneous. From the rareness of the collisions it is highly improbable that a single atom undergoes both types of disintegration.

Since the particles ejected from O and N are not produced at the source, but along the path of the α-particles, it is difficult to determine their mass and velocity with the precision desired. To overcome this drawback, attempts were made to determine the deflection of O atoms released from a mica plate placed over the source. In consequence of hydrogen in combination in the mica, the H atoms falling on the screen were so numerous compared with the O particles, and their deflexion under the experimental conditions so nearly alike, that it was difficult to distinguish between them.

21.2.2   Energy Considerations

In close collisions between an α-particle and an atom, the laws of conservation of energy and of momentum appear to hold,[3] but, in cases where the atoms are

---

[3] Rutherford, 'Phil. Mag.,' vol. 37, p. 562 (1919).

disintegrated, we should not necessarily expect these laws to be valid, unless we are able to take into account the change of energy and momentum of the atom in consequence of its disintegration.

In the case of the ejection of a hydrogen atom from the nitrogen nucleus, the data available are insufficient, for we do not know with certainty either the velocity of the H atom or the velocity of the α-particle after the collision.

If we are correct in supposing that atoms of mass 3 are liberated from O and N atoms, it can be easily calculated that there is a slight gain of energy as a result of the disintegration. If the mass is 3 exactly, the velocity of escape of the atom is 1.20 V, where V is the velocity of the impinging α-particle.

Thus $\quad\quad \dfrac{\text{energy of liberated atom}}{\text{energy of } \alpha\text{-particle}} = \dfrac{3 \times 1.44}{4} = 1.08,$

or there is a gain of 8 % in energy of motion, even though we disregard entirely the subsequent motion of the disintegrated nucleus or of the colliding α-particle. This extra energy must be derived from the nitrogen or oxygen nucleus in the same way that the α-particle gains energy of motion in escaping from the radio-active atom.

For the purpose of calculation, consider a direct collision between an α-particle and an atom of mass 3. The velocity of the latter is $\frac{8}{7}V$, where V is the velocity of the α-particle, and its energy is 0.96 of the initial energy of the α-particle. No doubt, in the actual case of a collision with the O or N atom, in which the atom of mass 3 is liberated, the α-particle comes under the influence of the main field of the nucleus, as well as of that of the part mass 3 immediately in its path. Under such conditions, it is not to be expected that the α-particle can give 0.96 of its energy to the escaping atom, but the latter acquires additional energy due to the repulsive field of the nucleus.

In our ignorance of the constitution of the nuclei and the nature of the forces in their immediate neighbourhood, it is not desirable to enter into speculations as to the mechanism of the collision at this stage, but it may be possible to obtain further information by a study of the trails of α-particles through oxygen or nitrogen by the well-known expansion method of C.T.R. Wilson. In a previous paper,[4] I discussed the photograph obtained by Mr. Wilson, in which there is a sudden change of 43° in the direction of the trail, with the appearance of a short spur at the fork. Evidence was given that the relative length of the tracks of the α-particle and of the spur were in rough accord with the view that the spur was due to the recoiling oxygen atom. This is quite probably the case, for the general evidence shows that the atoms of mass 3, after liberation, travel nearly in the direction of the α-particle, and an oblique collision may not result in the disintegration of the atom.

Recently, Dr. Shimizu, of the Cavendish Laboratory, has devised a modification of the Wilson expansion apparatus, in which expansions can be periodically produced several times a second, so that the trails of many particles can be inspected in a reasonable time. Under these conditions, both Shimizu and myself saw on several

---

[4] Rutherford, 'Phil. Mag.,' vol. 37, p. 577 (1919).

occasions what appeared to be branching trails of an α-particle in which the lengths of the two tracks were comparable. Eye observations of this kind are too uncertain to regard them with much confidence, so arrangements are being made by Mr. Shimizu to obtain photographs, so that the tracks can be examined in detail at leisure. In this way we may hope to obtain valuable information as to the conditions which determine the disintegration of the atoms, and on the relative energy communicated to the three systems involved, *viz.*, the α-particle, the escaping atom, and the residual nucleus.

So far no definite information is available as to the energy of the α-particle required to produce disintegration, but the general evidence indicates that fast α-particles, of range about 7 cm in air, are more effective than α-particles of range about 4 cm. This may not be connected directly with the actual energy required to effect the disintegration of the atom itself, but rather to the inability of the slower α-particle under the repulsive field to approach close enough to the nucleus to be effective in disrupting it. Possibly the actual energy required to disintegrate the atom is small compared with the energy of the α-particle.

If this be the case, it may be possible for other agents of less energy than the α-particle to effect the disintegration. For example, a swift electron may reach the nucleus with sufficient energy to cause its disintegration, for it moves in an attractive and not a repulsive field as in the case of the α-particle. Similarly, a penetrating γ-ray may have sufficient energy to cause disintegration. It is thus of great importance to test whether oxygen or nitrogen or other elements can he disintegrated under the action of swift cathode rays generated in a vacuum tube. In the case of oxygen and nitrogen, this could be tested simply by observing whether a spectrum closely resembling helium is given by the gas in the tube, after an intense bombardment of a suitable substance, by electrons. Experiments of this type are being undertaken by Dr. Ishida in the Cavendish Laboratory, every precaution being taken by the heating of the vacuum tube of special glass and electrodes to a high temperature to ensure the removal of any occluded helium which may be initially in the material. Helium has previously been observed by several investigators in vacuum tubes and is known to be released from substances by bombardment with cathode rays. The proof of the actual production of helium in such cases is exceedingly difficult, but the recent improvements in vacuum tube technique may make it easier to give a decisive answer to this important question.

21.2.3   Properties of the New Atom

We have shown that atoms of mass about 3 carrying two positive charges are liberated by α-particles both from nitrogen and oxygen, and it is natural to suppose that these atoms are independent units in the structure of both gases. Since probably the charged atom during its flight is the nucleus of a new atom without any external electrons, we should anticipate that the new atom when it has gained two negative electrons should have physical and chemical properties very nearly identical with those of helium, but with a mass 3 instead of 4. We should anticipate that the spectrum of helium and this isotope should be nearly the same, but on account of the marked difference in the relative masses of the nuclei, the displacement of the lines should be much greater than in the case of isotopes of heavy elements like lead.

It will be remembered that Bourget, Fabry, and Buisson,[5] from an examination of the width of the lines in the spectrum of nebulæ, conclude that the spectrum arises from an element of atomic mass about 2.7 or 3 in round numbers. It is difficult, however, on modern views to suppose that the spectrum of the so-called "nebulium" can be due to an element of nuclear charge 2 unless the spectrum under the conditions existing in nebulæ are very different from those observed in the laboratory. The possible origin of the spectrum of nebulium has been discussed at length by Nicholson[6] on quite other lines, and it is not easy at the moment to see how the new atoms from oxygen or nitrogen can be connected with the nebular material.

Since probably most of the helium in use is derived, either directly or indirectly, from the transformation of radio-active materials, and these, as far as we know, always give rise to helium of mass 4, the presence of an isotope of helium of mass 3 is not likely to be detected in such sources. It would, however, be of great interest to examine whether the isotope may be present in cases where the apparent presence of helium is difficult to connect with radio-active material; for example, in beryl, drawn attention to by Strutt.[7] This is based on the assumption that the atom of mass 3 is stable. The fact that it survives the intense disturbance of its structure due to a close collision with an α-particle is an indication that it is a structure difficult to disintegrate by external forces.

21.2.4 Constitution of Nuclei and Isotopes

In considering the possible constitution of the elements, it is natural to suppose that they are built up ultimately of hydrogen nuclei and electrons. On this view the helium nucleus is composed of four hydrogen nuclei and two negative electrons with a resultant charge of two. The fact that the mass of the helium atom 3.997 in terms of oxygen 16 is less than the mass of four hydrogen atoms, viz., 4.032, has been generally supposed to be due to the close interaction of the fields in the nucleus resulting in a smaller electromagnetic mass than the sum of the masses of the individual components. Sommerfeld[8] has concluded from this fact that the helium nucleus must be a very stable structure which would require intense forces to disrupt it. Such a conclusion is in agreement with experiment, for no evidence has been obtained to

[5] Bourget, Fabry and Buisson, 'C.R.,' April 6, May 18 (1914).

[6] Nicholson, 'Roy. Ast. Soc.,' vol. 72, No. 1, p. 49 (1911)'; vol. 74, No. 7, p. 623 (1914).

[7] Strutt, 'Roy. Soc. Proc.,' A, vol. 80, p. 572 (1908).

[8] Sommerfeld, 'Atombau und Spektrallinien,' p. 538. Vieweg and Son, 1919.

show that helium can be disintegrated by the swift α-particles which are able to disrupt the nuclei of nitrogen and oxygen. In his recent experiments on the isotopes of ordinary elements Aston[9] has shown that within the limit of experimental accuracy the masses of all the isotopes examined are given by whole numbers when oxygen is taken as 16. The only exception is hydrogen, which has a mass 1.008 in agreement with chemical observations. This does not exclude the probability that hydrogen is the ultimate constituent or which nuclei are composed, but indicates that either the grouping or the hydrogen nuclei and electrons is such that the average electromagnetic mass is nearly 1, or, what is more probable, that the secondary units, or which the atom is mainly built up, e.g., helium or its isotope, have a mass given nearly by a whole number when O is 16.

The experimental observations made so far are unable to settle whether the new atom has a mass exactly 3, but from the analogy with helium we may expect the nucleus of the new atom to consist of three H nuclei and one electron, and to have a mass more nearly 3 than the sum of the individual masses in the free state.

If we are correct in this assumption it seems very likely that one electron can also bind two H nuclei and possibly also one H nucleus. In the one case, this entails the possible existence of an atom of mass nearly 2 carrying one charge, which is to be regarded as an isotope of hydrogen. In the other case, it involves the idea of the possible existence of an atom of mass 1 which has zero nucleus charge. Such an atomic structure seems by no means impossible. On present views, the neutral hydrogen atom is regarded as a nucleus or unit charge with an electron attached at a distance, and the spectrum of hydrogen is ascribed to the movements or this distant electron. Under some conditions, however, it may be possible for an electron to combine much more closely with the H nucleus, forming a kind of neutral doublet. Such an atom would have very novel properties. Its external field would be practically zero, except very close to the nucleus, and in consequence it should be able to move freely through matter. Its presence would probably be difficult to detect by the spectroscope, and it may be impossible to contain it in a sealed vessel. On the other hand, it should enter readily the structure of atoms, and may either unite with the nucleus or be disintegrated by its intense field, resulting possibly in the escape of a charged H atom or an electron or both.

If the existence of such atoms be possible, it is to be expected that they may be produced, but probably only in very small numbers, in the electric discharge through hydrogen, where both electrons and H nuclei are present in considerable numbers. It is the intention of the writer to make experiments to test whether any indication of the production of such atoms can be obtained under these conditions.

The existence of such nuclei may not be confined to mass 1 but may be possible for masses 2, 3, or 4, or more, depending on the possibility of combination between the doublets. The existence of such atoms seems almost necessary to explain the building up of the nuclei of heavy elements; for unless we suppose the production of charged particles of very high velocities it is difficult to see how any positively

---

[9] Aston, 'Phil. Mag.,' December, 1919: April and May, 1920.

Fig. 21.1 Three possible isotopes of lithium built up from
electrons and light nuclei.—
[*K.K.*]

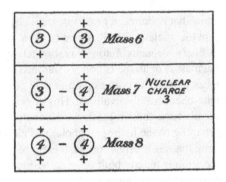

charged particle can reach the nucleus of a heavy atom against its intense repulsive
field.

We have seen that so far the nuclei of three light atoms have been recognised
experimentally as probable units of atomic structure, *viz.*,

$$\overset{+}{H_1}, \quad \overset{++}{X_3}, \quad \overset{++}{He_4}$$

where the subscript represents the mass of the element.

In considering the possible ways in which nuclei can be built up, difficulties at
once arise, for many combinations of these units with negative electrons are possible to give an element of the required nuclear charge and mass. In our complete
ignorance of the laws of force close to the nuclei, no criterion is available as to
the stability or relative probability of the theoretical systems. With the exception
of a few elements which can exist in the gaseous state, the possible isotopes of the
elements have not yet been settled. When further information is available as to the
products of the disintegration of other elements than the two so far examined, and
more complete data have been obtained as to the number and mass of the isotopes,
it may be possible to deduce approximate rules which may serve as a guide to the
mode in which the nuclei are built up from the simpler units. For these reasons it
seems premature at this stage to attempt to discuss with any detail the possible structure of even the lighter and presumably less complex atoms. It may, however, be of
some interest to give an example to illustrate a possible method of the formation of
isotopes in the case of the lighter elements. This is based on the view that probably
in many cases a helium nucleus of mass 4 may be substituted in the complex structure for the corresponding nucleus of mass 3 without seriously interfering with the
stability of the system. In such a case, the nuclear charge remains unchanged but the
masses differ by unity.

For example, take the case of lithium of nuclear charge 3 and atomic mass about
7. It is natural to suppose that the nucleus is composed of helium or its isotope of
mass 3 with one binding electron. The three possible combinations are shown in
Fig. 21.1.

On this view, at least three isotopes of lithium of mass 6, 7, and 8 are theoretically
probable, but even if the combinations were equally stable, the question of their

relative abundance in the element lithium on the earth will be dependent on many factors of which we know nothing; for example, the mode of actual formation of such nuclei, the relative amount of the combining units present, and the probability of their combinations.

The experimental results given in the paper support, as far as they go, the view that the atoms of hydrogen and of mass 3 are important units in the nuclear structure of nitrogen and oxygen. In the latter case, one could a priori have supposed that oxygen was in some way a combination of four helium nuclei of mass 4. It seems probable that the mass 3 is an important unit of the nuclei of light atoms in general, but it is not unlikely, with increasing complexity of the nuclei and corresponding increase of the electric field, the structures of mass 3 suffer a rearrangement and tend to revert to the presumably more stable nucleus of mass 4. This may be the reason why helium of mass 4 always appears to be expelled from the radio-active atoms, while the isotope of mass 3 arises in the artificial disintegration of lighter atoms like oxygen and nitrogen. It has long been known that for many of the elements the atomic weights can be expressed by the formula $4n$ or $4n + 3$, where n is an integer, suggesting that atoms of mass 3 and 4 are important units of the structure of nuclei.[10]

21.2.5   Structure of Carbon, Oxygen, and Nitrogen Nuclei

In the light of the present experiments, it may be of interest to give some idea, however crude, of the possible formation of the above atoms to account for the experimental facts. It will be remembered that nitrogen alone gives rise to H atoms while carbon and oxygen do not. Both nitrogen and oxygen give rise to atoms of mass 3, while carbon has not yet been investigated from this point of view. A possible structure is shown in Fig. 21.2 when the masses and charges of the combining units are indicated. Negative electrons are represented by the symbol $-$.

The carbon nucleus is taken to consist of four atoms of mass 3 and charge 2, and two binding electrons. The change to nitrogen is represented by the addition of two H atoms with a binding electron and an oxygen nucleus by the substitution of a helium nucleus in place of the two H atoms.

We can see from this type of structure that the chance of a direct collision with one of the four atoms of mass 3 in nitrogen is much greater than the chance of removing an H atom, for it is to be anticipated that the main nucleus would screen the H atom from a direct collision except in restricted regions facing the H atoms. This serves to illustrate why the number of H atoms of mass 3 liberated

---

[10] From these and other considerations, Harkins ('Phys. Rev.,' vol. 15, p. 73 (1920)) has proposed a constitutional formula for all the elements. The combining units employed by him are electrons and atoms of mass 1, 3, and 4 of nuclear charges 1, 1 and 2, respectively. The unit of mass 3 is taken by him to have a nucleus charge of 1 and not 2, and is thus to be regarded as an isotope of hydrogen and not an isotope of helium.

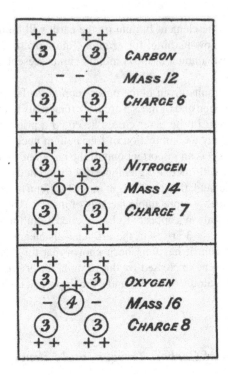

Fig. 21.2 Hypothetical structure of carbon, nitrogen and oxygen nuclei.—[*K.K.*]

from nitrogen should be much greater than the number of *H* atoms released under corresponding conditions. It should be borne in mind that the structures outlined are purely illustrative and no importance is attached to the particular arrangement employed.

It is natural to inquire as to the nature of the residual atoms after the disintegration of oxygen and nitrogen, supposing that they survive the collision and sink into a new stage of temporary or permanent equilibrium.

The expulsion of an *H* atom carrying one charge from nitrogen should lower the mass by 1 and the nuclear charge by 1. The residual nucleus should thus have a nuclear charge 6 and mass 13, and should be an isotope of carbon. If a negative electron is released at the same time, the residual atom becomes an isotope of nitrogen.

The expulsion of a mass 3 carrying two charges from nitrogen, probably quite independent of the release of the H atom, lowers the nuclear charge by 2 and the mass by 3. The residual atom should thus be an isotope of boron of nuclear charge 5 and mass 11. If an electron escapes as well, there remains an isotope of carbon of mass 11. The expulsion of a mass 3 from oxygen gives rise to a mass 13 of nuclear charge 6, which should be an isotope of carbon. In case of the loss of an electron as well, there remains an isotope of nitrogen of mass 13. The data at present available are quite insufficient to distinguish between these alternatives.

It is intended to continue experiments, to test whether any evidence can be obtained of the disintegration of other light atoms besides nitrogen and oxygen. The problem is more difficult in the case of elements which cannot be conveniently obtained in the gaseous state, since it is not an easy matter to ensure the absence of hydrogen or to prepare uniform thin films of such substances. For these reasons, and the strain involved in counting scintillations under difficult conditions, further progress is not likely to be rapid.

I am indebted to my assistant, G.A.R. Crowe, for the preparation of the radio-active sources and his help in counting; also to Mr. J. Chadwick and Dr. Ishida for assistance in counting scintillations in some of the later experiments.

21.3 Study Questions

QUES. 21.1. What are the properties of the new short-range particles liberated from nitrogen and oxygen?

a) How were the newly discovered particles produced? How are they different than the previously discovered "long-range" particles from nitrogen?
b) How did Rutherford measure the mass and charge of these new short-range particles? In particular, how did he prove that they were not singly (or doubly) charged oxygen atoms, helium atoms, or hydrogen atoms?
c) Are both momentum and mechanical energy conserved during every atomic collision? Why might some α-particles be more effective in producing nuclear disintegrations than others?
d) Is the newly discovered short-range particle stable? Is it as abundant as the helium-4 isotope? Does Rutherford anticipate that its spectrum will be the same as that of helium-4?

QUES. 21.2. What are atomic nuclei made of?

a) Can two isotopes of different masses have the same nuclear charge? If so, how? If atomic nuclei contain (positively charged) hydrogen nuclei, then what holds them together?
b) Is the atomic mass of every isotope an integral multiple of the the mass of a hydrogen nucleus? More generally, is the mass of an atomic nucleus equal to the sum of the masses of its components?
c) What new type of particle does Rutherford postulate? What is his motivation for considering the existence of such a particle? Why might such a particle be very difficult to detect?
d) What difficulty arises when trying to ascertain the units from which atomic nuclei might be built up? What additional information is required? What strategy does Rutherford (tentatively) suggest?
e) How, according to Rutherford's strategy, might one form a lithium atom? How many isotopes does it have? And what dictates the relative abundance of the various isotopes which might be formed?

f) How might one form a nitrogen atom? What are the structures of carbon, oxygen and nitrogen nuclei? What considerations go into constructing hypothetical
models such for these?

g) What renders the study of artificial disintegration more difficult in heavy
elements than in, say nitrogen and oxygen gas?

21.4   Exercises

Ex. 21.1 (DISINTEGRATION OF NITROGEN). Suppose an α-particle moving at speed
V strikes a nitrogen nucleus, causing it to eject a particle of mass 3 at speed $1.20\,V$.
Is kinetic energy conserved during this collision? If not, why not? Disregarding the
subsequent motion of the residual nucleus and of the α-particle, does your result
agree with Rutherford's calculation on p. 300?

Ex. 21.2 (RUTHERFORD'S NUCLEAR MODEL). Rutherford considers the possibility
that all atomic nuclei are built up of $^4_2$He, $^3_2$He and $^1_1$H particles. The masses of
these are found (by mass spectrometry) to be $4.00260\,u$, $3.0160\,u$ and $1.007823\,u$,
respectively. The *atomic mass unit*, u is defined as $\frac{1}{12}$ the atomic mass of a $^{12}_6$C atom.
According to this model, nuclear electrons serve both to bind the nucleus together
and to maintain the correct atomic number. The binding energy, ΔE, of the atomic
nucleus can be computed from the mass-deficit—the difference between the mass of
the nucleus and the mass of its constituents. What, then, would be the binding energy
of a carbon-12 isotope if the nucleus were comprised of (a) twelve $^1_1$H particles,
(b) three $^4_2$He particles, or (c) four $^3_2$He particles. Which of these compositions would
have the highest energy per nuclear bond between constituent particles?

21.5   Vocabulary

1. Provisional
2. Disintegrate
3. γ-ray
4. Cathode ray
5. Spectrum
6. Occlude
7. Isotope
8. Nebulium
9. Doublet
10. Spectroscope
11. Unity
12. Lithium

Chapter 22
The Discovery of the Neutron

> In order to explain the great penetrating power of the radiation
> we must further assume that the particle has no net charge.
> —James Chadwick

22.1 Introduction

In the course of Rutherford's effort to develop a plausible model of the atomic
nucleus, he postulated the existence of a particle having the same mass as the
hydrogen nucleus but having no electrical charge whatsoever.[1] He conceived of this
neutral particle—essentially a bound-state of a proton and an electron—in order to
account for isotopes having identical atomic charges but different atomic weights
(*e.g.* helium-3 and helium-4). Just a few years later, Rutherford's hypothetical particle was discovered by one of his students. James Chadwick (1891–1974) was born
in Cheshire, England. He attended the Manchester Municipal Secondary School
before entering Manchester University in 1908. He graduated from the honors program in physics in 1911, but remained at the university to study radioactivity under
Ernest Rutherford. His early work focused on the absorption of gamma-rays—
highly penetrating radiation which was produced during certain radioactive decays
and which could be detected using a Geiger counter. He earned his M.Sc. degree
in 1913, then traveled to Berlin to work under Hans Geiger. Professor Geiger had
just returned to Germany from Manchester to accept a position at the Physikalisch-
Technische Reichsanstalt. With the outbreak of the First World War, Chadwick was
interned in Germany. After the war, he return to England in 1918 to work with
Rutherford, who had moved to the Cavendish Laboratory in Cambridge. Chadwick
earned his Ph.D. in 1921, and in 1923 was appointed Assistant Director of Research
under Rutherford at the Cavendish Laboratory. It was around this time that Rutherford succeeded in artificially disintegrating nitrogen nuclei by bombarding them
with high-speed α-particles emitted by radioactive substances. Chadwick assisted
Rutherford in studying the artificial transmutation of light elements for several

---

[1] See Rutherford's *Bakerian Lecture* on the *Nuclear Constitution of Atoms* in Chap. 21 of the
present volume.

© Springer International Publishing Switzerland 2016
K. Kuehn, *A Student's Guide Through the Great Physics Texts*,
Undergraduate Lecture Notes in Physics, DOI 10.1007/978-3-319-21828-1_22

years, and in 1930 the two co-authored a book entitled *Radiations from Radioactive Substances*. During the second world war, Chadwick led a team of British scientists which collaborated with the Americans on the Manhattan project in Los Alamos. Chadwick went on to become a leading advocate for the development of Britain's very own atomic weapons program. He was subsequently knighted by King George VI in 1945.

In the reading selection that follows, Chadwick describes the 1932 discovery for which he is most famous, and for which he was awarded the Hughes Medal from the Royal Society in 1932 and the Nobel Prize for physics in 1935. He begins by describing the work of Bothe and Becker, who had recently detected very penetrating radiation emitted when certain light elements, such as beryllium and boron, were bombarded with α-rays from a polonium source. This radiation was initially thought to be gamma rays.

22.2 Reading: Chadwick, *The Existence of a Neutron*

Chadwick, J., The Existence of a Neutron, *Proceedings of the Royal Society of London. Series A, Containing Papers of a Mathematical and Physical Character*, 136(830), 692–708, 1932.

It was shown by Bothe and Becker[2] that some light elements when bombarded by α-particles of polonium emit radiations which appear to be of the γ-ray type. The element beryllium gave a particularly marked effect of this kind, and later observations by Bothe, by Mme. Curie-Joliot[3] and by Webster[4] showed that the radiation excited in beryllium possessed a penetrating power distinctly greater than that of any γ-radiation yet found from the radioactive elements. In Webster's experiments the intensity of the radiation was measured both by means of the Geiger-Müller tube counter and in a high pressure ionisation chamber. He found that the beryllium radiation had an absorption coefficient in lead of about 0.22 cm^{-1} as measured under his experimental conditions. Making the necessary corrections for these conditions, and using the results of Gray and Tarrant to estimate the relative contributions of scattering, photoelectric absorption, and nuclear absorption in the absorption of such penetrating radiation, Webster concluded that the radiation had a quantum energy of about 7×10^6 electron volts. Similarly he found that the radiation from boron bombarded by α-particles of polonium consisted in part of a radiation rather more penetrating than that from beryllium, and he estimated the quantum energy of this component as about 10×10^6 electron volts. These conclusions agree quite well

---

[2] 'Z. Physik,' vol. 66, p. 289 (1930).

[3] I. Curie. 'C. R. Acad. Sci. Paris,' vol. 193, p. 1412 (1931).

[4] 'Proc. Roy. Soc.,' A, vol. 136, p. 428 (1932).

with the supposition that the radiations arise by the capture of the α-particle into the beryllium (or boron) nucleus and the emission of the surplus energy as a quantum of radiation.

The radiations showed, however, certain peculiarities, and at my request the beryllium radiation was passed into an expansion chamber and several photographs were taken. No unexpected phenomena were observed though, as will be seen later, similar experiments have now revealed some rather striking events. The failure of these early experiments was partly due to the weakness of the available source of polonium, and partly to the experimental arrangement, which, as it now appears, was not very suitable.

Quite recently, Mme. Curie-Joliot and M. Joliot[5] made the very striking observation that these radiations from beryllium and from boron were able to eject protons with considerable velocities from matter containing hydrogen. In their experiments the radiation from beryllium was passed through a thin window into an ionisation vessel containing air at room pressure. When paraffin wax, or other matter containing hydrogen, was placed in front of the window, the ionisation in the vessel was increased, in some cases as much as doubled. The effect appeared to be due to the ejection of protons, and from further experiment they showed that the protons had ranges in air up to about 26 cm, corresponding to a velocity of nearly 3×10^9 cm/s. They suggested that energy was transferred from the beryllium radiation to the proton by a process similar to the Compton effect with electrons, and they estimated that the beryllium radiation had a quantum energy of about 50×10^6 electron volts. The range of the protons ejected by the boron radiation was estimated to be about 8 cm in air, giving on a Compton process an energy of about 35×10^6 electron volts for the effective quantum.[6]

There are two grave difficulties in such an explanation of this phenomenon. Firstly, it is now well established that the frequency of scattering of high energy quanta by electrons is given with fair accuracy by the Klein-Nishina formula, and this formula should also apply to the scattering of quanta by a proton. The observed frequency of the proton scattering is, however, many thousand times greater than that predicted by this formula. Secondly, it is difficult to account for the production of a quantum of 50×10^6 electron volts from the interaction of a beryllium nucleus and an α-particle of kinetic energy of 5×10^6 electron volts. The process which will give the greatest amount of energy available for radiation is the capture of the α-particle by the beryllium nucleus, Be^{11}, and its incorporation in the nuclear structure to form a carbon nucleus C^{13}. The mass defect of the O^{13} nucleus is known both from data supplied by measurements of the artificial disintegration of boron B^{10} and from observations of the band spectrum of carbon; it is about 10×10^6 electron volts. The mass defect of Be^9 is not known, but the assumption that it is zero will give a maximum value for the possible change of energy in the reaction

---

[5] Curie and Joliot, 'O. R. Acad. Sci. Paris,' vol. 194, p. 273 (1932).

[6] Many of the arguments of the subsequent discussion apply equally to both radiations, and the term "beryllium radiation" may often be taken to include the boron radiation.

$Be^9 + \alpha \rightarrow C^{13} +$ quantum. On this assumption it follows that the energy of the quantum emitted in such a reaction cannot be greater than about 14×10^6 electron volts. It must, of course, be admitted that this argument from mass defects is based on the hypothesis that the nuclei are made as far as possible of α-particles; that the Be^9 nucleus consists of 2 α-particles + 1 proton + 1 electron and the C^{13} nucleus of 3 α-particles + 1 proton + 1 electron. So far as the lighter nuclei are concerned, this assumption is supported by the evidence from experiments on artificial disintegration, but there is no general proof.

Accordingly, I made further experiments to examine the properties of the radiation excited in beryllium. It was found that the radiation ejects particles not only from hydrogen but from all other light elements which were examined. The experimental results were very difficult to explain on the hypothesis that the beryllium radiation was a quantum radiation, but followed immediately if it were supposed that the radiation consisted of particles of mass nearly equal to that of a proton and with no net charge, or neutrons. A short statement of some of these observations was published in 'Nature.'[7] This paper contains a fuller description of the experiments, which suggest the existence of neutrons and from which some of the properties of these particles can be inferred. In the succeeding paper Dr. Feather will give an account of some observations by means of the expansion chamber of the collisions between the beryllium radiation and nitrogen nuclei, and this is followed by an account by Mr. Dee of experiments to observe the collisions with electrons.

22.2.1   Observations of Recoil Atoms

The properties of the beryllium radiation were first examined by means of the valve counter used in the work[8] on the artificial disintegration by α-particles and described fully there. Briefly, it consists of a small ionisation chamber connected to a valve amplifier. The sudden production of ions in the chamber by the entry of an ionising particle is detected by means of an oscillograph connected in the output circuit of the amplifier. The deflections of the oscillograph were recorded photographically on a film of bromide paper.

The source of polonium was prepared from a solution of radium $(D + E + F)^9$ by deposition on a disc of silver. The disc had a diameter of 1 cm and was placed close to a disc of pure beryllium of 2 cm diameter, and both were enclosed in a small vessel which could be evacuated, Fig. 22.1. The first ionisation chamber used had an opening of 13 mm covered with aluminium foil of 4.5 cm air equivalent, and a

---

[7] 'Nature,' vol. 129, p. 312 (1932).

[8] Chadwick, Constable and Pollard, 'Proc. Roy. Soc.,' A, vol. 130, p. 463 (1931).

[9] The radium D was obtained from old radon tubes generously presented by Dr. C. F. Burnam and Dr. F. West, of the Kelly Hospital, Baltimore.

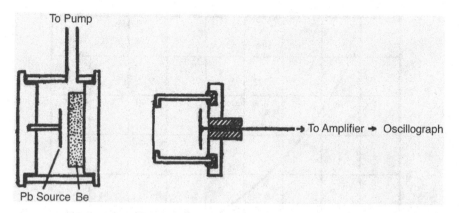

Fig. 22.1 A beryllium radiation source (*left*) and an ionization detector (*right*).—[*K.K.*]

depth of 15 mm. This chamber had a very low natural effect, giving on the average only about seven deflections per hour.

When the source vessel was placed in front of the ionisation chamber, the number of deflections immediately increased. For a distance of 3 cm between the beryllium and the counter the number of deflections was nearly 4 per minute. Since the number of deflections remained sensibly the same when thick metal sheets, even as much as 2 cm of lead, were interposed between the source vessel and the counter, it was clear that these deflections were due to a penetrating radiation emitted from the beryllium. It will be shown later that the deflections were due to atoms of nitrogen set in motion by the impact of the beryllium radiation.

When a sheet of paraffin wax about 2 mm thick was interposed in the path of the radiation just in front of the counter, the number of deflections recorded by the oscillograph increased markedly. This increase was due to particles ejected from the paraffin wax so as to pass into the counter. By placing absorbing screens of aluminium between the wax and the counter the absorption curve shown in Fig. 22.2, curve *A*, was obtained. From this curve it appears that the particles have a maximum range of just over 40 cm of air, assuming that an Al foil of 1.64 mg/cm^2 is equivalent to 1 cm of air. By comparing the sizes of the deflections (proportional to the number of ions produced in the chamber) due to these particles with those due to protons of about the same range it was obvious that the particles were protons. From the range-velocity curve for protons we deduce therefore that the maximum velocity imparted to a proton by the beryllium radiation is about 3.3×10^9 cm/s, corresponding to an energy of about 5.7×10^6 electron volts.

The effect of exposing other elements to the beryllium radiation was then investigated. An ionisation chamber was used with an opening covered with a gold foil of 0.5 mm air equivalent. The element to be examined was fixed on a clean brass plate and placed very close to the counter opening. In this way lithium, beryllium, boron, carbon and nitrogen, as paracyanogen, were tested. In each case the number of deflections observed in the counter increased when the element was bombarded

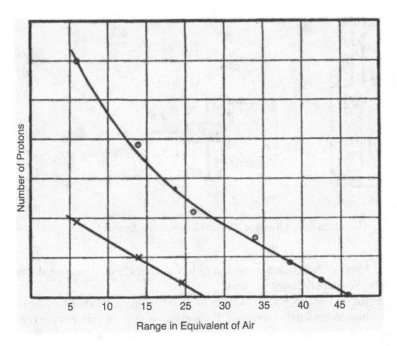

Fig. 22.2 Absorption curves for protons ejected from paraffin wax by beryllium radiation in the same (*A*) and in the opposite (*B*) direction as the incident α-particles from the polonium source.— [*K.K.*]

by the beryllium radiation. The ranges of the particles ejected from these elements were quite short, of the order of some millimetres in air. The deflections produced by them were of different sizes, but many of them were large compared with the deflection produced even by a slow proton. The particles therefore have a large ionising power and are probably in each case recoil atoms of the elements. Gases were investigated by filling the ionisation chamber with the required gas by circulation for several minutes. Hydrogen, helium, nitrogen, oxygen, and argon were examined in this way. Again, in each case deflections were observed which were attributed to the production of recoil atoms in the different gases. For a given position of the beryllium source relative to the counter, the number of recoil atoms was roughly the same for each gas. This point will be referred to later. It appears then that the beryllium radiation can impart energy to the atoms of matter through which it passes and that the chance of an energy transfer does not vary widely from one element to another.

It has been shown that protons are ejected from paraffin wax with energies up to a maximum of about 5.7×10^6 electron volts. If the ejection be ascribed to a Compton recoil from a quantum of radiation, then the energy of the quantum must be about 55×10^6 electron volts, for the maximum energy which can be given to a mass m by a quantum $h\nu$ is $\frac{2}{2+mc^2/h\nu} \cdot h\nu$. The energies of the recoil atoms produced by this radiation by the same process in other elements can be readily calculated.

For example, the nitrogen recoil atoms should have energies up to a maximum of 450,000 electron volts. Taking the energy necessary to form a pair of ions in air as 35 electron volts, the recoil atoms of nitrogen should produce not more than about 13,000 pairs of ions. Many of the deflections observed with nitrogen, however, corresponded to far more ions than this; some of the recoil atoms produced from 30,000 to 40,000 ion pairs. In the case of the other elements a similar discrepancy was noted between the observed energies and ranges of the recoil atoms and the values calculated on the assumption that the atoms were set in motion by recoil from a quantum of 55×10^6 electron volts. The energies of the recoil atoms were estimated from the number of ions produced in the counter, as given by the size of the oscillograph deflections. A sufficiently good measurement of the ranges could be made either by varying the distance between the element and the counter or by interposing thin screens of gold between the element and the counter.

The nitrogen recoil atoms were also examined, in collaboration with Dr. N. Feather, by means of the expansion chamber. The source vessel was placed immediately above an expansion chamber of the Shimizu type, so that a large proportion of the beryllium radiation traversed the chamber. A large number of recoil tracks was observed in the course of a few hours. Their range, estimated by eye, was sometimes as much as 5 or 6 mm in the chamber, or, correcting for the expansion, about 3 mm in standard air. These visual estimates were confirmed by a preliminary series of experiments by Dr. Feather with a large automatic expansion chamber, in which photographs of the recoil tracks in nitrogen were obtained. Now the ranges of recoil atoms of nitrogen of different velocities have been measured by Blackett and Lees. Using their results we find that the nitrogen recoil atoms produced by the beryllium radiation may have a velocity of at least 4×10^8 cm/s, corresponding to an energy of about 1.2×10^6 electron volts. In order that the nitrogen nucleus should acquire such an energy in a collision with a quantum of radiation, it is necessary to assume that the energy of the quantum should be about 90×10^6 electron volts, if energy and momentum are conserved in the collision. It has been shown that a quantum of 55×10^6 electron volts is sufficient to explain the hydrogen collisions. In general, the experimental results show that if the recoil atoms are to be explained by collision with a quantum, we must assume a larger and larger energy for the quantum as the mass of the struck atom increases.

22.2.2   The Neutron Hypothesis

It is evident that we must either relinquish the application of the conservation of energy and momentum in these collisions or adopt another hypothesis about the nature of the radiation. If we suppose that the radiation is not a quantum radiation, but consists of particles of mass very nearly equal to that of the proton, all the difficulties connected with the collisions disappear, both with regard to their frequency and to the energy transfer to different masses. In order to explain the great penetrating power of the radiation we must further assume that the particle has no net charge.

We may suppose it to consist of a proton and an electron in close combination, the "neutron" discussed by Rutherford[10] in his Bakerian Lecture of 1920.

When such neutrons pass through matter they suffer occasionally close collisions with the atomic nuclei and so give rise to the recoil atoms which are observed. Since the mass of the neutron is equal to that of the proton, the recoil atoms produced when the neutrons pass through matter containing hydrogen will have all velocities up to a maximum which is the same as the maximum velocity of the neutrons. The experiments showed that the maximum velocity of the protons ejected from paraffin wax was about 3.3×10^9 cm/s. This is therefore the maximum velocity of the neutrons emitted from beryllium bombarded by α-particles of polonium. From this we can now calculate the maximum energy which can be given by a colliding neutron to other atoms, and we find that the results are in fair agreement with the energies observed in the experiments. For example, a nitrogen atom will acquire in a head-on collision with the neutron of mass 1 and velocity 3.3×10^9 cm/s a velocity of 4.4×10^8 cm/s, corresponding to an energy of 1.4×10^6 electron volts, a range of about 3.3 mm in air, and a production of ions of about 40,000 pairs. Similarly, an argon atom may acquire an energy of 0.54×10^6 electron volts, and produce about 15,000 ion pairs. Both these values are in good accord with experiment.[11]

It is possible to prove that the mass of the neutron is roughly equal to that of the proton, by combining the evidence from the hydrogen collisions with that from the nitrogen collisions. In the succeeding paper, Feather records experiments in which about 100 tracks of nitrogen recoil atoms have been photographed in the expansion chamber. The measurement of the tracks shows that the maximum range of the recoil atoms is 3.5 mm in air at 15 °C and 760 mm pressure, corresponding to a velocity of 4.7×10^8 cm/s according to Blackett and Lees. If M, V be the mass and velocity of the neutron then the maximum velocity given to a hydrogen atom is

$$u_p = \frac{2M}{M+1} \cdot V, \tag{22.1}$$

and the maximum velocity given to a nitrogen atom is

$$u_n = \frac{2M}{M+14} \cdot V, \tag{22.2}$$

whence

$$\frac{M+14}{M+1} = \frac{u_p}{u_n} = \frac{3.3 \times 10^9}{4.7 \times 10^8}, \tag{22.3}$$

[10] Rutherford, 'Proc. Roy. Soc.,' A, vol. 97, p. 374 (1920). Experiments to detect the formation of neutrons in a hydrogen discharge tube were made by J. L. Glasson, 'Phil. Mag.,' vol. 42, p. 596 (1921), and by J. K. Roberts, 'Proc. Roy. Soc.,' A, vol. 102, p. 72 (1922). Since 1920 many experiments in search of these neutrons have been made in .this laboratory.
[11] It was noted that a few of the nitrogen recoil atoms produced about 50–60,000 ion pairs. These probably correspond to the oases of disintegration found by Feather and described in his paper.

and

$$M = 1.15 \qquad (22.4)$$

The total error in the estimation of the velocity of the nitrogen recoil atom may easily be about 10 %., and it is legitimate to conclude that the mass of the neutron is very nearly the same as the mass of the proton.

We have now to consider the production of the neutrons from beryllium by the bombardment of the α-particles. We must suppose that an α-particle is captured by a Be^9 nucleus with the formation of a carbon C^{12} nucleus and the emission of a neutron. The process is analogous to the well-known artificial disintegrations, but a neutron is emitted instead of a proton. The energy relations of this process cannot be exactly deduced, for the masses of the Be^9 nucleus and the neutron are not known accurately. It is, however, easy to show that such a process fits the experimental facts. We have

$$Be^9 + He^4 + \text{kinetic energy of } \alpha = C^{12} + n^1 + \text{kinetic energy of } C^{12}$$
$$+ \text{kinetic energy of } n^1.$$

If we assume that the beryllium nucleus consists of two α-particles and a neutron, then its mass cannot be greater than the sum of the masses of these particles, for the binding energy corresponds to a defect of mass. The energy equation becomes

$$(8.00212 + n^1) + 4.00106 + \text{K.E. of } \alpha > 12.0003 + n^1 + \text{K.E. of } n^1 \qquad (22.5)$$

or

$$\text{K.E. of } n^1 < \text{K.E. of } \alpha + 0.003 - \text{K.E. of } C^{12}. \qquad (22.6)$$

Since the kinetic energy of the α-particle of polonium is 5.25×10^6 electron volts, it follows that the energy of emission of the neutron cannot be greater than about 8×10^6 electron volts. The velocity of the neutron must therefore be less than 3.9×10^9 cm/s. We have seen that the actual maximum velocity of the neutron is about 3.3×10^9 cm/s, so that the proposed disintegration process is compatible with observation.

A further test of the neutron hypothesis was obtained by examining the radiation emitted from beryllium in the opposite direction to the bombarding α-particles. The source vessel, Fig. 22.1, was reversed so that a sheet of paraffin wax in front of the counter was exposed to the "backward" radiation from the beryllium. The maximum range of the protons ejected from the wax was determined as before, by counting the numbers of protons observed through different thicknesses of aluminium interposed between the wax and the counter.

The absorption curve obtained is shown in curve B, Fig. 22.2. The maximum range of the protons was about 22 cm in air, corresponding to a velocity of about 2.74×10^9 cm/s. Since the polonium source was only about 2 mm away from the beryllium, this velocity should be compared with that of the neutrons emitted not at 180° but at an angle not much greater than 90° to the direction of the incident α-particles. A simple calculation shows that the velocity of the neutron emitted at

90° when an α-particle of full range is captured by a beryllium nucleus should be 2.77 $\times$ 10^9 cm/s, taking the velocity of the neutron emitted at 0° in the same process as 3.3 $\times$ 10^9 cm/s. The velocity found in the above experiment should be less than this, for the angle of emission is slightly greater than 90°. The agreement with calculation is as good as can be expected from such measurements.

22.3   Study Questions

QUES. 22.1. Can beryllium radiation be identified with a quantum of light?

a) Why did Bothe and Becker initially conclude that γ-rays were produced when beryllium and boron were bombarded with α-particles? What was the problem with this conclusion?

b) What notable observation was made by Curie and Joliot when studying the properties of such beryllium radiation? What was their experimental apparatus? How did Curie and Joliot explain the ejection of protons from paraffin wax by beryllium radiation?

c) Why does Chadwick find "grave difficulties" with this conclusion? In particular, does the observed frequency of proton ejection from paraffin by the beryllium radiation match the theoretical predictions for the scattering of protons by light quanta? Also, can α-particles produce light quanta of sufficiently high-energy when striking a beryllium target?

d) Describe Chadwick's experimental apparatus in detail. How was beryllium radiation produced? How was it detected? Could it be blocked by thick metal sheets?

e) What happened when a sheet of paraffin wax was placed in front of the beryllium radiation source? Could recoil atoms be produced when other materials were subjected to the beryllium radiation? How were the energies of the recoil atoms measured?

f) If the recoiling hydrogen atoms from the paraffin have an energy of 5.7 $\times$ 10^6 eV, then how much energy would the beryllium radiation need if it is, in fact, a quantum of light? Is identifying beryllium radiation with a light quantum plausible?

g) What alternative explanation did Chadwick offer for the identity of the beryllium radiation? Why did he find this to be more plausible than the light-quantum explanation given by Curie and Joliot? Does it preserve the laws of conservation of energy and momentum?

QUES. 22.2. What is the charge of the (hypothetical) neutron? What is the mass of the neutron, and how can it be calculated from the speeds of recoiling hydrogen or nitrogen atoms?

QUES. 22.3. How might neutrons be produced when an α-particle from polonium strikes a beryllium target? What is the energy and speed of the ejected neutron? Are these values consistent with observations?

22.4 Exercises

EX. 22.1 (NEUTRON PRODUCTION).Consider the following nuclear reaction which occurs when an α-particle $(4.00260\,u)$ strikes a stationary beryllium-9 atom $(9.01219\,u)$.

$$^9_4\text{Be} + {}^4_2\text{He} \longrightarrow {}^{12}_6\text{C} + {}^1_0\text{n}$$

a) What is the maximum kinetic energy, and the velocity, of the ejected neutron $(1.00866\,u)$ if a 5.25 MeV α-particle from a polonium source is used? What is the minimum kinetic energy and velocity?
b) In which direction is the neutron ejected in each of these cases? Does one need to know the kinetic energy of the carbon-12 atom which is produced in order to solve this problem?
c) Finally, do your calculations agree with Chadwick's measurements for the velocity of forward-scattered and "backward"-scattered beryllium radiation? (HINT: see the discussion of nuclear reaction Q-values in Ex. 20.2.)

22.5 Vocabulary

1. Quantum
2. Paraffin
3. Deposition
4. Interpose
5. Relinquish
6. Neutron

Chapter 23
Neutron Scattering

The neutron should be able to penetrate the nucleus easily.
—James Chadwick

23.1 Introduction

In the previous reading selection, taken from the first half of his 1932 article on *The Existence of a Neutron*, Chadwick described his experiments with the recently discovered highly-penetrating beryllium radiation. First, a beryllium target was bombarded with α-particles from a silver disk coated with polonium. When a sheet of paraffin wax was held near the beryllium, protons were ejected from the paraffin. These protons had high enough energies that they could travel significant distances through air and still be detected by an ionization counter. Earlier investigators (Curie and Joliot) had supposed that the protons were being ejected from the paraffin by γ-particles—high-frequency light quanta—which had been emitted by the beryllium target. Chadwick rejected this hypothesis for two reasons. First, the scattering of protons from paraffin by light quanta should, according to theoretical considerations, produce far fewer protons than were actually being observed. Second, the hypothesized light quanta produced by the bombarding α-particles would not have the requisite energy to eject protons from the paraffin which could survive such long flights though air and still be detected by an ionization counter. So Chadwick instead proposed that the protons were being ejected from the paraffin by close collisions with uncharged particles whose masses are the same as a proton. These uncharged particles from the beryllium, he argued, were quite possibly the long-sought neutrons which Chadwick's mentor, Ernest Rutherford, had predicted years before. In the reading selection that follows (the second half of his 1932 article) Chadwick continues to explore the nature of the neutron and how it interacts with other materials. Neutron scattering would later become one of the most important techniques for condensed matter physics research and radioactive isotope production for both medical and industrial purposes.

© Springer International Publishing Switzerland 2016
K. Kuehn, *A Student's Guide Through the Great Physics Texts,*
Undergraduate Lecture Notes in Physics, DOI 10.1007/978-3-319-21828-1_23

23.2   Reading: Chadwick, *The Existence of a Neutron*

Chadwick, J., The Existence of a Neutron, *Proceedings of the Royal Society of London. Series A, Containing Papers of a Mathematical and Physical Character*, *136*(830), 692–708, 1932.

23.2.1   *The Nature of the Neutron*

It has been shown that the origin of the radiation from beryllium bombarded by α-particles and the behaviour of the radiation, so far as its interaction with atomic nuclei is concerned, receive a simple explanation on the assumption that the radiation consists of particles of mass nearly equal to that of the proton which have no charge. The simplest hypothesis one can make about the nature of the particle is to suppose that it consists of a proton and an electron in close combination, giving a net charge 0 and a mass which should be slightly less than the mass of the hydrogen atom. This hypothesis is supported by an examination of the evidence which can be obtained about the mass of the neutron.

As we have seen, a rough estimate of the mass of the neutron was obtained from measurements of its collisions with hydrogen and nitrogen atoms, but such measurements cannot be made with sufficient accuracy for the present purpose. We must turn to a consideration of the energy relations in a process in which a neutron is liberated from an atomic nucleus; if the masses of the atomic nuclei concerned in the process are accurately known, a good estimate of the mass of the neutron can be deduced. The mass of the beryllium nucleus has, however, not yet been measured, and, as was shown in Sect. 22.2.2, only general conclusions can be drawn from this reaction. Fortunately, there remains the case of boron. It was stated in Sect. 22.2 that boron bombarded by α-particles of polonium also emits a radiation which ejects protons from materials containing hydrogen. Further examination showed that this radiation behaves in all respects like that from beryllium, and it must therefore be assumed to consist of neutrons. It is probable that the neutrons are emitted from the isotope B^{11}, for we know that the isotope B^{10} disintegrates with the emission of a proton.[1] The process of disintegration will then be

$$B^{11} + He^4 \longrightarrow N^{14} + n^1.$$

The masses of B^{11} and N^{14} are known from Aston's measurements, and the further data required for the deduction of the mass of the neutron can be obtained by experiment.

In the source vessel of Fig. 22.1 the beryllium was replaced by a target of powdered boron, deposited on a graphite plate. The range of the protons ejected by the

---

[1] Chadwick, Constable and Pollard, *loc. cit.*

boron radiation was measured in the same way as with the beryllium radiation. The effects observed were much smaller than with beryllium, and it was difficult to measure the range of the protons accurately. The maximum range was about 16 cm in air, corresponding to a velocity of 2.5×10^9 cm/s. This then is the maximum velocity of the neutron liberated from boron by an α-particle of polonium of velocity 1.59×10^9 cm/s. Assuming that momentum is conserved in the collision, the velocity of the recoiling N^{14} nucleus can be calculated, and we then know the kinetic energies of all the particles concerned in the disintegration process. The energy equation of the process is

$$\text{Mass of } B^{11} + \text{mass of } He^4 + \text{K.E. of } He^4 = \text{mass of } N^{14} + \text{mass of } n^1$$
$$= \text{K.E. of } N^{14} + \text{K.E. of } n^1.$$

The masses are $B^{11} = 11.00825 \pm 0.0016$; $He^4 = 4.00106 \pm 0.0006$; $N^{14} = 14.0042 \pm 0.0028$. The kinetic energies in mass units are α-particle $= 0.00565$; neutron $= 0.0035$; and nitrogen nucleus $= 0.00061$. We find therefore that the mass of the neutron is 1.0067. The errors quoted for the mass measurements are those given by Aston. They are the maximum errors which can be allowed in his measurements, and the probable error may be taken as about one-quarter of these.[2] Allowing for the errors in the mass measurements it appears that the mass of the neutron cannot be less than 1.003, and that it probably lies between 1.005 and 1.008.

Such a value for the mass of the neutron is to be expected if the neutron consists of a proton and an electron, and it lends strong support to this view. Since the sum of the masses of the proton and electron is 1.0078, the binding energy, or mass defect, of the neutron is about 1 to 2 million electron volts. This is quite a reasonable value. We may suppose that the proton and electron form a small dipole, or we may take the more attractive picture of a proton embedded in an electron. On either view, we may expect the "radius" of the neutron to be a few times 10^{-13} cm.

23.2.2   The Passage of the Neutron Through Matter

The electrical field of a neutron of this kind will clearly be extremely small except at very small distances of the order of 10^{-12} cm. In its passage through matter the neutron will not be deflected unless it suffers an intimate collision with a nucleus. The potential of a neutron in the field of a nucleus may be represented roughly by Fig. 23.1. The radius of the collision area for sensible deflection of the neutron will be little greater than the radius of the nucleus. Further, the neutron should be able to penetrate the nucleus easily, and it may be that the scattering of the neutrons

---

[2] The mass of B^{11} relative to B^{10} has been checked by optical methods by Jenkins and McKellar ('Phys. Rev.,' vol. 39, p. 549 (1932)). Their value agrees with Aston's to 1 part in 10. This suggests that great confidence may be put in Aston's measurements.

Fig. 23.1 The potential of
a neutron in the field of a
nucleus.—[*K.K.*]

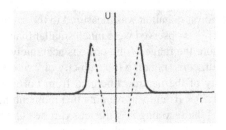

will be largely due to the internal field of the nucleus, or, in other words, that the
scattered neutrons are mainly those which have penetrated the potential barrier. On
these views we should expect the collisions of a neutron with a nucleus to occur
very seldom, and that the scattering will be roughly equal in all directions, at least
as compared with the Coulomb scattering of a charged particle.

These conclusions were confirmed in the following way. The source vessel, with
Be target, was placed rather more than 1 in. from the face of a closed counter filled
with air, Fig. 22.1. The number of deflections, or the number of nitrogen recoil atoms
produced in the chamber, was observed for a certain time. The number observed
was 190/h, after allowing for the natural effect. A block of lead 1 in. thick was then
introduced between the source vessel and the counter. The number of deflections
fell to 166/h. Since the number of recoil atoms produced must be proportional to the
number of neutrons passing through the counter, these observations show that 13 %
of the neutrons had been absorbed or scattered in passing through 1 in. of lead.

Suppose that a neutron which passes within a distance p from the centre of the
lead nucleus is scattered and removed from the beam. Then the fraction removed
from the beam in passing through a thickness t of lead will be $\pi p^2 nt$, where n is the
number of lead atoms per unit volume. Hence $\pi p^2 nt = 0.13$, and $p = 7 \times 10^{-13}$ cm.
This value for the collision radius with lead seems perhaps rather small, but it is
not unreasonable. We may compare it with the radii of the radioactive nuclei calculated from the disintegration constants by Gamow and Houtermans,[3] *viz.*, about
7×10^{-13} cm. Similar experiments were made in which the neutron radiation was
passed through blocks of brass and carbon. The values of p deduced in the same
way were 6×10^{-13} cm and 3.5×10^{-13} cm respectively.

The target areas for collision for some light elements were compared by another
method. The second ionisation chamber was used, which could be filled with different gases by circulation. The position of the source vessel was kept fixed relative to
the counter, and the number of deflections was observed when the counter was filled
in turn with hydrogen, nitrogen, oxygen, and argon. Since the number of neutrons
passing through the counter was the same in each case, the number of deflections
should be proportional to the target area for collision, neglecting the effect of the
material of the counter, and allowing for the fact that argon is monatomic. It was
found that nitrogen, oxygen, and argon gave about the same number of deflections;

---

[3] 'Z. Physik,' vol. 52, p. 453 (1928).

the target areas of nitrogen and oxygen are thus roughly equal, and the target area of argon is nearly twice that of these. With hydrogen the measurements were very difficult, for many of the deflections were very small owing to the low ionising power of the proton and the low density of the gas. It seems probable from the results that the target area of hydrogen is about two-thirds that of nitrogen or oxygen, but it may be rather greater than this.

There is as yet little information about the angular distribution of the scattered neutrons. In some experiments kindly made for me by Dr. Gray and Mr. Lea, the scattering by lead was compared in the backward and forward directions, using the ionisation in a high pressure chamber to measure the neutrons. They found that the amount of scattering was about that to be expected from the measurements quoted above, and that the intensity per unit solid angle was about the same between 300 and 900 in the forward direction as between 900 and 1500 in the backward direction. The scattering by lead is therefore not markedly anisotropic.

Two types of collision may prove to be of peculiar interest, the collision of a neutron with a proton and the collision with an electron. A detailed study of these collisions with an elementary particle is of special interest, for it should provide information about the structure and field of the neutron, whereas the other collisions will depend mainly on the structure of the atomic nuclei. Some preliminary experiments by Mr. Lea, using the pressure chamber to measure the scattering of neutrons by paraffin wax and by liquid hydrogen, suggest that the collision with a proton is more frequent than with other light atoms. This is not in accord with the experiments described above, but the results are at present indecisive. These collisions can be more directly investigated by means of the expansion chamber or by counting methods, and it is hoped to do so shortly.

The collision of a neutron with an electron has been examined in two ways, by the expansion chamber and by the counter. An account of the expansion chamber experiments is given by Mr. Dee in the third paper of this series. Mr. Dee has looked for the general ionisation produced by a large number of neutrons in passing through the expansion chamber, and also for the short electron tracks which should be the result of a very close collision between a neutron and an electron. His results show that collisions with electrons are extremely rare compared even with those with nitrogen nuclei, and he estimates that a neutron can produce on the average not more than 1 ion pair in passing through 3 m of air.

In the counter experiments a beam of neutrons was passed through a block of brass, 1 in. thick, and the maximum range of the protons ejected from paraffin wax by the emergent beam was measured. From this range the maximum velocity of the neutrons after travelling through the brass is obtained and it can be compared with the maximum velocity in the incident beam. No change in the velocity of the neutrons due to their passage through the brass could be detected. The accuracy of the experiment is not high, for the estimation of the end of the range of the protons was rather difficult. The results show that the loss of energy of a neutron in passing through 1 in. of brass is not more than about 0.4×10^6 electron volts. A path of 1 in. in brass corresponds as regards electron collisions to a path of nearly 2×10^4 cm of air, so that this result would suggest that a neutron loses less than 20 V/cm path in air in electron collisions. This experiment thus lends general support to those with the expansion chamber, though it is of far inferior accuracy. We conclude that the transfer of energy from the neutron to electrons is of very rare occurrence. This is not unexpected. Bohr[4] has shown on quite general ideas that collisions of a neutron with an electron should be very few compared with nuclear collisions. Massey,[5] on plausible assumptions about the field of the neutron, has made a detailed calculation of the loss of energy to electrons, and finds also that it should be small, not more than 1 ion pair per metre in air.

23.2.3 General Remarks

It is of interest to examine whether other elements, besides beryllium and boron, emit neutrons when bombarded by α-particles. So far as experiments have been made, no case comparable with these two has been found. Some evidence was obtained of the emission of neutrons from fluorine and magnesium, but the effects were very small, rather less than 1 % of the effect obtained from beryllium under the same conditions. There is also the possibility that some elements may emit neutrons spontaneously, e.g., potassium, which is known to emit a nuclear β-radiation accompanied by a more penetrating radiation. Again no evidence was found of the presence of neutrons, and it seems fairly certain that the penetrating type is, as has been assumed, a γ-radiation.

Although there is certain evidence for the emission of neutrons only in two cases of nuclear transformations, we must nevertheless suppose that the neutron is a common constituent of atomic nuclei. We may then proceed to build up nuclei out of α-particles, neutrons and protons, and we are able to avoid the presence of uncombined electrons in a nucleus. This has certain advantages for, as is well known, the electrons in a nucleus have lost some of the properties which they have outside, e.g., their spin and magnetic moment. If the α-particle, the neutron, and the proton are the only units of nuclear structure, we can proceed to calculate the mass defect or binding energy of a nucleus as the difference between the mass of the nucleus and the sum of the masses of the constituent particles. It is, however, by no means certain that the α-particle and the neutron are the only complex particles in the nuclear structure, and therefore the mass defects calculated in this way may not be the true binding energies of the nuclei. In this connection it may be noted that the examples of disintegration discussed by Dr. Feather in the next paper are not all of one type, and he suggests that in some cases a particle of mass 2 and charge 1, the hydrogen isotope recently reported by Urey, Brickwedde and Murphy, may be emitted. It is indeed possible that this particle also occurs as a unit of nuclear structure.

---

[4] Bohr, Copenhagen discussions, unpublished.

[5] Massey, 'Nature,' vol. 129, p. 469, corrected p. 691 (1932).

It has so far been assumed that the neutron is a complex particle consisting of a proton and an electron. This is the simplest assumption and it is supported by the evidence that the mass of the neutron is about 1.006, just a little less than the sum of the masses of a proton and an electron. Such a neutron would appear to be the first step in the combination of the elementary particles towards the formation of a nucleus. It is obvious that this neutron may help us to visualise the building up of more complex structures, but the discussion of these matters will not be pursued further for such speculations, though not idle, are not at the moment very fruitful. It is, of course, possible to suppose that the neutron may be an elementary particle. This view has little to recommend it at present, except the possibility of explaining the statistics of such nuclei as N^{14}.

There remains to discuss the transformations which take place when an α-particle is captured by a beryllium nucleus, Be^9. The evidence given here indicates that the main type of transformation is the formation of a C^{12} nucleus and the emission of a neutron. The experiments of Curie-Joliot and Joliot,[6] of Auger,[7] and of Dee show quite definitely that there is some radiation emitted by beryllium which is able to eject fast electrons in passing through matter. I have made experiments using the Geiger point counter to investigate this radiation and the results suggest that the electrons are produced by a γ-radiation. There are two distinct processes which may give rise to such a radiation. In the first place, we may suppose that the transformation of Be^9 to C^{12} takes place sometimes with the formation of an excited C^{12} nucleus which goes to the ground state with the emission of γ-radiation. This is similar to the transformations which are supposed to occur in some cases of disintegration with proton emission, e.g., B^{10}, F^{19}, Al^{27}; the majority of transformations occur with the formation of an excited nucleus, only in about one-quarter is the final state of the residual nucleus reached in one step. We should then have two groups of neutrons of different energies and a γ-radiation of quantum energy equal to the difference in energy of the neutron groups. The quantum energy of this radiation must be less than the maximum energy of the neutrons emitted, about 5.7×10^6 electron volts. In the second place, we may suppose that occasionally the beryllium nucleus changes to a C^{13} nucleus and that all the surplus energy is emitted as radiation. In this case the quantum energy of the radiation may be about 10×10^6 electron volts.

It is of interest to note that Webster has observed a soft radiation from beryllium bombarded by polonium α-particles, of energy about 5×10^5 electron volts. This radiation may well be ascribed to the first of the two processes just discussed, and its intensity is of the right order. On the other hand, some of the electrons observed by Curie-Joliot and Joliot had energies of the order of 2–10×10^6 V, and Auger recorded one example of an electron of energy about 6.5×10^6 V. These electrons may be due to a hard γ-radiation produced by the second type of transformation.[8]

---

[6] 'C. R. Acad. Sci. Paris,' vol. 194, p. 708 and p. 876 (1932).

[7] 'C. R. Acad. Sci. Paris,' vol. 194, p. 877 (1932).

[8] Although the presence of fast electrons can be easily explained in this way, the possibility that some may be due to secondary effects of the neutrons must not be lost sight of.

It may be remarked that no electrons of greater energy than the above appear to be present. This is confirmed by an experiment[9] made in this laboratory by Dr. Occhialini. Two tube counters were placed in a horizontal plane and the number of coincidences recorded by them was observed by means of the method devised by Rossi. The beryllium source was then brought up in the plane of the counters so that the radiation passed through both counters in turn. No increase in the number of coincidences could be detected. It follows that there are few, if any, β-rays produced with energies sufficient to pass through the walls of both counters, a total of 4 mm brass; that is, with energies greater than about 6×10^6 V. This experiment further shows that the neutrons very rarely produce coincidences in tube counters under the usual conditions of experiment.

In conclusion, I may restate briefly the case for supposing that the radiation the effects of which have been examined in this paper consists of neutral particles rather than of radiation quanta. Firstly, there is no evidence from electron collisions of the presence of a radiation of such a quantum energy as is necessary to account for the nuclear collisions. Secondly, the quantum hypothesis can be sustained only by relinquishing the conservation of energy and momentum. On the other hand, the neutron hypothesis gives an immediate and simple explanation of the experimental facts; it is consistent in itself and it throws new light on the problem of nuclear structure.

23.2.4 Summary

The properties of the penetrating radiation emitted from beryllium (and boron) when bombarded by the α-particles of polonium have been examined. It is concluded that the radiation consists, not of quanta as hitherto supposed, but of neutrons, particles of mass 1, and charge 0. Evidence is given to show that the mass of the neutron is probably between 1.005 and 1.008. This suggests that the neutron consists of a proton and an electron in close combination, the binding energy being about $1–2 \times 10^6$ electron volts. From experiments on the passage of the neutrons through matter the frequency of their collisions with atomic nuclei and with electrons is discussed.

I wish to express my thanks to Mr. H. Nutt for his help in carrying out the experiments.

---

[9] *Cf.* also Rasetti, 'Naturwiss.,' vol. 20, p. 252 (1932).

23.3 Study Questions

QUES. 23.1. Is the neutron an elementary particle, a composite particle, or a quantum of radiation?

a) Do any other elements, besides beryllium and boron, produce neutrons when bombarded with α-particles?
b) What is the mass of the neutron? How can the mass of the neutron be measured, and with what precision?
c) What, according to Chadwick, is the size and the internal structure of the neutron? What experiments are particularly useful in measuring its internal structure?

QUES. 23.2. Can neutrons be effectively used to probe the structure of atomic nuclei?

a) Which are able to more closely approach an atomic nucleus before scattering, a proton, a neutron, or an electron? Why is this?
b) Is the range of protons ejected from paraffin wax by a beam of neutrons reduced by passing the beam through a block of lead? How about through a block of brass? What does this imply?
c) How can the size of a lead nucleus be estimated from neutron scattering data? How does the size of the lead nucleus compare to that of other atomic nuclei? And what renders a comparison of the sizes of hydrogen, nitrogen, oxygen and argon nuclei difficult?
d) Are neutrons more likely to collide with an atomic nucleus or with an electron? How can one study the collision of neutrons with hydrogen atoms? with electrons? Does this accord with experimental predictions?

QUES. 23.3. What are the elementary units of structure of an atomic nucleus? Are there any advantages of avoiding the possibility of uncombined electrons in the nucleus?

23.4 Exercises

EX. 23.1 (NEUTRON SHIELDING). Suppose a beam of neutrons emitted by a beryllium target (as described by Chadwick) strikes a 1 in. thick sheet of lead. What fraction of the incident neutrons is transmitted through the lead? How would this fraction change if the lead was 2 in. thick instead? How thick would the lead need to be so that no more than 1 % of the incident neutrons are transmitted?

EX. 23.2 (NUCLEAR CROSS-SECTION). As explained by Chadwick, the fraction, f, of incident particles scattered or absorbed by a target may be expressed in terms of the thickness of the target, t, the number of target atoms per unit volume, n, and the radius, p, of the target atoms:

$$f = \pi p^2 nt. \qquad (23.1)$$

The quantity πp^2 is the *cross section* for an interaction between the incident particle and a target nucleus. What is the cross-section for neutrons striking a lead target, as described in Ex. 23.1? Why do you suppose σ is typically reported in *barns*? How big is a barn?

Ex. 23.3 (URANIUM FISSION). The *cross-section* for a particular nuclear reaction may be defined in terms of the nuclear reaction rate, R, and the incident particle flux, I:

$$\sigma_s = \frac{R}{I} \qquad\qquad (23.2)$$

The incident flux is the number of particles striking the target per second per unit area. The nuclear reaction rate is the number of induced reactions per unit time per target nucleus. Generally speaking, the cross-section depends not only on the properties of the target nuclei, but also on the speed of the incident particles. For example, by slowing a beam of high-energy neutrons down to thermal velocities— velocities at which their kinetic energies are of order $k_B T$ (Boltzmann's constant times the temperature)—the likelihood that they will induce the fission of uranium-235 is greatly increased. This is partly due to the fact that slow neutrons spend more time in the vicinity of each uranium-235 nucleus, increasing the likelihood that they will be captured, producing an unstable uranium-236 nucleus. This heavier uranium isotope then breaks apart into two lighter nuclei, and also produced a (7 MeV) γ-particle and two (2 MeV) neutrons:

$$n + {}^{235}_{92}\text{U} \longrightarrow {}^{236}_{92}\text{U}^* \longrightarrow {}^{89}_{36}\text{Kr} + {}^{144}_{56}\text{Ba} + \gamma + 2n$$

Suppose that the cross-section for uranium-235 fission by (low-energy) thermal neutrons is about 580 barns. What incident neutron flux is necessary to produce one fission per second? If the kinetic energy of the neutron beam is about 1 electron-volt, then what density of neutrons (neutrons per cubic centimeter) must be provided in the beam to produce this reaction rate? Notably, if one or more of the neutrons produced by the fission reaction described above can be made to induce another fission reaction, then one might produce a fission "chain reaction." What do you suppose makes it difficult to sustain such a chain reaction?

23.5   Vocabulary

1. Monatomic
2. Anisotropic
3. Speculation
4. Ascribe

Chapter 24
X-Ray Diffraction

*Whatever we may find regarding the nature of x-rays, it would
take a bold man indeed to suggest, in light of these experiments,
that they differ in nature from ordinary light.*

—Arthur Holly Compton

24.1 Introduction

Arthur Holly Compton (1892–1962) was born in Wooster, Ohio. He earned his
Bachelor of Science degree in 1913 from the College of Wooster, where is father
served as Dean and Professor of Philosophy. He earned his Master of Arts degree in
1914, and his Ph.D. in 1916, both from Princeton University. He went on to serve
as an instructor of physics for a year at the University of Minnesota, then for a short
time as a research engineer at Westinghouse Lamp Company in Pittsburgh. In 1919
he was appointed a National Research Council Fellow and studied at Cambridge
University. His subsequent academic appointments included positions at Washington University as the head of the physics department and later as Chancellor, and at
the University of Chicago as Professor of Physics.

Compton's academic career was largely devoted to the study of x-rays and their
interaction with matter. X-rays had been discovered in 1895 by William Röntgen
while passing an electrical current through an evacuated glass *Crookes tube.* The
tube, he surmised, was emitting invisible rays which could penetrate a heavy black
cardboard shield and cause a distant screen to glow with a fluorescent light.[1] Almost
immediately, the usefulness of these new highly-penetrating rays was recognized,
and they were soon employed for medical diagnostic purposes. In fact, the earliest
x-ray images are of Röntgen's wife's hand—revealing her skeleton beneath a thin
shadow of flesh. The true nature of x-rays, however, remained a mystery until 1912,
when Max von Laue discovered that they exhibit a diffraction pattern when passed
through a copper sulfate crystal. This confirmed that the mysterious x-rays were
indeed electromagnetic waves with lengths comparable to the atomic spacing of the

---

[1] An English translation by Arthur Stanton of Röntgen's 1895 German publication can be found in
Röntgen, W. C., On a New Kind of Rays, *Nature*, *3*(59), 277–231, 1896.

© Springer International Publishing Switzerland 2016
K. Kuehn, *A Student's Guide Through the Great Physics Texts,*
Undergraduate Lecture Notes in Physics, DOI 10.1007/978-3-319-21828-1_24

crystal.[2] Inspired by von Laue's work, William Henry Bragg and his son, William Lawrence Bragg, carried out a systematic investigation of crystal structures using the new technique of x-ray diffraction.[3]

But are these x-rays truly waves? Since Einstein's 1905 publication, theoretical and experimental evidence was beginning to suggest that light behaves—at least in certain situations—like discrete packets of energy.[4] Could this quantum theory of light be extended to x-rays? And if so, how could it be experimentally verified? These are the issues which Compton addresses in the reading selection contained in the next two chapters.

24.2   Reading: Compton, *X-Rays as a Branch of Optics*

Compton, A. H., X-Rays as a Branch of Optics, *Journal of the Optical Society of America, 16*(2), 71–86, 1928.

One of the most fascinating aspects of recent physics research has been the gradual extension of the familiar laws of optics to the very high frequencies of x-rays, until at present there is hardly a phenomenon in the realm of light whose parallel is not found in the realm of x-rays. Reflection, refraction, diffuse scattering, polarization, diffraction, emission and absorption spectra, photoelectric effect, all of the essential characteristics of light have been found also to be characteristic of x-rays. At the same time it has been found that some of these phenomena undergo a gradual change as we proceed to the extreme frequencies of x-rays, and as a result of these interesting changes in the laws of optics we have gained new information regarding the nature of light.

It has not always been recognized that x-rays is a branch of optics. As a result of the early studies of Röntgen and his followers it was concluded that x-rays could not be reflected or refracted, that they were not polarized on traversing crystals, and that they showed no signs of diffraction on passing through narrow slits. In fact, about the only property which they were found to possess in common with light was that of propagation in straight lines. Many will recall also the heated debate between Barkla and Bragg, as late as 1910, one defending the idea that x-rays are waves like light, the other that they consist of streams of little bullets called "neutrons." It is a debate on which the last word has not yet been said!

---

[2] Laue, M., Concerning the detection of X-ray interferences, in *Nobel Lectures, Physics 1901–1921*, Elsevier Publishing Company, 1914.

[3] Lawrence, B. W., The diffraction of X-rays by crystals, in *Nobel Lectures, Physics 1901–1921*, Elsevier Publishing Company, 1915.

[4] For Einstein's introduction of the photon concept, see Chap. 16 of the present volume.

24.2.1   The Refraction and Reflection of X-Rays

We should consider the phenomena of refraction and reflection as one problem, since it is a well known law of optics that reflection can occur only from a boundary surface between two media of different indices of refraction. If one is found, the other must be present.

In his original examination of the properties of x-rays, Röntgen[5] tried unsuccessfully to obtain refraction by means of prisms of a variety of materials such as ebonite, aluminium and water. Perhaps the experiment of this type most favorable for detecting refraction was one by Barkla.[6] In this work x-rays of a wave length which excited strongly the characteristic K radiation from bromine were passed through a crystal of potassium bromide. The precision of his experiment was such that he was able to conclude that the refractive index for a wave length of 0.5A probably differed from unity by less than 5 parts in a million.

Although these direct tests for refraction of x-rays were unsuccessful, Stenström observed[7] that for x-rays whose wave lengths are greater than about 3A, reflected from crystals of sugar and gypsum, Bragg's law, $n\lambda = 2D \sin \theta$, does not give accurately the angles of reflection. He interpreted the difference as due to an appreciable refraction of the x-rays as they enter the crystal. Measurements by Duane and Siegbahn and their collaboraters[8] showed that discrepancies of the same type occur, though they are very small indeed, when ordinary x-rays are reflected from calcite.

The direction of the deviations in Stenström's experiments indicated that the index of refraction of the crystals employed was less than unity. If this is the case also for other substances, total reflection should occur when x-rays in air strike a polished surface at a sufficiently sharp glancing angle, just as light in a glass prism is totally reflected from a surface between the glass and air if the light strikes the surface at a sufficiently sharp angle. From a measurement of this critical angle for total reflection, it should be possible to determine the index of refraction of the x-rays.

When the experiment was tried,[9] the results were strictly in accord with these predictions. The apparatus was set up as shown in Fig. 24.1, reflecting a very narrow sheet of x-rays from a polished mirror onto the crystal of a Bragg spectrometer. It was found that the beam could be reflected from surfaces of polished glass and silver through several minutes of arc. By studying the spectrum of the reflected beam, the critical glancing angle was found to be approximately proportional to the wave length. For ordinary x-rays whose wave length is half an Angström, the critical

---

[5] W. Röntgen, Sitzungber. der Wurzburger Phys. Med. Ges. Jahrg. 1895. These papers are reprinted in German in Ann. d. Phys., *64*, p. 1; 1898, and in English translation by A. Stanton in Science, *3*, p. 227; 1896.

[6] C. G. Barkla, Phil. Mag., *31*, p. 257; 1916.

[7] W. Stenstrom, Dissertation, Lund, 1919.

[8] Duane and Patterson, Phys. Rev., *16*, p. 532; 1920. M. Siegbahn, C. R., *173*, p. 1350; 1921; *174*, p. 745; 1922.

[9] A. H. Compton , Phil. Mag., *45*. p. 1121; 1923.

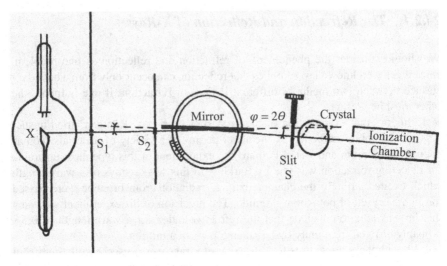

Fig. 24.1 Apparatus for studying the total reflection of x-rays

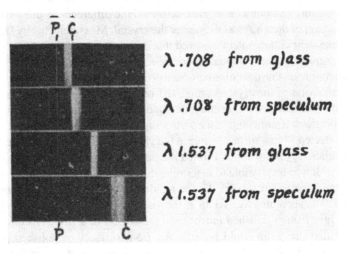

Fig. 24.2 Total reflection of x-rays from polished glass and speculum metal (Doan). *P* directed beam, *C* critical angle of the totally reflected beam

glancing angle from crown glass was found to be about 4.5 minutes of arc, which means a refractive index differing from unity by a little less than 1 part in a million. Figure 24.2 shows some photographs of the totally reflected beam and the critical angle for total reflection taken recently by Dr. Doan[10] working at Chicago. From the sharpness of the critical angles shown in this figure, it is evident that a precise determination of the refractive index can thus be made.

---

[10] R. L. Doan, Phil. Mag., *20*, p. 100; 1927.

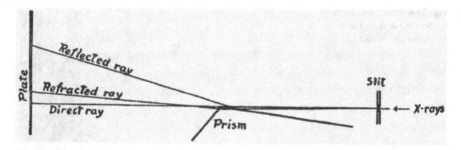

Fig. 24.3 Refraction of x-rays by a glass prism. Arrangement by Larsson, Siegbahn and Waller

You will recall that when one measures the index of refraction of a beam of light in a glass prism it is customary to set the prism at the angle for minimum deviation. This is done primarily because it simplifies the calculation of the refractive index from the measured angles. It is an interesting comment on the psychology of habit that most of the earlier investigators of the refraction of x-rays by prisms also used their prisms set at the angle for minimum deviation. Of course, since the effect to be measured was very small indeed, the adjustments should have been made to secure not the minimum deviation but the maximum deviation possible. After almost 30 years of attempts to refract x-rays by prisms, experiments under the conditions to secure maximum refraction were first performed by Larsson, Siegbahn and Waller,[11] using the arrangement shown diagrammatically in Fig. 24.3. The x-rays struck the face of the prism at a fine glancing angle, just greater than the critical angle for the rays which are refracted. Thus the direct rays, the refracted rays, and the totally reflected rays of greater wave length were all recorded on the same plate.

Figure 24.4 shows one of the resulting photographs. Here we see a complete dispersion spectrum of the refracted x-rays, precisely similar to the spectrum obtained when light is refracted by a prism of glass. The presence of the direct ray and the totally reflected ray on the same plate make possible all the angle measurements necessary for a precise determination of the refractive index for each spectrum line.

For a generation we have been trying to obtain a quantitative test of Drude and Lorentz's dispersion theory in the ordinary optical region. But our ignorance regarding the number and the natural frequency of the electron oscillators in the refractive medium has foiled all such attempts. For the extreme frequencies of x-rays, however, the problem becomes greatly simplified. In the case of substances such as glass, the x-ray frequencies are much higher than the natural frequencies of the oscillators in the medium, and the only knowledge which the theory requires is that of the number of electrons per unit volume in the dispersive medium. If we assume the number of electrons per atom to be equal to the atomic number, we are thus able to calculate at once the refractive index of the medium for x-rays. In the case of glass this calculation gives agreement with experiment within the experimental error, which is in

---

[11] Larsson, Siegbahn and Waller, Naturwiss, 1924.

Fig. 24.4 Prism spectrum of
x-rays obtained by Larsson,
Siegbahn and Waller

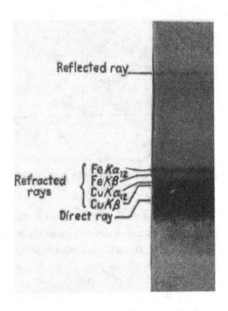

some cases less than 1 %. So we may say that the laws of optical dispersion given by
the electron theory are first established on a quantitative basis by these experiments
on the refraction of x-rays.

Another way of looking at the problem is to assume the validity of the dispersion
equation developed from the electron theory, and to use these measurements of the
refraction of x-rays to calculate the number of electrons in each atom of the refracting material. This affords us what is probably our most direct as well as our most
precise means of determining this number. The precision of the experiments is now
such that we can say that the number of electrons per atom effective in refracting
x-rays is within less than one half of 1 % equal to the atomic number of the atom.

Thus optical refraction and reflection are extended to the region of x-rays, and
this extension has brought with it more exact knowledge not only of the laws of
optics but also of the structure of the atom.

24.2.2   The Diffraction of X-Rays

Early in the history of x-rays it was recognized that most of the properties of these
rays might be explained if, as suggested by Wiechert,[12] they consist of electromagnetic waves much shorter than those of light. Haga and Wind performed a careful
series of experiments[13] to detect any possible diffraction by a wedge shaped slit a

---

[12] E. Wiechert, Sitz. d. Phys-okon Ges. zu Konigsberg, 1894.

[13] Haga and Wind, Wied. Ann., *68*, p. 884; 1899.

few thousandths of an inch broad at its widest part. The magnitude of the broadening was about that which would result[14] from rays of 1.3A wave length. The experiments were repeated by yet more refined methods by Walter and Pohl,[15] who came to the conclusion that if any diffraction effects were present, they were considerably smaller than Raga and Wind had estimated. But on the basis of photometric measurements of Walter and Pohl's plates by Koch,[16] using his new photoelectric microphotometer, Sommerfeld found[17] that their photographs indicated an effective wave length for hard x-rays of .4A, and for soft x-rays a wave length measurably greater.

It may have been because of their difficulty that these experiments did not carry as great conviction as their accuracy would seem to have warranted. Nevertheless it was this work perhaps more than any other which encouraged Laue to undertake his remarkable experiments on the diffraction of x-rays by crystals.

Within the last few years Walter has repeated these slit diffraction experiments, making use of the K_α line of copper, and has obtained perfectly convincing diffraction effects.[18] Because of the difficulty in determining the width of the slit where the diffraction occurs, it was possible to make from his photographs only a rough estimate of the wave length of the x-rays. But within this rather large probable error the wave length agreed with that determined by crystal spectrometry.

While these slit diffraction experiments were being developed, and long before they were brought to a successful conclusion, Laue and his collaborators discovered the remarkable fact that crystals act as suitable gratings for diffracting x-rays. You are all acquainted with the history of this discovery. The identity in nature of x-rays and light could no longer be doubted. It gave a tool which enabled the Braggs to determine with a definiteness previously almost unthinkable the manner in which crystals are constructed of their elementary components. By its help Moseley and Siegbahn have studied the spectra of x-rays, we have learned to count one by one the electrons in the different atoms, and we have found out something regarding the arrangement of these electrons. The measurement of x-ray wave lengths thus made possible gave Duane the means of making his precise determination of Planck's radiation constant. By showing the change of wave length when x-rays are scattered, it has helped us to find the quanta of momentum of radiation which had previously been only vaguely suspected. Thus in the two great fields of modern physical inquiry, the structure of matter and the nature of radiation, the discovery of the diffraction of x-rays by crystals has opened the gateway to many new and fruitful paths of investigation. As the Duc de Broglie has remarked, "if the value of a discovery is to be measured by the fruitfulness of its consequences, the work of

---

[14] A. Sommerfeld, Phys. ZS., *2*, p. 59; 1900.

[15] Walter and Pohl, Ann. der Phys., *29*, p. 331; 1909.

[16] P. P. Koch, Ann. der Phys., 38, p. 507; 1912.

[17] A. Sommerfeld, Ann. der Phys., *38*, p. 473; 1912.

[18] B. Walter, Ann. der Phys., *74*, p. 661; 1924; *75*, Sept. 1924.

Laue and his collaborators should be considered as perhaps the most important in modern physics."

These are some of the consequences of extending the optical phenomenon of diffraction into the realm of x-rays.

There is, however, another aspect of the extension of optical diffraction into the x-ray region, which has also led to interesting results. It is the use of ruled diffraction gratings for studies of spectra. By a series of brilliant investigations, Schumann, Lyman and Millikan, using vacuum spectrographs, have pushed the optical spectra by successive stages far into the ultraviolet. Using a concave reflection grating at nearly normal incidence, Millikan and his collaborators[19] found a line, probably belonging to the L series of aluminium, of a wave length as short as 136.6A, only a twenty-fifth that of yellow light. Why his spectra stopped here, whether because of failure of his gratings to reflect shorter wave lengths, or because of lack of sensitiveness of the plates, or because his hot sparks gave no rays of shorter wave length, was hard to say.

Röntgen had tried to get x-ray spectra by reflection from a ruled grating, but the task seemed hopeless. How could one get spectra from a reflection grating if the grating would not reflect? But when it was found that x-rays could be totally reflected at fine glancing angles, hope for the success of such an experiment was revived. Carrara,[20] working at Pisa, tried one of Rowland's optical gratings, but without success. Fortunately we at Chicago did not know of this failure, and with one of Michelson's gratings ruled specially for the purpose, Doan found that he could get diffraction spectra of the K series radiations from both copper and molybdenum.[21] Figure 24.5 shows one of our diffraction spectra, giving several orders of the K_{α_1} line of molybdenum, obtained by reflection at a small glancing angle. This work was quickly followed by Thibaud,[22] who photographed a beautiful spectrum of the K series lines of copper from a grating of only a few hundred lines ruled on glass. That x-ray spectra could be obtained from the same type of ruled reflection gratings as those used with light was now established.

The race to complete the spectrum between the extreme ultraviolet of Millikan and the soft x-ray spectra of Siegbahn began again with renewed enthusiasm. It had seemed that the work of Millikan and his coworkers had carried the ultraviolet spectra to as short wave lengths as it was possible to go. On the x-ray side, the long wave length limit was placed, theoretically at least, by the spacing of the reflecting layers in the crystal used as a natural grating. De Broglie, W. H. Bragg, Siegbahn and their collaborators were finding suitable crystals of greater and greater spacing, until Thoraeus and Siegbahn,[23] using crystals of palmitic acid, measured the L_α line of

---

[19] Millikan, Bowen, Sawyer, Shallenberger, Proc. Nat. Acad., 7, p. 289; 1921; Phys. Rev., 23, p. 1; 1924.

[20] N. Carrara, N. Cimento, 1, p. 107; 1924.

[21] A. H. Compton and R. L. Doan, Proc. Nat. Acad., 11, p. 598; 1925.

[22] J. Thibaud, C. R., Jan. 4,1926.

[23] Siegbahn and Thoraeus, Arkiv f M. o F., 19, p. 1; 1925.

Fig. 24.5 Spectrum of the K_{α_1} line of molybdenum, $\lambda = 0.708A$, from a grating ruled on speculum metal (Compton and Doan). *D* marks the direct beam, and *O* the directly reflected beam

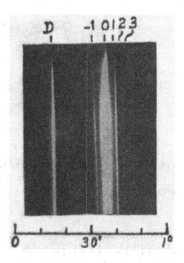

chromium, with a wave length 21.69A. But there still remained a gap of almost three octaves between these x-rays and the shortest ultraviolet in which, though radiation had been detected by photoelectric methods, no spectral measurements has been made.

Thibaud, working in de Broglie's laboratory at Paris, made a determined effort to extend the limit of the ultraviolet spectrum, using his glass grating at glancing incidence.[24] His spectra, however, stopped at 144A, a little greater than the shortest wave length observed in Millikan's experiments.

But meanwhile Dauvillier, also working with de Broglie, was making rapid strides working from the soft x-ray side of the gap. First,[25] using a grating of palmitic acid, he found the K_{α} line of carbon of wave length 45A. Then[26] using for a grating a crystal of the lead salt of mellissic acid, with the remarkable grating space of 87.5A, he measured a spectrum line of thorium as long as 121A, leaving only a small fraction of an octave between his longest x-ray spectrum lines and Millikan's shortest ultraviolet lines. The credit for filling in the greater part of the remaining gap must thus be given to Dauvillier.

The final bridge between the x-ray and the ultraviolet spectra has however been laid by Osgood,[27] a young Scotchman working with me at Chicago. He also used soft x-rays as did Dauvillier, but instead of a crystal grating, he did his experiments with a concave glass grating in a Rowland mounting, but with the rays at glancing incidence. Figure 24.6 shows a series of Osgood's spectra. The shortest wave length here shown is the K_{α} line of carbon, 45A, and we see a series of lines up to 211A. An

---

[24] J. Thibaud, J. de Phys. et Rad., *8*, p. 15; 1927.

[25] A. Dauvillier, C. R., *182*, p. 1083; 1926.

[26] A. Dauvillier, J. de Phys. et Rad., 6, p. 1; Jan. 1927.

[27] T. H. Osgood, Nature, *119*, p. 817; June 4,1927; Phys. Rev., November, 1927.

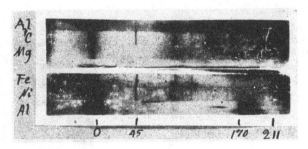

Fig. 24.6 Osgood's grating spectra of soft x-rays from Al, C, Mg, Fe and Ni, showing lines from $\lambda = 45A$ to $\lambda = 211A$. These are the first spectra bridging the gap between the soft x-rays and the ultraviolet

interesting feature of these spectra is an emission band in the aluminium spectrum at about 170A, which is probably in some way associated with the L series spectrum of aluminium. These spectra overlap, on the short wavelength side, Dauvillier's crystal measurements, and on the side of the great wave lengths, Millikan's ultraviolet spectra.

In the September number of the Physical Review, Hunt[28] describes similar experiments, using however a plane ruled grating at glancing incidence, in which he has measured lines from 2A down to the carbon line at 45A, thus meeting the shortest of Osgood's measurements. On the other hand, Fig. 24.7 shows some beautiful spectra of the extreme ultraviolet obtained recently by Dr. Hoag, working with Professor Gale at Chicago, using a concave grating at grazing incidence. These spectra extend from 200A to 1760A, overlapping Osgood's x-ray spectra on the short wave length side, and reaching the ordinary ultraviolet region on the side of the great wave lengths. Thus from the extreme infrared to the region of ordinary x-rays we now have a continuous series of spectra from ruled gratings.

Whatever we may find regarding the nature of x-rays, it would take a bold man indeed to suggest, in light of these experiments, that they differ in nature from ordinary light.

It is too early to predict what may be the consequences of these grating measurements of x-rays. It seems clear, however, that they must lead to a new and more precise knowledge of the absolute wave length of x-rays, and thus to direct determinations of the grating spaces of crystals. This will in turn afford a new means of determining Avogadro's number and the electronic charge, which should be of precision comparable with that of Millikan's oil drops.

---

[28] F. L. Hunt, Phys. Rev., Sept. 1927.

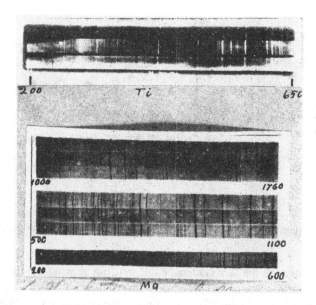

Fig. 24.7 Spectra of the extreme ultraviolet, from Mg and Ti, 200A to 1760A (Hoag)

24.3   Study Questions

QUES. 24.1. In what way are x-rays similar to visible light? Do x-rays exhibit reflection? Refraction? Total internal reflection? Why was it not immediately obvious that x-rays are waves, like light?

QUES. 24.2. How does the Drude-Lorentz theory allow one to theoretically calculate the refractive index of a material? Why were x-rays particularly suitable for testing this theory? And how did such tests, in turn, provide a more exact knowledge of the structure of the atom?

QUES. 24.3. Under what conditions, if any, do x-rays exhibit diffraction? What does this suggest?

QUES. 24.4. Where do x-rays fit in the electromagnetic spectrum? How much of the spectrum had been explored at the time of Compton's writing? And what techniques are used to measure the wavelengths of various regions the spectrum?

24.4   Exercises

EX. 24.1 (X-RAY CRYSTALLOGRAPHY AND THE BRAGG CONDITION). By carefully examining the scattering of x-rays from various crystals, William Bragg was able to deduce and catalog the arrangement and spacing of their constituent atoms. To understand how he did this, consider a simple cubic crystal, such as NaCl, whose

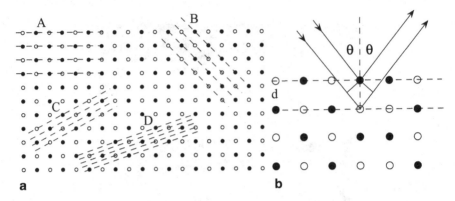

Fig. 24.8 (a) Four sets of Bragg planes constructed in different regions of a crystal lattice. (b) Incident x-rays reflecting from adjacent Bragg planes which are separated by distance d

atoms are arranged in the *face centered cubic* structure drawn (two-dimensionally) in Fig. 24.8a. Various sets of parallel equally-spaced planes—called *Bragg planes*—can be drawn so as to contain all of the atoms in the crystal. In Fig. 24.8a, the four sets of Bragg planes which have been drawn appear as dashed lines labeled A, B, C and D. When incident x-rays strike the atoms lying in a *single* Bragg plane, the outgoing x-rays obey the law of reflection; but when incident x-rays strike atoms lying in two *adjacent* Bragg planes, constructive interference only occurs for certain incidence angles. In this problem we will explore the conditions under which such constructive interference is exhibited by the reflected x-rays.

a) Consider the set of Bragg planes which are separated by distance d in Fig. 24.8b. Prove that constructive interference between x-rays reflecting from adjacent Bragg planes occurs only for incidence angles θ_n that satisfy the *Bragg condition*:

$$n\lambda = 2d \sin \theta_n \qquad\qquad n = 0, 1, 2 \ldots \qquad (24.1)$$

b) If $d = 5\text{Å}$ and $\lambda = 0.2$ nm, find all of the incident angles, θ_n which satisfy the Bragg condition for the Bragg planes shown in Fig. 24.8b. Conversely, by observing the angles at which constructive interference occurs for x-rays of a known wavelength λ, the spacing of a particular set of Bragg planes can be deduced.

24.5   Vocabulary

1. Reflection
2. Refraction
3. Diffraction
4. Neutron

 5. Ebonite
 6. Unity
 7. Calcite
 8. Dispersion
 9. Molybdenum
10. Grating
11. Octave
12. Photoelectric
13. Ultraviolet
14. Infrared

Chapter 25
Compton Scattering

We are thus confronted with the dilemma of having before us convincing evidence that radiation consists of waves, and at the same time that it consists of corpuscles.

—Arthur Holly Compton

25.1 Introduction

In the first half of his 1928 publication entitled *X-rays as a Branch of Optics*, Compton describes numerous experiments which reveal that (i) x-rays obey the same laws of reflection and refraction as ordinary (visible) light rays, (ii) the angle by which x-rays bend when entering a body can be computed from the body's electron density by using the optical dispersion theory of Drude and Lorentz, and (iii) x-rays fit nicely into the high-frequency region of the electromagnetic spectrum, since they have a wavelength on the order of a few angstroms. Now, in the second half of his *X-rays as a Branch of Optics*, Compton explains the surprising results of certain x-ray scattering experiments. To set the stage, Compton briefly reminds the reader of the main features of light scattering by turbid media. A turbid medium (such as milk) consists of a fluid in which tiny impurities are suspended; these impurities, in turn, are capable of scattering incident light.[1] What makes Compton's novel x-ray scattering experiments difficult—if not impossible—to reconcile with the classical theory of light scattering?

25.2 Reading: Compton, *X-Rays as a Branch of Optics*

Compton, A. H., X-Rays as a Branch of Optics, *Journal of the Optical Society of America*, 16(2), 71–86, 1928.

---

[1] For a more comprehensive discussion of the scattering of light by turbid media (such as milk) see the final section of Tyndall's fourth lecture on light, included in Chap. 24 of volume III.

© Springer International Publishing Switzerland 2016
K. Kuehn, *A Student's Guide Through the Great Physics Texts,*
Undergraduate Lecture Notes in Physics, DOI 10.1007/978-3-319-21828-1_25

25.2.1   The Scattering of X-rays and Light

The phenomena that we have been considering are ones in which the laws which have been found to hold in the optical region apply equally well in the x-ray region. This is not the case, however, for all optical phenomena.

The theory of the diffuse scattering of light by turbid media has been examined by Drude, Lord Rayleigh, Raman and others, and an essentially similar theory of the diffuse scattering of x-rays has been developed by Thomson, Debye and others. Two important consequences of these theories are, (1) that the scattered radiation shall be of the same wave length as the primary rays, and (2) that the rays scattered at 90° with the primary rays shall be plane polarized. The experimental tests of these two predictions have led to interesting results.

A series of experiments performed during the last few years[2] have shown that secondary x-rays are of greater wave length than the primary rays which produce them. This work is too well known to require description. On the other hand, careful experiments to find a similar increase in wave length in light diffusely scattered by a turbid medium have failed to show any such effect.[3] An examination of the spectrum of the secondary x-rays shows that the primary beam has been split into two parts, as shown in Fig. 25.1, one of the same wave length and the other of increased wave length. When different primary wave lengths are used, we find always the same difference in wave length between these two components; but the relative intensity of the two components changes. For the longer wave lengths the unmodified ray has the greater energy, while for the shorter wave lengths the modified ray is predominant. In fact when hard γ-rays are employed, it is not possible to find any radiation of the original wave length.

Thus in the wave length of secondary radiation we have a gradually increasing departure from the classical electron theory of scattering as we go from the optical region to the region of x-rays and γ-rays.

The question arises, are these secondary x-rays of increased wavelength to be classed as scattered x-rays or as fluorescent rays? An important fact bearing on this point is the intensity of the secondary rays. From the theories of Thomson, Debye and others it is possible to calculate the absolute intensity of the scattered rays. It is found that this calculated intensity agrees very nearly with the total intensity of the modified and unmodified rays, but that in many cases the observed intensity of the unmodified ray taken alone is very small compared with the calculated intensity. If the electron theory of the intensity of scattering is even approximately correct, we must thus include the modified with the unmodified rays as scattered rays.

Information regarding the origin of these secondary rays is also given by a study of their state of polarization. We have called attention to the fact that the electron theory demands that the x-rays scattered at 90° should be completely plane polarized.

---

[2] For an account of this work, see *e.g.* the writer's "X-Rays and Electrons," Chap. IX, Van Nostrand, 1926.

[3] E. g., P. A. Ross, Proc. Nat. Acad., *9*, p. 246; 1923.

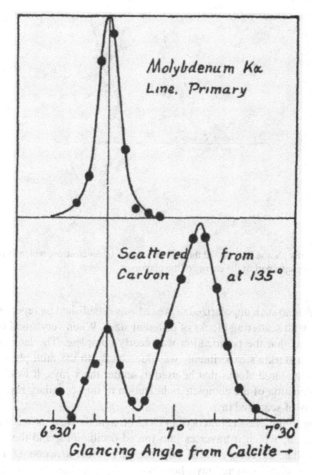

Molybdenum Kα
Line, Primary

Scattered from
Carbon at 135°

6°30'   7°   7°30'
Glancing Angle from Calcite →

Fig. 25.1 A typical spectrum of scattered x-rays, showing the splitting of the primary ray into a modified and an unmodified ray

If the rays of increased wave-length are fluorescent, however, we should not expect them to be strongly polarized. You will remember the experiments performed by Barkla[4] some 20 years ago in which he observed strong polarization in x-rays scattered at right angles. It was this experiment which gave us our first strong evidence of the similar character of x-rays and light. But in this work the polarization was far from complete. In fact the intensity of the secondary rays at 90° dropped only to one third of its maximum value, whereas for complete polarization it should have fallen to zero. It might have seemed that the remaining third was due to really unpolarized rays of a fluorescent type.

---

[4] C. G. Barkla, Proc. Roy. Soc. A., *77*, p. 247; 1906.

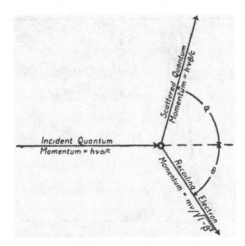

Fig. 25.2 An x-ray photon is deflected through an angle ϕ by an electron, which in turn recoils at an angle θ, taking up a part of the energy of the photon

The fact that no such unpolarized rays exist was established by repeating Barkla's experiment[5] with scattering blocks of different sizes. When very small blocks were used, we found that the polarization was nearly complete. The lack of complete polarization in Barkla's experiments was due chiefly to the multiple scattering of the x-rays in the large blocks that he used to scatter the x-rays. It would seem that the only explanation of the complete polarization of the secondary rays is that they consist wholly of scattered rays.

According to the classical theory, an electromagnetic wave is scattered when it sets the electrons which it traverses into forced oscillations, and these oscillating electrons reradiate the energy which they receive. In order to account for the change in wave-length of the scattered rays, however, we have had to adopt a wholly differ-ent picture of the scattering process—that shown in Fig. 25.2. Here we do not think of the x-rays as waves, but as light corpuscles, quanta, or, as we may call them, photons. Moreover, there is nothing here of the forced oscillation pictured on the classical view, but a sort of elastic collision, in which the energy and momentum are conserved.

This new picture of the scattering process leads at once to three consequences that can be tested by experiment. There is a change of wave-length

$$\delta\lambda = \frac{h}{mc}(1 - \cos\phi) \tag{25.1}$$

which accounts for the modified line in the spectra of scattered x-rays. Experiment has shown that this formula is correct within the precision of our knowledge of h, m

[5] A. H. Compton and C. F. Hagenow, J.O.S.A. and R.S.I., *8*, p. 487; 1924.

and c. The electron which recoils from the scattered x-ray should have the kinetic energy,

$$E_{\text{kin}} = h\nu \cdot \frac{h\nu}{mc^2} \cos^2 \theta \qquad (25.2)$$

approximately. When this theory was first proposed, no electrons of this type were known; but they were discovered by Wilson[6] and Bothe[7] within a few months after their prediction. Now we know that the number, energy and spatial distribution of these recoil electrons are in accord with the predictions of the photon theory. Finally, whenever a photon is deflected at an angle ϕ, the electron should recoil at an angle θ given by the relation,

$$\cot \frac{1}{2}\phi = \tan \theta \qquad (25.3)$$

approximately.

This relation we have tested[8] using the apparatus shown diagrammatically in Fig. 25.3. A narrow beam of x-rays enters a Wilson expansion chamber. Here it produces a recoil electron. If the photon theory is correct, associated with this recoil electron, a photon is scattered in the direction ϕ. If it should happen to eject a β-ray, the origin of this β-ray tells the direction in which the photon was scattered. Figure 25.4 shows a typical photograph of the process. A measurement of the angle θ at which the recoil electron on this plate is ejected and the angle ϕ of the origin of the secondary β-particle, shows close agreement with the photon formula. This experiment is of especial significance, since it shows that for each recoil electron there is a scattered photon, and that the energy and momentum of the system photon plus electron are conserved in the scattering process.

The evidence for the existence of directed quanta of radiation afforded by this experiment is very direct. The experiment shows that associated with each recoil electron there is scattered x-ray energy enough to produce a secondary beta ray, and that this energy proceeds in a direction determined at the moment of ejection of the recoil electron. Unless the experiment is subject to improbably large experimental errors, therefore, *the scattered x-rays proceed in the form of photons.*

Thus we see that as a study of the scattering of radiation is extended into the very high frequencies of x-rays, the manner of scattering changes. For the lower frequencies the phenomena could be accounted for in terms of waves. For these higher frequencies we can find no interpretation of the scattering except in terms of the deflection of corpuscles or photons of radiation. Yet it is certain that the two types of radiation, light and x-rays, are essentially the same kind of thing. We are thus confronted with the dilemma of having before us convincing evidence that radiation consists of waves, and at the same time that it consists of corpuscles.

---

[6] C. T. R. Wilson, Proc. Roy. Soc., *109*, p. 1; 1923.

[7] W. Bothe, ZS. f. Phys., *16*, p. 319; 1923; 20, p. 237; 1923.

[8] A. H. Compton and A. W. Simon, Phys. Rev., *26*, p. 289; 1925.

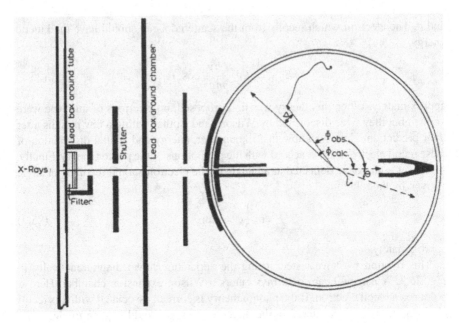

Fig. 25.3 An electron recoiling at an angle θ should be associated with a photon deflected through an angle ϕ

Fig. 25.4 Photograph showing recoil electron and associated secondary β-ray. The upper photograph is retouched

It would seem that this dilemma is being solved by the new wave mechanics. De Broglie[9] has assumed that associated with every particle of matter in motion there

---

[9] L. de Broglie, Thèse, Paris, 1924.

is a wave whose wave length is given by the relation,

$$mv = h/\lambda, \tag{25.4}$$

where mv is the momentum of the particle. A very similar assumption was made at about the same time by Duane,[10] to account for the diffraction of x-ray photons. As applied to the motion of electrons, Schrödinger has shown the great power of this conception in studying atomic structure.[11] It now seems, through the efforts of Heisenberg, Bohr and others, that this conception of the relation between corpuscles and waves is capable of giving us a unified view of the diffraction and interference of light, and at the same time of its diffuse scattering and of the photoelectric effect. It would however take too long to describe these new developments in detail.

We have thus seen how the essentially optical properties of radiation have been recognized and studied in the realm of x-rays. A study of the refraction and specular reflection of x-rays has given an important confirmation of the electron theory of dispersion, and has enabled us to count with high precision the number of electrons in the atom. The diffraction of x-rays by crystals has given wonderfully exact information regarding the structure of crystals, and has greatly extended our knowledge of spectra. When x-rays were diffracted by ruled gratings, it made possible the study of the complete spectrum from the longest to the shortest waves. In the diffuse scattering of radiation, we have found a gradual change from the scattering of waves to the scattering of corpuscles.

Thus by a study of x-rays as a branch of optics we have found in x-rays all of the well known wave characteristics of light, but we have found also that we must consider these rays as moving in directed quanta. It is these changes in the laws of optics when extended to the realm of x-rays which have been in large measure responsible for the recent revision of our ideas regarding the nature of the atom and of radiation.

University of Chicago,
Chicago, Illinois,
October 10, 1927.

25.3 Study Questions

QUES. 25.1. Which is better, the classical wave theory of scattering or the new photon theory of scattering?

a) What are the two important consequences of the classical theory of diffuse scattering of light by turbid media? Do experiments confirm these predictions?

---

[10] W. Duane, Proc. Nat. Acad., 1924.

[11] E. Shrödinger, Ann. der Phys., *79*, pp. 361, 489, 734; *80*, 437; *81*,109; 1926; Phys. Rev., *28*, p. 1051; 1926.

b) In what way did the scattering of very high frequency rays contradict the classical electron theory of scattering? In particular, what happens to the wavelength of the scattered ray?

c) How does the classical theory depict the scattering process of electromagnetic waves from electrons? How, by contrast, does the light-corpuscle theory depict the same process?

d) According to the light-corpuscle theory, what is the wavelength (and the direction) of a photon scattered by an electron? How are these actually measured?

e) Are x-rays then best understood as waves or as particles? Is there a way out of this dilemma?

QUES. 25.2. How has the study of x-rays contributed to our ideas regarding the nature of the atom and radiation?

25.4   Exercises

EX. 25.1 (COMPTON SCATTERING). In this problem, we will explore the scattering of light from an electron, as depicted in Fig. 25.2. Let us assume that the electron is initially stationary and that the incoming quantum of light (the photon) carries a momentum p_0.

a) First, use the principle of conservation of momentum to demonstrate that the momenta of the scattered photon, p_1, and of the scattered electron, p, are related by

$$p_0^2 + p_1^2 - 2\,p_0\,p_1 \cos\phi = p^2 \tag{25.5}$$

b) According to the theory of special relativity, the energy, E, of a particle of rest mass m_0 is related to its momentum, p, by[12]

$$E^2 = (pc)^2 + (m_0 c^2)^2 \tag{25.6}$$

Assuming that the photon has zero rest mass, use the conservation of energy to demonstrate that the kinetic energy of the outgoing electron, T, can be written as

$$T = (p_0 - p_1)c \tag{25.7}$$

and that the electron's kinetic energy, rest mass and momentum are related by

$$T^2 + 2\,T\,m_0\,c^2 = (p\,c)^2 \tag{25.8}$$

---

[12] See the discussion of Einstein's theory of relativity, and especially Ex. 32.2, in Chap. 32 of volume II.

c) Combine Eqs. 25.5, 25.7, and 25.8 to show that the momenta of the incoming and outgoing photons are related by

$$(p_0 - p_1)^2 = p_0\, p_1(1 - \cos\phi) \qquad (25.9)$$

d) Assuming that the photon obeys the de Broglie relation, show that the change of wavelength of the scattered photon is given by Compton's scattering formula, Eq. 25.1.

e) What is the change in wavelength of a photon scattering from an electron in a carbon sample through an angle $\phi = 135°$, as depicted in Fig. 25.1? At what scattering angle, ϕ, would the photon momentum change by the *Compton wavelength*, which is defined in Eq. 25.10?

$$\lambda_c = \frac{h}{m_0 c} \qquad (25.10)$$

f) Finally, how would the preceding analysis change if the photon were to scatter from a hydrogen nucleus rather than from an electron? Would the compton wavelength remain the same?

25.5 Vocabulary

1. Turbid
2. Photon

Chapter 26
Electron Scattering and Diffraction

> Electron scattering is not, it would seem, the mildly interesting
> matter of flying particles and central fields that we supposed,
> but is instead a much more interesting phenomenon in which
> electrons exhibit the properties of waves.
>
> —Clinton Davisson

26.1 Introduction

Clinton Joseph Davisson (1881–1958) was born in Bloomington, Illinois.[1] He enrolled at the University of Chicago in 1902 after graduating from the public high school. Before finishing his Bachelor of Science degree, however, he was hired as a part-time physics instructor by Princeton University. During his time at Princeton, he was able to complete the requirements for his undergraduate degree by returning to Chicago during summer sessions. He eventually earned his Ph.D. from Princeton in 1911, writing a dissertation under the guidance of O.W. Richardson *On The Thermal Emission of Positive Ions From Alkaline Earth Salts*. Davisson then went on to serve as an instructor at the Carnegie Institute of Technology in Pittsburg until 1917, when he moved to the Western Electric Company (later Bell Telephone Laboratories) in New York City. This industrial research position, which initially consisted in war-related work, eventually provided Davisson with the freedom to do scientific research which he lacked in his previous academic appointment. While at Western Electric, Davisson studied the ejection of electrons from metals by heat (thermionic emission) and by bombardment with other electrons. This latter research program eventually led to his famous work, in collaboration with Lester Germer, for which Davisson was awarded the Nobel Prize in physics in 1937. The famous Davisson-Germer experiment is described in the reading selection below. It begins with a short introduction by the editor of the *Bell System Technical Journal*, in which the article was published in 1928.

---

[1] Much of the information in this introduction is from the Davisson entry in the Complete Dictionary of Scientific Biography. 2008. Retrieved January 04, 2015 from *Encyclopedia.com*.

© Springer International Publishing Switzerland 2016
K. Kuehn, *A Student's Guide Through the Great Physics Texts*,
Undergraduate Lecture Notes in Physics, DOI 10.1007/978-3-319-21828-1_26

26.2   Reading: Davisson, *The Diffraction of Electrons by a Crystal of Nickel*

Davisson, C., The Diffraction of Electrons by a Crystal of Nickel, *Bell System Technical Journal*, 7(1), 90–105, 1928.

This article is taken from the manuscript prepared by the author for his address at the joint meeting of Section B of the American Association for the Advancement of Science and the American Physical Society on December 28, 1927, at Nashville, Tennessee. An account of this work giving fuller experimental details is given by Davisson and Germer in the December, 1927, issue of the *Physical Review*.

These experiments are fundamental to some of the newer theories in physics. Until they were performed, it could be said that all experimental facts about the electron could be explained by regarding it as a particle of negative electricity. It now appears that in some way a "wavelength" is connected with the electron's behavior. The work thus shows an interesting contrast with the discovery of A.H. Compton that a ray of light (a light pulse) suffers a change of wave-length upon impact with an electron, the change of wave-length corresponding exactly to the momentum gained by the electron. Until Compton's work, all the known facts about light could be explained by thinking of light as a wave motion. The Compton effect seems to prove the existence of particles of light.

Physics is thus faced with a double duality. Compton showed that light is in some sense *both* a wave motion and a stream of particles. Davisson and Germer have now shown that a beam of electrons is in some sense *both* a stream of particles and a wave motion.

At the same time, theoretical advances have been made which seem to pave the way for an understanding of this curious situation. A general account of these new developments was given by K.K. Darrow in his series "Contemporary Advances in Physics" in the *Bell System Technical Journal* for October, 1927. Some remarks on the relation of the Davisson and Germer experiments to the new mechanics were given in this article, p. 692 *et seq.*— EDITOR.

The experiments which I have been asked to describe are the most recent of an investigation of the scattering of electrons by metals on which we have been engaged in the Bell Telephone Laboratories for the last 7 or 8 years.

The investigation had its inception in a simple but significant observation. We observed some time in the year 1919 that when a beam of electrons is directed against a metal target, electrons having the same speed as those in the incident beam stream out in all directions from the bombarded area. It seemed to us at the time that these could be no other than particular electrons from the incident beam that had suffered large deflections in simple elastic encounters with single atoms of the target. The mechanism of scattering, as we pictured it, was similar to that of alpha ray scattering. There was a certain probability that an incident electron would be caught in the field of an atom, turned through a large angle, and sent on its way without loss of energy. If this were the nature of electron scattering it would be possible, we thought, to deduce from a statistical study of the deflections some information in regard to the field of the deflecting atom. It was with these ideas in mind that the investigation was begun. What we were attempting, it will be seen, were atomic explorations similar to those of Sir Ernest Rutherford and his collaborators but explorations in which the probe should be an electron instead of an alpha particle. I shall not stop to recount the earlier experiments of this investigation, but shall pass at once to the most recent ones—those in which Dr. Germer and I have studied the scattering of electrons by a single crystal of nickel.

The unusual interest that attaches to these experiments is due to their revealing the phenomenon of electron scattering in a new and, I may say, fashionable role. Electron scattering is not, it would seem, the mildly interesting matter of flying particles and central fields that we supposed, but is instead a much more interesting phenomenon in which electrons exhibit the properties of waves. The experiments reveal that the way in which electrons are scattered by a crystal is very similar to the way in which x-rays are scattered by a crystal. The analogy is not so much with the alpha ray experiments of Sir Ernest Rutherford, as with the x-ray diffraction experiments of Professor von Laue.

My task of describing these experiments is much simplified by the fact that the experiments of Professor von Laue are so well known and so thoroughly comprehended. I remind you very briefly that in the original Laue experiment a beam of x-rays was directed against a crystal of zincblende, that about the transmitted beam was found an array of regularly disposed subsidiary beams proceeding outward from the irradiated portion of the crystal, and that these subsidiary beams could be interpreted completely and precisely in terms of the then already popular wave theory of x-radiation. They could indeed be explained as diffraction beams that resulted from the superposition of secondary wave trains expanding from the regularly arranged atoms of the crystal lattice.

There are two features of the Laue experiment which we shall need particularly to remember. The first is that diffraction beams issue not only from the far side of the crystal along with the transmitted beam, but also from the near or incidence side of the crystal—these latter being disposed in a regular array about the incident beam. The second is that each diffraction beam is characterized by a particular wavelength, and that a given beam appears in the diffraction pattern if the incident beam contains radiation of its characteristic wavelength, or of some submultiple value of this wave-length, but not otherwise. If the incident beam is monochromatic, no diffraction beams appear at all unless the wave-length of the incident beam happens to coincide with a wave-length of one or more of the diffraction beams. In that case the favored beams appear but no others.

With this picture of x-ray scattering in mind one sees at once the significance of the main results of the present experiments. A homogeneous beam of electrons is directed against a crystal of nickel, and at certain critical speeds of bombardment full speed scattered electrons issue from the incidence side of the crystal in sharply defined beams—a few beams at each of the critical speeds—the totality of such beams making up a regularly disposed array similar to the array of Laue beams that would issue from the same side of the same crystal if the incident beam were a beam of x-rays.

The electron beams are not identical in disposition with the Laue beams, and yet it is possible to treat them as diffraction beams, and from their position and from the geometry and scale of the crystal to calculate "wave-lengths" of the incident beam—just as we might do if we were dealing with x-rays or with any other wave radiation. When this is done we arrive at a definite and simple relation between the speed of the electron beam and its apparent wave-length—the wave-length is inversely proportional to the speed.

Surprising as it is to find a beam of electrons exhibiting thus the properties of a beam of waves, the phenomenon is less surprising today than it would have been a few years ago. We have been prepared, to a certain extent, by recent developments in the theory of mechanics for surprises of just this sort—for the discovery of circumstances in which particles exhibit the properties of waves. We have witnessed, during the last 3 years, the inception and development of the idea that all mechanical phenomena are in some sense wave phenomena—that the rigorous solution of every problem in mechanics must concern itself with the propagation and interference of waves. The wave nature of mechanical phenomena is not ordinarily apparent, we are told, because the length of the waves involved is ordinarily small compared to the dimensions of the system. It is only in such small scale phenomena as the intimate reactions between atoms and electrons that the wave-lengths are comparable with the dimensions of the system. Here only are we to expect notable departures from classical mechanics, and here only are we to find evidence of a more comprehensive wave mechanics.[2] The success of this new theory has been confined, up to the present time, to explanations of certain of the data of spectroscopy. In this field the theory has appealed very strongly to all of us because of the elegance of its methods and because of its remarkable facility in accounting for various of the inhibitions with which the radiating atom is afflicted. We have been prepared by these successes to view with not too great Surprise—or alarm—evidence for the wave nature of phenomena involving freely moving electrons. And any reluctance we may feel in treating electron scattering as a wave phenomenon is apt to be dispelled when we find that the value calculated for the wave-length of the equivalent radiation is in acceptable agreement with that which L. de Broglie assigned to the waves which he associated with a freely moving particle—that is to say, the value h/mv (Planck's constant divided by the momentum of the particle).

In this account of the experiments I will describe the general method of the measurements and the general character of the results rather than attempt to go into these matters in detail.

Nickel forms crystals of the face centered cubic type. In Fig. 26.1a the crystal which we had at our disposal is represented by a block of unit cubes of this type.

Our first step in preparing the crystal for bombardment was to cut through this structure at right angles to one of the cube diagonals. The appearance of the crystal after the cut was made, and the corner of the cube removed, is indicated in Fig. 26.1b. It is this newly formed triangular surface that was exposed to electron bombardment. The bombardment was at normal incidence as indicated in Fig. 26.1c. We are to think of electrons raining down normally upon this triangular surface, and

---

[2] It was predicted by W. Elsasser in 1925 (Naturwiss., 13, 711 (1925)) that evidence for the wave mechanics would be found in the interaction between a beam of electrons and a crystal.

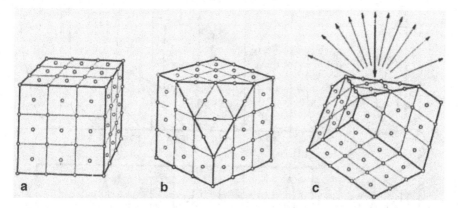

Fig. 26.1 Diagrams of nickel lattice, of cut lattice, and of lattice with incident and scattered beams

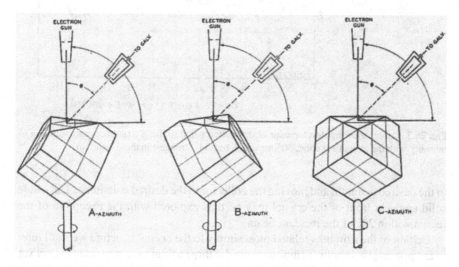

Fig. 26.2 Showing the three principal azimuths

of some of these emerging from the crystal without loss of energy, and proceeding from it in various directions.

What is measured is the current density of these full speed scattered electrons as a function of direction and of bombarding potential. The way in which the measurements are made is illustrated in Fig. 26.2. The electrons proceeding in a given direction from the crystal enter the inner box of a double Faraday collector and a galvanometer of high sensitivity is used to measure the current to which they give rise. An appropriate retarding potential between the parts of the collector excludes from the inner box all but full speed electrons.

The collector may be moved over an arc of a circle in the plane of the drawing as indicated, and the crystal may be rotated about an axis which coincides with the axis of the incident beam of electrons. Thus the collector may be set for measuring the intensity of scattering in any direction relative to the crystal—by turning the crystal

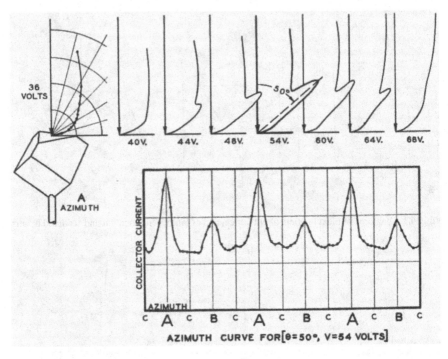

Fig. 26.3 Curves showing development of diffraction beam in the *A*-azimuth...and variation of intensity with the azimuth at colat. 50° for which beam is strongest in the *A*-azimuth

to the desired azimuth, and moving the collector to the desired colatitude. The whole solid angle in front of the crystal may be thus explored with the exception of the region within 20° of the incident beam.

Certain of the azimuths related most simply to the crystal structure we shall refer to as "principal azimuths." Thus there are the three azimuths that include the apexes of the triangle. If we find the intensity of scattering depending on colatitude in a certain way in one of these azimuths, we expect, of course, to find it depending upon colatitude in the same way in each of the other two. We shall call these the "*A*-azimuths." On the left in Fig. 26.2 the crystal has been turned to bring one of the *A*-azimuths into the plane of rotation of the collector.

Another triad of principal azimuths consists of the three which include the mid-points of the sides of the triangle. These we shall call the "*B*-azimuths." The next most important family of azimuths comprises those which are parallel to the sides of the triangle; of these there are six, the "*C*-azimuths."

If we turn the crystal to any arbitrarily chosen azimuth, set the bombarding potential at any arbitrarily chosen value, and measure the intensity of scattering as a function of colatitude, what we find ordinarily is the type of relation represented by the curve on the left in Fig. 26.3.

This curve is actually one found for scattering in the *A*-azimuth when the bombarding potential is 36 V. It is typical, however, of the curves that are obtained when no diffraction beam is showing. The intensity of scattering in a given direction is indicated by the length of the vector from the point of bombardment to the curve. The intensity is zero in the plane of the crystal surface, and increases regularly as the colatitude angle is decreased. This type of scattering forms a background upon which the diffraction beams are superposed.

The occurrence of a diffraction beam is illustrated in the series of curves to the right in Fig. 26.3. When the bombarding potential is increased from 36 to 40 V, the curve is characterized by a slight hump at colatitude 60°. With further increase in bombarding potential this hump moves upward, and at the same time develops into a strong spur. The spur reaches its maximum development at 54 V in colatitude 50°, then decreases in intensity, and finally vanishes from the curve at about 70 V in colatitude 40°.

We next make an exploration in azimuth through this spur at its maximum; we adjust the bombarding potential to 54 V, set the collector in colatitude 50°, and make measurements of the intensity of scattering as the crystal is rotated. The results of this exploration are exhibited by the curve at the bottom of Fig. 26.3, in which current to the collector is plotted against azimuth. We find that the spur is sharp in azimuth as well as in latitude and that it is one of a set of three spurs as required by the symmetry of the crystal.

We observe also that there are small spurs showing in the *B*-azimuths. We turn the crystal to bring the *B*-azimuth under observation, and again make explorations in latitude for various speeds of bombardment. We find that the spur in the *B*-azimuth is similar to the "54 volt" spur in the *A*-azimuth, but that it attains its maximum development at a higher voltage and at a higher angle. Curves exhibiting its growth and decay are shown in Fig. 26.4. Maximum development is attained at 65 V in colatitude 44°. At the bottom of the figure we show the intensity-azimuth curve through this spur at its maximum. The small maxima in the *A*-azimuths represent the remnants of the "54-volt" spurs.

We have thus a set of spurs at colatitude 50° in the *A*-azimuths when the bombarding potential is 54 V and a set of 44° in the *B*-azimuths when the bombarding potential is 65 V. These spurs are due to beams of full speed scattered electrons which are comparable in sharpness and definition with the beam of incident electrons. This is inferred from the widths of the spurs and the resolving power of the apparatus.

It is hardly necessary to point out that these sharply defined beams of scattered electrons are similar in their behavior to x-ray diffraction beams. If the incident beam were a beam of monochromatic x-rays of adjustable wave-length instead of a homogeneous beam of electrons of adjustable speed, quite similar effects could be produced. If the wave-length of the x-ray beam were varied, critical values would be found at which intense diffraction beams would issue from the crystal in its *A*-azimuths and others at which such beams would issue in the *B*-azimuths. The x-ray diffraction beams would indeed be more sharply defined in wave-length than the electron beams defined in voltage. No diffraction beam would be observed until the wave-length of the incident x-rays were very close indeed to its critical value, and the beam would disappear again when the wave-length had passed only very

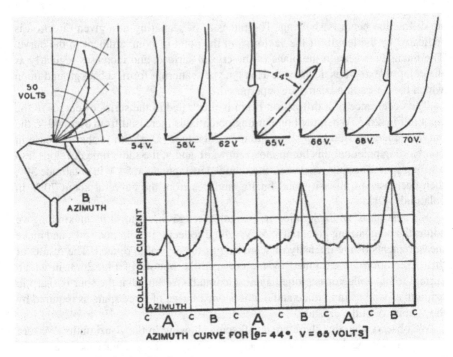

Fig. 26.4 Similar for the *B*-azimuth

slightly beyond the critical value. This "wave-length sharpness" or "wave-length resolving power" is dependent, however, upon the number and disposition of the atoms involved in the diffraction. If the crystal were only a few atom layers in thickness, or if the x-rays were extinguished on penetrating through only a few atom layers of the crystal, then the x-ray diffraction beams would be much less sharply defined in wave-length; they would behave more like the electron beams. We may say then that the electron beams exhibit the general behavior of diffraction beams resulting from the scattering of a beam of very soft wave radiation—radiation that is very rapidly extinguished in the crystal.

26.3   Study Questions

QUES. 26.1. To whose work did Davisson originally compare his own studies of the scattering of a beam of electrons from a metal target? What information did he hope to obtain from such measurements? What surprising observation did he make when scattering electrons from a nickel crystal?

QUES. 26.2. What are the key features of Laue's x-ray diffraction experiments? In particular, are the x-rays scattered with equal intensity in all directions? If not, then how does the spacing between the crystal's atoms (along lines of different orientations) relate to the scattering of x-rays? Are diffraction beams of every conceivable wavelength emitted when a single wavelength x-ray beam strikes the crystal?

QUES. 26.3. Was the similarity between Davisson's experiments with electrons and Laue's experiments with x-rays entirely surprising? Why had the wave-like character of electrons not been observed previously? What properties of an electron determines its (De Broglie) wavelength?

QUES. 26.4. Describe the experimental apparatus of Davisson and Germer. What is the nature of the nickel target? In which direction was it oriented? How did they control the speed of the electron beam? And how did they detect the scattered electrons?

QUES. 26.5. What notable pattern did the scattered electrons display? In particular, how did the scattering intensity depend upon (i) the accelerating voltage, (ii) the co-latitude angle (for a fixed azimuthal angle), and (iii) the azimuthal angle (for a fixed co-latitude angle)?

QUES. 26.6. In what way was the scattering of electron beams qualitatively similar to the scattering of x-ray beams? In what way(s) was it different? Do electrons scatter from crystals like particles or like waves?

26.4 Exercises

EX. 26.1 (DE BROGLIE WAVELENGTH). According to Louis de Broglie, the momentum of any particle, whether an electron or a sand grain, is given by

$$p = \frac{h}{\lambda} \qquad (26.1)$$

a) Assuming this to be true, what is the momentum of a single photon of red light ($\lambda = 633$ nm)?
b) What (average) force would be produced on a black (absorbing) plate if it were struck by a beam of red light photons at a rate of 1000 photons per second.
c) Would the same beam exert the same force on a perfectly reflecting mirror? If not, by how much would it differ?

EX. 26.2 (ELECTRON SCATTERING FROM THE SURFACE OF A CRYSTAL). Suppose that a beam of electrons is accelerated through a potential difference of 100 V. The beam strikes the surface of a crystal whose atoms are arranged in a simple cubic lattice with atomic spacing 4 Å. Assuming that the (mutually perpendicular) rows of surface atoms act as a diffraction grating, at what angle with respect to the normal is the first-order ($n = 1$) interference maximum observed? Does the diffraction pattern appear as a series of linear fringes? To answer these questions, you might recognize that when waves strike a diffraction grating, the angular locations of the principle maxima of the scattered waves occur whenever the partial waves from the grating lines constructively interfere with one another.[3] (ANSWER: 18°.)

26.5   Vocabulary

1. Duality
2. Zincblende
3. Subsidiary
4. Superposition
5. Monochromatic
6. Diffraction
7. Homogeneous
8. Inception
9. Spectroscopy
10. Apt
11. Inhibition
12. Galvanometer
13. Azimuth
14. Colatitude
15. Apex
16. Triad

---

[3] See Ex. 20.4, and especially Eq. 20.2, in volume III.

Chapter 27
Matter Waves

The electron as a particle is too well established to be
discredited by a few experiments with a nickel crystal.

—Clinton Davisson

27.1 Introduction

In the previous reading selection, Davisson interpreted his experiments on the scattering of electrons from a nickel crystal using De Broglie's recently developed theory of matter waves. According to this novel theory of mechanics, the momentum and wavelength of any particle are related by

$$p = \frac{h}{\lambda}. \tag{26.1}$$

This peculiar relationship had been devised by Prince Louis de Broglie in his 1924 doctoral thesis in an attempt to generalize Einstein's theory of light quanta.[1] De Broglie claimed that just as light waves could exhibit particle-like properties (in the form of photons), so too, particles (such as electrons) could exhibit wave-like properties. This counter-intuitive idea of *wave-particle duality* had been recently employed by Compton in order to make sense of the scattering of photons from electrons,[2] and it would soon form the basis of Schrödinger's wave-mechanical formulation of quantum theory.[3] In the reading selection below, Davisson continues to discuss his famous electron scattering experiments. You will notice that he treats the top layer of atoms in the nickel crystal as a diffraction grating whose spacing depends on the orientation of the crystal.[4] Do his results provide quantitative

---

[1] Einstein's interpretation of the photoelectric effect is described in Sect. 16.2.8 of the present volume.

[2] The phenomenon of Compton scattering is discussed in Chap. 25 of the present volume.

[3] See Schrödinger's 1933 nobel lecture on *The Fundamental Idea of Wave Mechanics*, contained in Chap. 31 of the present volume.

[4] For a discussion of diffraction gratings and their effect on incident waves, refer to Ex. 20.4 in volume III.

© Springer International Publishing Switzerland 2016
K. Kuehn, *A Student's Guide Through the Great Physics Texts,*
Undergraduate Lecture Notes in Physics, DOI 10.1007/978-3-319-21828-1_27

(as opposed to merely qualitative)support for De Broglie's theory of matter waves? What conclusion does he finally draw from his data?

27.2   Reading: Davisson, *The Diffraction of Electrons by a Crystal of Nickel*

Davisson, C., The Diffraction of Electrons by a Crystal of Nickel, *Bell System Technical Journal*, 7(1), 90–105, 1928.

Let us try now to forget that what we are measuring in these experiments is a current of discrete electrons arriving one by one at our collector. Let us imagine that what we are dealing with is indeed a monochromatic wave radiation, and that our Faraday box and galvanometer are instruments suitable for measuring the intensity of this radiation. We are to think of the incident electron beam as a beam of monochromatic waves, and of the "54-volt beam" in the A-azimuth and the "65-volt beam" in the B-azimuth as diffraction beams that owe their intensities, in the usual way, to constructive interference among elements of the incident beam scattered by the atoms of the crystal. With this picture in mind we try next to calculate wave-lengths of this electron radiation from the data of these beams and from the geometry and scale of the crystal.

To begin with, we shall need to look more closely into our crystal. The atoms in the triangular face of the crystal may be regarded as arranged in lines or files at right angles to the plane of the A- and B-azimuths (Fig. 27.1). If a beam of radiation were scattered by this single layer of atoms, these lines of atoms would function as the lines of an ordinary line grating. In particular, if the beam met the plane of atoms at normal incidence, diffraction beams would appear in the A- and B-azimuths, and the wave-lengths and inclinations of these beams would be related to one another and to the grating constant d by the well-known formula, $n\lambda = d \sin\theta$, as illustrated at the top of the figure.

In the actual experiments the diffracting system is not quite so simple. It comprises not a single layer of atoms, but many layers; it is equivalent not to a single line grating, but to many line gratings piled one above the other, as shown graphically at the bottom of the figure. What diffraction beams will issue from this pile of similar and similarly oriented plane gratings?

The answer to this question is twofold. In respect of position all the beams which appear will coincide with beams which would issue from a single grating. We get no additional beams by adding extra layers to the lattice. In respect of intensity, however, the results are greatly changed. A given beam may be accentuated or it may be diminished, both absolutely and relatively to the other beams; it may in fact be blotted out completely, or reduced to such an extent that it can no longer be perceived. These are effects of interference among the similar beams proceeding from the various plane gratings that make up the pile. Later we shall consider under what conditions these component beams combine to produce a resultant beam of

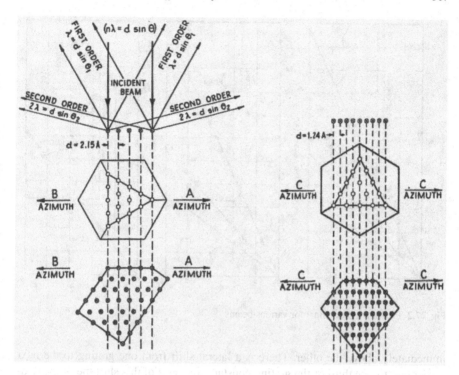

Fig. 27.1 Showing $n\lambda = d \sin \theta$ relation in the A-, B- and C-azimuths

maximum intensity; for the present, however, I wish only to stress the fact that whenever and wherever a space lattice beam appears its wave-length and colatitude angle θ will be related to the constant d of the plane grating through the ordinary plane grating formula. We therefore apply this formula to the 54- and 65-V beams that have been described. The grating constant d has the value 2.15 Å., the 54-V beam occurs at $\theta = 50°$ so that $n\lambda$ for this beam should have the value $2.15 \times \sin 50°$, or 1.65 Å. For the 65-V beam we obtain for $n\lambda$ the value 1.50 Å.

We now compare these wave-lengths with the wave-lengths associated with freely moving electrons of these speeds in the theory of wave mechanics. Translated into bombarding potentials, de Broglie's relation

$$\lambda = \frac{h}{mv} \qquad \text{becomes} \qquad \lambda = \sqrt{\frac{150}{V}}\,\text{Å,} \qquad (27.1)$$

where V represents the bombarding potential in volts. The length of the phase wave of a "54-volt electron" is $(150/54)^{1/2} = 1.67$Å, and for a 65-V electron 1.52Å. The 54- and 65-V electron beams do very well indeed as first order phase wave diffraction beams.

It may be mentioned that beams occur at different voltages in the A- and B-azimuths because the plane gratings that make up the crystal are not piled one

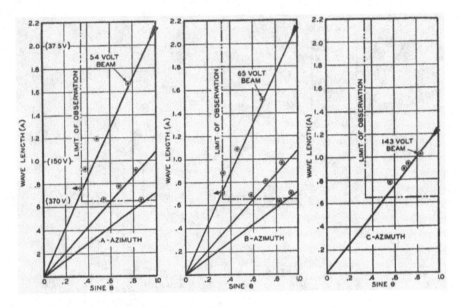

Fig. 27.2 Plot of λ against $\sin \theta$ for various beams

immediately above the other. There is a lateral shift from one grating to the next amounting to one third of the grating constant. Because of this shift the phase relation among the elementary beams emerging in the A-azimuth is not the same as that among those emerging in the B-azimuth—and coincidence of phase among these beams occurs at different voltages, or at different wave-lengths, in the two azimuths.

We next make similar calculations for a beam occurring in the C-azimuth. One such beam attains its maximum development in colatitude 56° when the bombarding potential is 143 V. For diffraction into the C-azimuth we must regard the atoms in the surface layer as arranged in lines normal to the plane of this azimuth as illustrated in Fig. 27.1. The grating constant is 1.24 Å, and the similar gratings that make up the whole crystal are piled up without lateral shift. For this reason the C-azimuth is six-fold instead of only three-fold. For a beam occurring in this azimuth in colatitude 56°, $n\lambda$ should be equal to $1.24 \times \sin 56°$ or 1.03 Å. The value of h/mv for electrons that have been accelerated from rest through 143 V is $(150/143)^{1/2}$ or 1.025 Å. Again the beam does very well as a first order diffraction beam.

The total number of such beams which we have observed in all azimuths in explorations up to 370 V is 24—nine in the A-azimuth, ten in the B-azimuth, and five in the C-azimuth. It would be possible to calculate an observed wave-length for each of these beams from $n\lambda = d \sin \theta$, and to compare this in each case with the theoretical wave-length calculated from $\lambda = h/mv$, just as we have done already for three of the beams. We have chosen, however, to display the results graphically rather than numerically.

The data for the 24 beams are exhibited in diagrams in Fig. 27.2, in which wavelength λ is plotted against the sine of the colatitude, θ. There is a separate diagram for each azimuth and in each the straight lines passing through the origin represent the plane grating formula $n\lambda = d \sin \theta$ in its different orders. Each of the 24 beams is represented by a point or by a wedge-shaped symbol in one of these diagrams. The quantities coordinated in each case are the wave-length of the incident beam as calculated from $\lambda = h/mv = (150/V)^{1/2}$ and the sine of the colatitude angle of the diffraction beam as observed. There are no points to the left of the line $\theta = 20°$ as this represents the lower limit of our colatitude range of observation, and none below the line $\lambda = 0.637$ Å as this corresponds to the upper limit of our voltage range, 370 V. The bombarding potentials corresponding to various wave-lengths are shown by figures enclosed in brackets.

When the data are exhibited in this fashion the question as to whether or not the observed wave-length of a beam agrees with its theoretical wave-length is answered by whether or not the point representing the beam falls on one or another of the lines representing the plane grating formula. If there were perfect agreement in all cases, each of the points would lie on some one of these lines.

It will be seen that the points all lie close to the lines, though not as a rule exactly on them. It is of course very important to decide whether the departures of the points from the lines are or are not too great to be attributed to uncertainties of measurements. It is our belief that they are in fact due to experimental error in the determination of the colatitude angles. If we accept the theoretical values of the wave-lengths as correct, and calculate the values of θ which we should have observed, we find that in no case do they deviate by more than 4° from the values of θ actually set down. Corrections of this magnitude do not seem excessive when it is considered that we are making measurements with what amounts to a rather crude spectrometer, that the arm of the spectrometer is but 11 mm in length, that the opening in the collector is 5° in width, and that the spectrometer itself is sealed into a glass bulb. We therefore assume that in every case the value of the wave-length assigned by de Broglie is the correct one.

I now direct your attention to a particular group of these beams—the group comprising the beam of greatest wave-length in each of the three azimuths, which are represented in the figure by wedge-shaped symbols. The interpretation of these three is quite simple. The radiation to which our electron beam is equivalent is extremely soft as already noted. Its intensity suffers a considerable decrement when the beam passes normally through only a single layer of atoms. This characteristic is inferred from the low resolving power of the crystal, and is consistent with what we know of the penetrating power of low speed electrons. When the beam passes through a layer of atoms at other than normal incidence the decrement in its intensity is greater still—and in the limit as the angle of incidence approaches grazing to the atom layer the intensity of the transmitted beam will approach zero. Thus we may expect that when a diffraction beam leaves the crystal at near grazing emergence the contributions to the resultant beam which come from the second and lower layers of atoms will be much less important than when the beam emerges from the crystal at a higher angle. Near grazing the radiation proceeding from the second and lower layers will be heavily absorbed in its passage through the overlying layers. Within a limited angular range near grazing the diffraction beam will be made up almost entirely of radiation scattered by the uppermost layer of atoms. The diffracting system becomes essentially a single plane grating and what we should observe is ordinary plane grating diffraction.

The first order diffraction beam from a line grating appears at grazing emergence when the wave-length of the incident radiation is equal to the grating constant. The grating constant for diffraction into the A- and B-azimuths is 2.15 Å and grazing beams should appear in both azimuths when the wave-length of the incident electron beam has this value. The bombarding potential corresponding to wave-length 2.15 Å is 32.5 V, and at just 32.5 V diffraction beams appear at grazing in both these azimuths. As the bombarding potential is increased the beams move up from the surface to satisfy the relation $\lambda = d \sin \theta$. Ten or fifteen degrees above the surface radiation from the second and lower layers escapes in sufficient amounts to reduce the intensity of the resultant beam through interference, and at a somewhat higher angle the beam disappears.

An exactly similar beam is found at grazing in the C-azimuth. The grating constant here is 1.24 Å and the bombarding potential corresponding to wave-length 1.24 Å is 97.5 V. The beam appears at grazing at just this voltage. These three beams occurring and behaving exactly as required by the theory constitute the strongest evidence we have in favor of the wave interpretation of electron scattering.

We have been less successful in trying to account for the occurrences of the remaining 21 sets of beams. We do not know why they occur where they do. The most we have been able to do is to relate their occurrences with those of the Laue beams that would issue from the same crystal if the incident beam were a beam of x-rays.

In Fig. 27.3 we indicate by crossed circles in a ($\lambda \sin \theta$) diagram the x-ray diffraction beams that would be observed in the B-azimuth. We show also again the electron beams as actually observed. It is obvious that the law of occurrence of electron beams is not the same as the law of occurrence of Laue beams, and yet we see that the occurrences of the two sets of beams have certain features in common. The dots representing electron beams occur along the plane grating lines at about the same intervals as the crossed circles representing the Laue beams. Other points of similarity are found with further study of the data and one is led finally to the conviction that each electron beam is the analogue of a particular Laue beam. The electron beam represented by a given dot appears to be the analogue of the B-azimuth Laue beam of the same order represented by the crossed circle occurring next above it in the diagram. This association of beams is indicated in the figure.

The occurrences of the Laue beams are determined in part by the separation between the atomic plane gratings that make up the crystal. If the separation between adjacent planes were increased the crossed circles representing the Laue beams would be moved upward along the plane grating lines; if the separation were decreased the crossed circles would be moved downward. Merely as a mode of description, then, we may say that a given electron beam has the wave-length and position that its Laue beam analogue would have if the separation between planes were decreased by a certain factor.

Fig. 27.3 $\lambda \sin \theta$ diagram for
B-azimuth

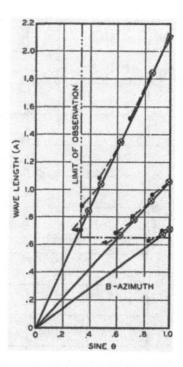

We have calculated this spacing factor for each of the 21 beams and the values
found are plotted in the upper part of Fig. 27.4 against the voltages of the beams. The
points form a very bad curve. They do indicate, however, that the factor increases
with the speed of the electron, and there is the suggestion that it approaches unity
as a limiting value. There is the suggestion, that is, that at high voltages the law of
occurrence of electron beams is the same as the law of occurrence of Laue beams.

It has been pointed out by Eckart that if the index of refraction of the crystal
for the electron radiation is other than unity diffraction beams will occur as if the
separation between atom planes were other than normal. We have computed the
indices of refraction that would give rise to the observed occurrence of beams and
these are plotted in the lower part of the diagram against bombarding potential.
Again the points fan very irregularly. While it cannot be said that there is at present
a satisfactory explanation of the peculiar occurrence of the space lattice electron
diffraction beams, it should be clearly understood that this deficiency in no way
affects either the wave-length measurements of these beams or the agreement of
these wave-lengths with the values of *h/mv*.

The electron diffraction beams which I have described are the only ones observed
when the surface of the crystal is free from gas. When the surface is not free from
gas still other beams appear. These beams are due to the scattering of electrons by
the adsorbed gas and therefore we shall not consider them at this time.

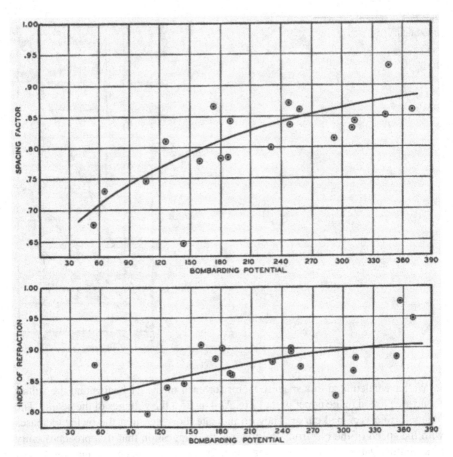

Fig. 27.4 Plot of values of spacing factor and associated values of refractive index for 21 beams

In closing I should like to say a few words about the conceptual difficulty in which these experiments involve us. When Laue and his collaborators investigated the scattering of x-rays by crystals the results of their observations were accepted at once as establishing the wave theory of x-rays. It was a very simple matter for W.H. Bragg and others to give up the corpuscular theory because of the hypothetical nature of the x-ray corpuscle. It was only necessary to recognize that Laue's results were contrary to hypothesis and the corpuscle disappeared.

If the electron were not the well-authenticated particle we know it to be, it is possible that the experiment I have described would cause it to vanish in like manner. We do not, however, anticipate any such event. The electron as a particle is too well established to be discredited by a few experiments with a nickel crystal. The most we are apt to allow is that there are circumstances in which it is more convenient to regard electrons as waves than as particles. We will allow perhaps that electrons have a dual nature—when they produce tracks in a C.T.R. Wilson cloud experiment they are particles, but when they are scattered by a crystal they are waves.

A quite similar situation exists, of course, in the case of x-rays. It has been evident for some years that the adherents of the corpuscular theory of x-rays were too enthusiastic in their recantations. X-rays also exhibit a dual nature—when they give rise to diffraction patterns they are waves, but when they exhibit the Compton effect or cause the emission of electrons from atoms they are particles—quanta or photons.

This state of affairs is one that should appeal to us as intolerable. There must, it would seem, be comprehensive modes of description applicable to all electron and x-ray phenomena, but what these are we do not yet know. We do not know whether we shall eventually believe with de Broglie and Schroedinger that electrons and x-rays are waves that sometimes masquerade as particles, with Duane that electrons and x-rays are particles that sometimes masquerade as waves, or whether eventually we shall believe with Born that we are dealing in both cases with actual particles and phantom waves.

I believe, however, that for the present and for a long time to come we shall, in describing experiments, worry but little about ultimate realities and logical consistency. We will describe each phenomenon in whatever terms we find most convenient.

27.3 Study Questions

QUES. 27.1. Do Davisson's experiments provide quantitative support for the theory of matter waves?

a) How does Davisson compute the wavelength of electrons scattered from a nickel crystal based on the spacing of the atoms at its surface? How do you suppose he knew (beforehand) the arrangement and spacing of the nickel atoms?
b) How does Davisson compute the wavelengths of the scattered electrons based on De Broglie's theory of matter waves? What were the shortest wavelength electrons which Davisson was able to produce?
c) Do the calculations based on De Broglie's theory of matter waves compare favorably with the calculations based on Laue's scattering of waves from surface atoms? To what does Davisson attribute any inconsistencies? Do you find his conclusions convincing?

QUES. 27.2. Is there a comprehensive (*i.e.* logically consistent) theory of both electrons and x-rays?

a) Why was the wave-theory of x-rays accepted more readily (when Laue performed his scattering experiments) than the wave-theory of electrons (when Davisson performed his scattering experiments)? And why does Davisson suggest that "the adherents of the corpuscular theory of x-rays were too enthusiastic in their recantations."

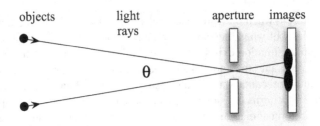

Fig. 27.5 Light reflected from two objects having an angular separation θ form two broadened images after passing through an aperture

b) How does the electron and x-ray theory of de Broglie and Schrödinger differ from that of Duane? How does the theory of Born differ from both of these? To which of these (if any) does Davisson subscribe? To which do you?

27.4   Exercises

Ex. 27.1 (DIFFRACTION LIMITS, ELECTRON IMAGING AND THE RAYLEIGH CRITERION). The resolution of any measurement apparatus—the size of the smallest discernible feature—is limited by the wavelength of the illumination source. This is essentially due to the diffraction of waves when passing through an aperture. In this exercise, we will explore how such wave diffraction imposes a limit on the resolution achievable using both optical and electron imaging techniques.

a) As a first example, consider an owl hunting for prey. Moonlight illuminates two mice on the ground far below. Light reflected from the mice (separated by 10 cm) enters the owl's iris (aperture diameter 5 mm) and forms two images on the owl's retina. Since light which passes through an aperture experiences diffraction, each of these images will be spread out a bit, as depicted schematically in Fig. 27.5. As long as the spreading is not too severe (so that the images do not overlap too much) the owl will still be able to resolve two distinct mice. According to the *Rayleigh criterion*, two images can be resolved so long as each of their central bright spots lie outside of its neighboring image's first diffraction minimum.[5] With this in mind, demonstrate that, according to the Rayleigh criterion, two images are resolvable if their angular separation is greater than θ_R, where

$$\sin \theta_R = \frac{\lambda}{d} \tag{27.2}$$

---

[5] Recall that when a light ray passes through an aperture of width d, the spreading of the ray is determined by the angular locations of the diffraction minima on either side of the central bright spot. See the treatment of single-slit diffraction in Ex. 14.2 of volume III.

At what maximum altitude can the owl still resolve the two mice?[6] (ANSWER: About 1 km.)

b) As a second example, consider a simplified electron microscope. It operates by aiming a beam of high-energy electrons (100 keV) at a sample having many tiny, detailed features. The electrons are (elastically) reflected from the sample. They travel 1 m, pass through a small circular aperture (diameter 1 mm), and are detected when they strike a small zinc-sulphide screen. What is the smallest feature that can be resolved using this electron imaging apparatus? How does this compare to the best resolution available for reflection of visible light from a sample?

EX. 27.2 (LOGICAL CONSISTENCY). Should scientists worry about "ultimate realities and logical consistency." Should anyone?

27.5   Vocabulary

1. Accentuate
2. Lateral
3. Resolve
4. Analogue
5. Unity
6. Adsorb
7. Corpuscular
8. Authenticate
9. Discredit
10. Recantation
11. Photon
12. Masquerade

---

[6] Strictly speaking, Eq. 27.2 is only valid for slit-shaped apertures. For circular apertures, the Rayleigh criterion becomes $\sin \theta_R = 1.22 \frac{\lambda}{d}$.

Chapter 28
Bohr's Atomic Model

> *A definite relation may be obtained between the spectra of the elements and the structure of their atoms on the basis of the postulates.*
>
> —Niels Bohr

28.1 Introduction

Niels Henrick David Bohr (1885–1962) was born in Copenhagen, Denmark. He received his Ph.D. from Copenhagen University in 1911. Afterwards, he moved to England, where he studied under J.J. Thomson in Cambridge and under Ernest Rutherford in Manchester. Bohr's model of the atom, based on Rutherford's planetary model, was published in 1913. He was awarded the Nobel Prize for physics in 1922 "for his services in the investigation of the structure of atoms and of the radiation emanating from them." He went on to serve as a professor at the University of Copenhagen and then as the director of the newly founded Institute of Theoretical Physics. He also served a minor role in the Manhattan project during World War II, after which he became a leading advocate for the peaceful use of nuclear energy. Among his most famous writings are a series of philosophical essays in which he clarified the meaning and the significance of the new atomic theory.[1]

The reading selections in the next few chapters make up Bohr's 1922 Noble lecture. He begins by providing a brief overview of the most important developments and discoveries which led up to his own model of the atom. How did Bohr's atomic model differ from that of Rutherford? What specific problems motivated Bohr's novel theory? And could Bohr's model account for the emission of light by atomic hydrogen?

---

[1] See, for example, Bohr, N., *Atomic Theory and the Description of Nature*, Cambridge University Press, Cambridge, 1934.

© Springer International Publishing Switzerland 2016
K. Kuehn, *A Student's Guide Through the Great Physics Texts*,
Undergraduate Lecture Notes in Physics, DOI 10.1007/978-3-319-21828-1_28

28.2   Reading: Bohr, *The Structure of the Atom*

Bohr, N., The Structure of the Atom, in *Nobel Lectures, Physics 1922–1941*, Elsevier Publishing Company, 1965. Lecture delivered by Niels Bohr on December 11, 1922.

Ladies and Gentlemen. Today, as a consequence of the great honour the Swedish Academy of Sciences has done me in awarding me this year's Nobel Prize for Physics for my work on the structure of the atom, it is my duty to give an account of the results of this work and I think that I shall be acting in accordance with the traditions of the Nobel Foundation if I give this report in the form of a survey of the development which has taken place in the last few years within the field of physics to which this work belongs.

28.2.1   The General Picture of the Atom

The present state of atomic theory is characterized by the fact that we not only believe the existence of atoms to be proved beyond a doubt, but also we even believe that we have an intimate knowledge of the constituents of the individual atoms. I cannot on this occasion give a survey of the scientific developments that have led to this result; I will only recall the discovery of the electron towards the close of the last century, which furnished the direct verification and led to a conclusive formulation of the conception of the atomic nature of electricity which had evolved since the discovery by Faraday of the fundamental laws of electrolysis and Berzelius's electrochemical theory, and had its greatest triumph in the electrolytic dissociation theory of Arrhenius. This discovery of the electron and elucidation of its properties was the result of the work of a large number of investigators, among whom Lenard and J.J. Thomson may be particularly mentioned. The latter especially has made very important contributions to our subject by his ingenious attempts to develop ideas about atomic constitution on the basis of the electron theory. The present state of our knowledge of the elements of atomic structure was reached, however, by the discovery of the atomic nucleus, which we owe to Rutherford, whose work on the radioactive substances discovered towards the close of the last century has much enriched physical and chemical science.

According to our present conceptions, an atom of an element is built up of a nucleus that has a positive electrical charge and is the seat of by far the greatest part of the atomic mass, together with a number of electrons, all having the same negative charge and mass, which move at distances from the nucleus that are very great compared to the dimensions of the nucleus or of the electrons themselves. In this picture we at once see a striking resemblance to a planetary system, such as we have in our own solar system. Just as the simplicity of the laws that govern the motions of the solar system is intimately connected with the circumstance that the dimensions of the moving bodies are small in relation to the orbits, so the corresponding relations in atomic structure provide us with an explanation of an essential feature of natural phenomena in so far as these depend on the properties of the elements. It makes clear at once that these properties can be divided into two sharply distinguished classes.

To the first class belong most of the ordinary physical and chemical properties of substances, such as their state of aggregation, colour, and chemical reactivity. These properties depend on the motion of the electron system and the way in which this motion changes under the influence of different external actions. On account of the large mass of the nucleus relative to that of the electrons and its smallness in comparison to the electron orbits, the electronic motion will depend only to a very small extent on the nuclear mass, and will be determined to a close approximation solely by the total electrical charge of the nucleus. Especially the inner structure of the nucleus and the way in which the charges and masses are distributed among its separate particles will have a vanishingly small influence on the motion of the electron system surrounding the nucleus. On the other hand, the structure of the nucleus will be responsible for the second class of properties that are shown in the radioactivity of substances. In the radioactive processes we meet with an explosion of the nucleus, whereby positive or negative particles, the so-called α- and β-particles, are expelled with very great velocities.

Our conceptions of atomic structure afford us, therefore, an immediate explanation of the complete lack of interdependence between the two classes of properties, which is most strikingly shown in the existence of substances which have to an extraordinarily close approximation the same ordinary physical and chemical properties, even though the atomic weights are not the same, and the radioactive properties are completely different. Such substances, of the existence of which the first evidence was found in the work of Soddy and other investigators on the chemical properties of the radioactive elements, are called isotopes, with reference to the classification of the elements according to ordinary physical and chemical properties. It is not necessary for me to state here how it has been shown in recent years that isotopes are found not only among the radioactive elements, but also among ordinary stable elements; in fact, a large number of the latter that were previously supposed simple have been shown by Aston's well-known investigations to consist of a mixture of isotopes with different atomic weights. The question of the inner structure of the nucleus is still but little understood, although a method of attack is afforded by Rutherford's experiments on the disintegration of atomic nuclei by bombardment with α-particles. Indeed, these experiments may be said to open up a new epoch in natural philosophy in that for the first time the artificial transformation of one element into another has been accomplished. In what follows, however, we shall confine ourselves to a consideration of the ordinary physical and chemical properties of the elements and the attempts which have been made to explain them on the basis of the concepts just outlined.

It is well known that the elements can be arranged as regards their ordinary physical and chemical properties in a *natural system* which displays most suggestively the peculiar relationships between the different elements. It was recognized for the

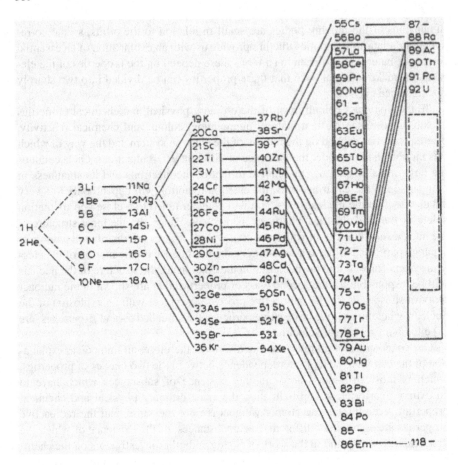

Fig. 28.1 Julius Thomsen's depiction of the periodic table.—[K.K.]

first time by Mendeleev and Lothar Meyer that when the elements are arranged in
an order which is practically that of their atomic weights, their chemical and physical properties show a pronounced periodicity. A diagrammatic representation of this
so-called Periodic Table is given in Fig. 28.1, where, however, the elements are not
arranged in the ordinary way but in a somewhat modified form of a table first given
by Julius Thomsen, who has also made important contributions to science in this
domain. In the figure the elements are denoted by their usual chemical symbols, and
the different vertical columns indicate the so-called periods. The elements in successive columns which possess homologous chemical and physical properties are
connected with lines. The meaning of the square brackets around certain series of
elements in the later periods, the properties of which exhibit typical deviations from
the simple periodicity in the first periods, will be discussed later.

In the development of the theory of atomic structure the characteristic features
of the natural system have found a surprisingly simple interpretation. Thus we are

led to assume that the ordinal number of an element in the Periodic Table, the so-called atomic number, is just equal to the number of electrons which move about the nucleus in the neutral atom. In an imperfect form, this law was first stated by Van den Broek; it was, however, foreshadowed by J.J. Thomson's investigations of the number of electrons in the atom, as well as by Rutherford's measurements of the charge on the atomic nucleus. As we shall see, convincing support for this law has since been obtained in various ways, especially by Moseley's famous investigations of the X-ray spectra of the elements. We may perhaps also point out, how the simple connexion between atomic number and nuclear charge offers an explanation of the laws governing the changes in chemical properties of the elements after expulsion of α- or β-particles, which found a simple formulation in the so-called radioactive displacement law.

28.2.2   Atomic Stability and Electrodynamic Theory

As soon as we try to trace a more intimate connexion between the properties of the elements and atomic structure, we encounter profound difficulties, in that essential differences between an atom and a planetary system show themselves here in spite of the analogy we have mentioned.

The motions of the bodies in a planetary system, even though they obey the general law of gravitation, will not be completely determined by this law alone, but will depend largely on the previous history of the system. Thus the length of the year is not determined by the masses of the sun and the earth alone, but depends also on the conditions that existed during the formation of the solar system, of which we have very little knowledge. Should a sufficiently large foreign body some day traverse our solar system, we might among other effects expect that from that day the length of the year would be different from its present value.

It is quite otherwise in the case of atoms. The definite and unchangeable properties of the elements demand that the state of an atom cannot undergo permanent changes due to external actions. As soon as the atom is left to itself again, its constituent particles must arrange their motions in a manner which is completely determined by the electric charges and masses of the particles. We have the most convincing evidence of this in spectra, that is, in the properties of the radiation emitted from substances in certain circumstances, which can be studied with such great precision. It is well known that the wavelengths of the spectral lines of a substance, which can in many cases be measured with an accuracy of more than one part in a million, are, in the same external circumstances, always exactly the same within the limit of error of the measurements, and quite independent of the previous treatment of this substance. It is just to this circumstance that we owe the great importance of spectral analysis, which has been such an invaluable aid to the chemist in the search for new elements, and has also shown us that even on the most distant bodies of the universe there occur elements with exactly the same properties as on the earth.

On the basis of our picture of the constitution of the atom it is thus impossible, so long as we restrict ourselves to the ordinary mechanical laws, to account for the characteristic atomic stability which is required for an explanation of the properties of the elements.

The situation is by no means improved if we also take into consideration the well-known electrodynamic laws which Maxwell succeeded in formulating on the basis of the great discoveries of Oersted and Faraday in the first half of the last century. Maxwell's theory has not only shown itself able to account for the already known electric and magnetic phenomena in all their details, but has also celebrated its greatest triumph in the prediction of the electromagnetic waves which were discovered by Hertz, and are now so extensively used in wireless telegraphy.

For a time it seemed as though this theory would also be able to furnish a basis for an explanation of the details of the properties of the elements, after it had been developed, chiefly by Lorentz and Larmor, into a form consistent with the atomistic conception of electricity. I need only remind you of the great interest that was aroused when Lorentz, shortly after the discovery by Zeeman of the characteristic changes that spectral lines undergo when the emitting substance is brought into a magnetic field, could give a natural and simple explanation of the main features of the phenomenon. Lorentz assumed that the radiation which we observe in a spectral line is sent out from an electron executing simple harmonic vibrations about a position of equilibrium, in precisely the same manner as the electromagnetic waves in radiotelegraphy are sent out by the electric oscillations in the antenna. He also pointed out how the alteration observed by Zeeman in the spectral lines corresponded exactly to the alteration in the motion of the vibrating electron which one would expect to be produced by the magnetic field.

It was, however, impossible on this basis to give a closer explanation of the spectra of the elements, or even of the general type of the laws holding with great exactness for the wavelengths of lines in these spectra, which had been established by Balmer, Rydberg, and Ritz. After we obtained details as to the constitution of the atom, this difficulty became still more manifest; in fact, so long as we confine ourselves to the classical electrodynamic theory we cannot even understand why we obtain spectra consisting of sharp lines at all. This theory can even be said to be incompatible with the assumption of the existence of atoms possessing the structure we have described, in that the motions of the electrons would claim a continuous radiation of energy from the atom, which would cease only when the electrons had fallen into the nucleus.

28.2.3   The Origin of the Quantum Theory

It has, however, been possible to avoid the various difficulties of the electrodynamic theory by introducing concepts borrowed from the so-called quantum theory, which marks a complete departure from the ideas that have hitherto been used for the explanation of natural phenomena. This theory was originated by Planck, in the year 1900, in his investigations on the law of heat radiation, which, because of its independence of the individual properties of substances, lent itself peculiarly well to a test of the applicability of the laws of classical physics to atomic processes.

Planck considered the equilibrium of radiation between a number of systems with the same properties as those on which Lorentz had based his theory of the Zeeman effect, but he could now show not only that classical physics could not account for the phenomena of heat radiation, but also that a complete agreement with the experimental law could be obtained if—in pronounced contradiction to classical theory—it were assumed that the energy of the vibrating electrons could not change continuously, but only in such a way that the energy of the system always remained equal to a whole number of so-called energy-quanta. The magnitude of this quantum was found to be proportional to the frequency of oscillation of the particle, which, in accordance with classical concepts, was supposed to be also the frequency of the emitted radiation. The proportionality factor had to be regarded as a new universal constant, since termed Planck's constant, similar to the velocity of light, and the charge and mass of the electron.

Planck's surprising result stood at first completely isolated in natural science, but with Einstein's significant contributions to this subject a few years after, a great variety of applications was found. In the first place, Einstein pointed out that the condition limiting the amount of vibrational energy of the particles could be tested by investigation of the specific heat of crystalline bodies, since in the case of these we have to do with similar vibrations, not of a single electron, but of whole atoms about positions of equilibrium in the crystal lattice. Einstein was able to show that the experiment confirmed Planck's theory, and through the work of later investigators this agreement has proved quite complete. Furthermore, Einstein emphasized another consequence of Planck's results, namely, that radiant energy could only be emitted or absorbed by the oscillating particle in so-called "quanta of radiation," the magnitude of each of which was equal to Planck's constant multiplied by the frequency.

In his attempts to give an interpretation of this result, Einstein was led to the formulation of the so-called "hypothesis of light-quanta", according to which the radiant energy, in contradiction to Maxwell's electromagnetic theory of light, would not be propagated as electromagnetic waves, but rather as concrete light atoms, each with an energy equal to that of a quantum of radiation. This concept led Einstein to his well-known theory of the photoelectric effect. This phenomenon, which had been entirely unexplainable on the classical theory, was thereby placed in a quite different light, and the predictions of Einstein's theory have received such exact experimental confirmation in recent years, that perhaps the most exact determination of Planck's constant is afforded by measurements on the photoelectric effect. In spite of its heuristic value, however, the hypothesis of light-quanta, which is quite irreconcilable with so-called interference phenomena, is not able to throw light on the nature of radiation. I need only recall that these interference phenomena constitute our only means of investigating the properties of radiation and therefore of assigning any closer meaning to the frequency which in Einstein's theory fixes the magnitude of the light-quantum.

In the following years many efforts were made to apply the concepts of the
quantum theory to the question of atomic structure, and the principal emphasis was
sometimes placed on one and sometimes on the other of the consequences deduced
by Einstein from Planck's result. As the best known of the attempts in this direction,
from which, however, no definite results were obtained, I may mention the work of
Stark, Sommerfeld, Hasenöhrl, Haas, and Nicholson.

From this period also dates an investigation by Bjerrum on infrared absorption
bands, which, although it had no direct bearing on atomic structure, proved significant for the development of the quantum theory. He directed attention to the fact
that the rotation of the molecules in a gas might be investigated by means of the
changes in certain absorption lines with temperature. At the same time he emphasized the fact that the effect should not consist of a continuous widening of the lines
such as might be expected from classical theory, which imposed no restrictions on
the molecular rotations, but in accordance with the quantum theory he predicted
that the lines should be split up into a number of components, corresponding to a
sequence of distinct possibilities of rotation. This prediction was confirmed a few
years later by Eva von Bahr, and the phenomenon may still be regarded as one of
the most striking evidences of the reality of the quantum theory, even though from
our present point of view the original explanation has undergone a modification in
essential details.

28.2.4   The Quantum Theory of Atomic Constitution

The question of further development of the quantum theory was in the meantime
placed in a new light by Rutherford's discovery of the atomic nucleus (1911). As
we have already seen, this discovery made it quite clear that by classical conceptions
alone it was quite impossible to understand the most essential properties of atoms.
One was therefore led to seek for a formulation of the principles of the quantum
theory that could immediately account for the stability in atomic structure and the
properties of the radiation sent out from atoms, of which the observed properties
of substances bear witness. Such a formulation was proposed (1913) by the present
lecturer in the form of two postulates, which may be stated as follows:

(1) Among the conceivably possible states of motion in an atomic system there
exist a number of so-called stationary states which, in spite of the fact that the
motion of the particles in these states obeys the laws of classical mechanics to
a considerable extent, possess a peculiar, mechanically unexplainable stability,
of such a sort that every permanent change in the motion of the system must
consist in a complete transition from one stationary state to another.

(2) While in contradiction to the classical electromagnetic theory no radiation takes
place from the atom in the stationary states themselves, a process of transition
between two stationary states can be accompanied by the emission of electromagnetic radiation, which will have the same properties as that which would be

sent out according to the classical theory from an electrified particle executing an harmonic vibration with constant frequency. This frequency v has, however, no simple relation to the motion of the particles of the atom, but is given by the relation

$$h v = E' - E'' \tag{28.1}$$

where h is Planck's constant, and E' and E'' are the values of the energy of the atom in the two stationary states that form the initial and final state of the radiation process. Conversely, irradiation of the atom with electromagnetic waves of this frequency can lead to an absorption process, whereby the atom is transformed back from the latter stationary state to the former.

While the first postulate has in view the general stability of the atom, the second postulate has chiefly in view the existence of spectra with sharp lines. Furthermore, the quantum-theory condition entering in the last postulate affords a starting-point for the interpretation of the laws of series spectra.

The most general of these laws, the combination principle enunciated by Ritz, states that the frequency v for each of the lines in the spectrum of an element can be represented by the formula

$$v = T'' - T', \tag{28.2}$$

where T'' and T' are two so-called "spectral terms" belonging to a manifold of such terms characteristic of the substance in question.

According to our postulates, this law finds an immediate interpretation in the assumption that the spectrum is emitted by transitions between a number of stationary states in which the numerical value of the energy of the atom is equal to the value of the spectral term multiplied by Planck's constant. This explanation of the combination principle is seen to differ fundamentally from the usual ideas of electrodynamics, as soon as we consider that there is no simple relation between the motion of the atom and the radiation sent out. The departure of our considerations from the ordinary ideas of natural philosophy becomes particularly evident, however, when we observe that the occurrence of two spectral lines, corresponding to combinations of the same spectral term with two other different terms, implies that the nature of the radiation sent out from the atom is not determined only by the motion of the atom at the beginning of the radiation process, but also depends on the state to which the atom is transferred by the process.

At first glance one might, therefore, think that it would scarcely be possible to bring our formal explanation of the combination principle into direct relation with our views regarding the constitution of the atom, which, indeed, are based on experimental evidence interpreted on classical mechanics and electrodynamics. A closer investigation, however, should make it clear that a definite relation may be obtained between the spectra of the elements and the structure of their atoms on the basis of the postulates.

Fig. 28.2 Selected transitions
(*arrows*) between stationary
states (*circles*).—[*K.K.*]

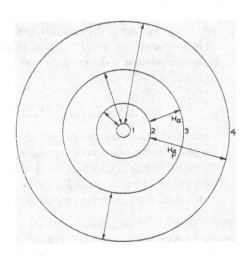

28.2.5 The Hydrogen Spectrum

The simplest spectrum we know is that of hydrogen. The frequencies of its lines
may be represented with great accuracy by means of Balmer's formula:

$$\nu = K\left(\frac{1}{n''^2} - \frac{1}{n'^2}\right) \tag{28.3}$$

where K is a constant and n' and n'' are two integers. In the spectrum we accordingly
meet a single series of spectral terms of the form K/n^2, which decrease regularly
with increasing term number n. In accordance with the postulates, we shall therefore assume that each of the hydrogen lines is emitted by a transition between two
states belonging to a series of stationary states of the hydrogen atom in which the
numerical value of the atom's energy is equal to hK/n^2.

Following our picture of atomic structure, a hydrogen atom consists of a positive nucleus and an electron which—so far as ordinary mechanical conceptions are
applicable—will with great approximation describe a periodic elliptical orbit with
the nucleus at one focus. The major axis of the orbit is inversely proportional to the
work necessary completely to remove the electron from the nucleus, and, in accordance with the above, this work in the stationary states is just equal to hK/n^2. We
thus arrive at a manifold of stationary states for which the major axis of the electron
orbit takes on a series of discrete values proportional to the squares of the whole
numbers. The accompanying Fig. 28.2 shows these relations diagrammatically. For
the sake of simplicity the electron orbits in the stationary states are represented by
circles, although in reality the theory places no restriction on the eccentricity of the
orbit, but only determines the length of the major axis. The arrows represent the
transition processes that correspond to the red and green hydrogen lines, H_α and
H_β, the frequency of which is given by means of the Balmer formula when we put
$n'' = 2$ and $n' = 3$ and 4 respectively. The transition processes are also represented

which correspond to the first three lines of the series of ultraviolet lines found by Lyman in 1914, of which the frequencies are given by the formula when n'' is put equal to 1, as well as to the first line of the infrared series discovered some years previously by Paschen, which are given by the formula if n'' is put equal to 3.

This explanation of the origin of the hydrogen spectrum leads us quite naturally to interpret this spectrum as the manifestation of a process whereby the electron is bound to the nucleus. While the largest spectral term with term number 1 corresponds to the final stage in the binding process, the small spectral terms that have larger values of the term number correspond to stationary states which represent the initial states of the binding process, where the electron orbits still have large dimensions, and where the work required to remove an electron from the nucleus is still small. The final stage in the binding process we may designate as the normal state of the atom, and it is distinguished from the other stationary states by the property that, in accordance with the postulates, the state of the atom can only be changed by the addition of energy whereby the electron is transferred to an orbit of larger dimensions corresponding to an earlier stage of the binding process.

The size of the electron orbit in the normal state calculated on the basis of the above interpretation of the spectrum agrees roughly with the value for the dimensions of the atoms of the elements that have been calculated by the kinetic theory of matter from the properties of gases. Since, however, as an immediate consequence of the stability of the stationary states that is claimed by the postulates, we must suppose that the interaction between two atoms during a collision cannot be completely described with the aid of the laws of classical mechanics, such a comparison as this cannot be carried further on the basis of such considerations as those just outlined.

A more intimate connexion between the spectra and the atomic model has been revealed, however, by an investigation of the motion in those stationary states where the term number is large, and where the dimensions of the electron orbit and the frequency of revolution in it vary relatively little when we go from one stationary state to the next following. It was possible to show that the frequency of the radiation sent out during the transition between two stationary states, the difference of the term numbers of which is small in comparison to these numbers themselves, tended to coincide in frequency with one of the harmonic components into which the electron motion could be resolved, and accordingly also with the frequency of one of the wave trains in the radiation which would be emitted according to the laws of ordinary electrodynamics.

The condition that such a coincidence should occur in this region where the stationary states differ but little from one another proves to be that the constant in the Balmer formula can be expressed by means of the relation

$$K = \frac{2\pi^2 e^4 m}{h^3} \tag{28.4}$$

where e and m are respectively the charge and mass of the electron, while h is Planck's constant. This relation has been shown to hold to within the considerable accuracy with which, especially through the beautiful investigations of Millikan, the quantities e, m, and h are known.

This result shows that there exists a connexion between the hydrogen spectrum and the model for the hydrogen atom which, on the whole, is as close as we might hope considering the departure of the postulates from the classical mechanical and electrodynamic laws. At the same time, it affords some indication of how we may perceive in the quantum theory, in spite of the fundamental character of this departure, a natural generalization of the fundamental concepts of the classical electrodynamic theory. To this most important question we shall return later, but first we will discuss how the interpretation of the hydrogen spectrum on the basis of the postulates has proved suitable in several ways, for elucidating the relation between the properties of the different elements.

28.3   Study Questions

QUES. 28.1. What are the strengths and weaknesses of the planetary model of the atom?

a) What is the planetary model of the atom? On what grounds was it accepted? In particular, what types of phenomena could it elucidate?
b) How are electron orbits *unlike* those of planets? And how is this connected to spectral analysis? To atomic stability?

QUES. 28.2. What is a light quantum?

a) What is the origin of idea of quantization of energy? What experiment did Einstein use the hypothesis of light quanta to explain?
b) With what phenomena is the concept of the light quantum irreconcilable?

QUES. 28.3. How is the Bohr model of the atom similar to, or different from, the planetary model?

a) What were the postulates which formed the basis of Bohr's quantum theory? In particular, what is a *stationary state*, and why is it so-called? How are stationary states related to electromagnetic radiation from atoms?
b) What do these two postulates aim to explain? Are they successful? In what sense do these postulates provide a departure from "the ordinary ideas of natural philosophy"?

QUES. 28.4. Can the Bohr model of the atom account for the observed emission spectrum of atomic hydrogen?

a) What is the Balmer formula? What does it mean?
b) Is the emission spectrum of hydrogen continuous or discreet? How does the Balmer formula account for this?
c) According to the Bohr model, how can one interpret, say, the red and green hydrogen emission lines? The ionization energy of hydrogen?
d) What fundamental constants appear in Bohr's model of the hydrogen atom? Do the measured values of these constants support Bohr's model? What does this suggest?

28.4 Exercises

EX. 28.1 (BOHR MODEL). According to the planetary model, a hydrogen atom consists of a positively charged proton orbited by a negatively charged electron. In this exercise, we will explore how Bohr's model is similar to, and different from, Rutherford's planetary model.

a) Using Newton's second law of motion to relate the coulomb force to the electron's centripetal acceleration, show that the speed of an electron orbiting the proton at distance r is given by

$$v(r) = \sqrt{\frac{kq^2}{mr}},$$

where k is coulomb's constant, q is the elementary charge, and m is the electron mass.

b) What is the angular momentum, l, of the electron in terms of m, v, and r?

c) Suppose that l were somehow "quantized" so that it could only have values $l = n\hbar$, where n is an integer and $\hbar$ is a constant with units of angular momentum. Show that the electron orbits are now also quantized, with orbital radii

$$r_n = \frac{n^2\hbar^2}{kq^2m}.$$

d) What is the kinetic energy of the orbiting electron? The kinetic energy should be positive, and should depend upon the orbital radius. What is the electrical potential energy of the electron-proton system? (HINT: The kinetic and potential energies should have opposite signs and should depend on the orbital radius.)

e) Show that the total energy of an orbiting electron (whose angular momentum is quantized) can be written as

$$E_n = -\frac{2k^2q^4m\pi^2}{h^2}\frac{1}{n^2}, \tag{28.5}$$

where $\hbar = h/2\pi$. Does this answer agree with Bohr's result? (HINT: What system of units is Bohr using?)

f) What is the frequency of a photon emitted during a transition between states E_4 and E_2. Does this lie in the visible range of the spectrum?

g) What is the shortest wavelength photon which can be absorbed by a hydrogen atom?

EX. 28.2 (BRACKETT SERIES PROBLEM). In the line spectrum of atomic hydrogen there is also a group of lines known as the Brackett series. These lines are produced when electrons, excited to high energy levels, make transitions to the $n = 4$ level. Find the longest and the shortest wavelengths of the Brackett series. Are these in the visible range of the electromagnetic spectrum?

Fig. 28.3 Clockwise from
left: a spectrometer (Sargent-
Welch model S-75903-80), a
5000 V discharge tube power
supply (model WL2393D)
with a hydrogen spectrum
tube (model WLS-68755-
30G), the spectrometer light
shields, and a 600 line/mm
diffraction grating (model
CP86763-01)

EX. 28.3 (HYDROGEN EMISSION SPECTRUM LABORATORY). In this laboratory exercise, you will observe the emission spectrum of hydrogen gas and thereby measure
the value of the Rydberg constant. You will need a grating spectrometer, a discharge
tube power supply, and both hydrogen and mercury spectrum tubes (Fig. 28.3). The
grating spectrometer consists of three main components: a collimator, a diffraction
grating, and a telescope. The collimator is an optical device which forms a beam
of parallel light rays from an initially divergent light source (such as a hydrogen
spectrum tube). The front of the collimator is fitted with an adjustable slit which
allows only a small ribbon of light from the source into the collimator tube. The
collimated light strikes the diffraction grating, which separates the light according
to the wavelengths of its constituent colors. By measuring the angular position of
each colored fringe using the telescope, the precise wavelength of each line in the
hydrogen emission spectrum can be readily determined.

The particular spectrometer that you use will likely include a detailed description
of its components and specific instructions for its use. Before measuring the emission lines of hydrogen, you will need to calibrate your spectrometer by using a light
source with a well-known emission line, such as the mercury green line. Set up your
spectrometer in a dark room and place the vertical entrance slit of the collimator
tube very near the mercury discharge tube. Adjust the spectrometer until you can
clearly see the mercury emission lines. One of these is a brilliant green line, which
has a wavelength of 546.1 nm. Measure the angles at which the first order ($m = 1$)
and second order ($m = 2$) mercury green lines occur with respect to the central
white fringe ($m = 0$). Then use the diffraction grating formula,[2]

$$m\lambda = d \sin \theta_m \qquad\qquad m = 0, 1, 2, \ldots \qquad (28.6)$$

---
[2] See the discussion of wave interference, and especially Ex. 20.4, in Chap. 20 of volume III.

to determine the grating spacing, d. After thus calibrating your diffraction grating, set up your spectrometer to measure the hydrogen emission lines. Depending on conditions, you should be able to observe first order and second order fringes of red, blue-green, violet, and perhaps even far-violet light by sweeping your telescope through a wide range of angles. It may take some time for your eyes to adjust to the dark; a small red LED light is useful for recording information in your lab book without disturbing your night-vision. Record the angular positions of each of these fringes, both to the left and to the right of the central white fringe. From your data, and the grating spacing which you previously determined, calculate the wavelength (and frequency) of each of the observed hydrogen lines.

Now, attempt to model your data using Balmer's formula (Eq. 28.3). Each of the visible lines in the hydrogen emission spectrum corresponds to a transition from a high energy state, n', to a low energy state, $n'' = 2$. You will need to assign to each of your emission lines a particular value of n'. (How should you do this?) Then, plot the frequency, ν, versus $1/n'$ and find the slope; this yields the value of the Rydberg constant, K. (By how much does your measured value differ from the accepted value?) Also, use your plot to determine the series limit: the frequency as $n' \to \infty$. (What does the series limit signify?) Finally, compute the value of Planck's constant, h, from your measured value of the Rydberg constant.

28.5 Vocabulary

1. Radioactive
2. Isotope
3. Element
4. Homologous
5. X-ray
6. Spectra
7. Radiotelegraphy
8. Quanta
9. Lattice
10. Photoelectric
11. Heuristic
12. Interference
13. Manifold
14. Eccentricity

Chapter 29
Atomic Spectra and Quantum Numbers

> The stationary states compose a more complex manifold, in
> which, according to these formal methods, each state is
> characterized by several whole numbers, the so-called quantum
> numbers.
>
> —Niels Bohr

29.1 Introduction

Since at least the thirteenth century, it had been known that sunlight could be divided
into a broad rainbow of colors by passing it through droplets of water or transparent crystals. This phenomenon of the *dispersion* of light was explored with great
care by Isaac Newton in the late seventeenth century by shining pencil-thin rays of
sunlight through triangular glass prisms.[1] But it was not until the mid-eighteenth
century—with the development of the spectroscope by Kirchhoff and Bunsen in
Heidelberg—that specific chemical elements could be identified by measuring their
unique emission spectra. Anders Ångström had recently found that hydrogen gas
emits a discrete set of spectral lines (rather than the continuous rainbow-like emission spectrum studied by Newton) when subjected to a high-voltage electrical
discharge. The wavelengths of these spectral lines could be measured using the
spectroscope. They were found to exhibit a clearly discernible pattern which was, in
turn, described mathematically in 1885 by the Swiss mathematician Johann Balmer.
Balmer's formula was later generalized by Johannes Rydberg so as to account more
completely for the emission spectrum of hydrogen.[2]

From the viewpoint of a planetary model of the atom, the discrete emission
spectrum of hydrogen (and the other elements) made no sense. After all, like planets orbiting the sun, electrons should be able to orbit an atomic nucleus at any
conceivable frequency. So according to the Maxwell's theory of electrodynamics,

---

[1] See, for example, Propositions II-V of Book I in Newton, I., *Opticks: or A Treatise of the
Reflections, Refractions, Inflections & Colours of Light*, 4th ed., William Innes at the West-End
of St. Pauls, London, 1730.

[2] See Eq. 28.3.

© Springer International Publishing Switzerland 2016

K. Kuehn, *A Student's Guide Through the Great Physics Texts*,
Undergraduate Lecture Notes in Physics, DOI 10.1007/978-3-319-21828-1_29

they should be capable of emitting every conceivable color of light. But atomic gases typically emit (and absorb) only particular colors. Moreover, orbiting electrons should *continually* emit radiation—and hence lose energy—spiraling into the atomic nucleus. The atom should thus be unstable.

In order to address these seemingly fatal flaws of the planetary model, Bohr made two novel postulates. First, he proposed that electrons can reside in so-called *stationary states*. When in one of these peculiar states, either the electron is not moving (hence, it is "stationary"), or the classical theory of radiation simply does not apply. How such a scenario might arise Bohr could only surmise. Second, he proposed that when an electron makes a transition between two stationary states, it emits a single photon—Einstein's proposed quantum of light—whose frequency (as it were) is determined strictly by the energy difference between the initial and final stationary states. Bohr's two postulates, when appended to the planetary model, were able to account (at least nominally) for the empirical formula developed by Balmer and Rydberg to describe the emission spectrum of hydrogen. In the reading selection that follows, which is a continuation of his 1922 Nobel lecture, Bohr explains how his model was expanded so as to account for the emission spectra of atoms having more complex electronic structures. In so doing, he introduces the reader to the concept of *quantum numbers*.

29.2   Reading: Bohr, *The Structure of the Atom*

Bohr, N., The Structure of the Atom, in *Nobel Lectures, Physics 1922-1941*, Elsevier Publishing Company, 1965. Lecture delivered by Niels Bohr on December 11, 1922.

29.2.1   *Relationships Between the Elements*

The discussion above can be applied immediately to the process whereby an electron is bound to a nucleus with any given charge. The calculations show that, in the stationary state corresponding to a given value of the number n, the size of the orbit will be inversely proportional to the nuclear charge, while the work necessary to remove an electron will be directly proportional to the square of the nuclear charge. The spectrum that is emitted during the binding of an electron by a nucleus with charge N times that of the hydrogen nucleus can therefore be represented by the formula:

$$v = N^2 K \left(\frac{1}{n''^2} - \frac{1}{n'^2} \right) \tag{29.1}$$

If in this formula we put $N = 2$, we get a spectrum which contains a set of lines in the visible region which was observed many years ago in the spectrum of certain stars. Rydberg assigned these lines to hydrogen because of the close analogy with the series of lines represented by the Balmer formula. It was never possible to produce these lines in pure hydrogen, but just before the theory for the hydrogen spectrum was put forward, Fowler succeeded in observing the series in question by sending a strong discharge through a mixture of hydrogen and helium. This investigator also assumed that the lines were hydrogen lines, because there existed no experimental evidence from which it might be inferred that two different substances could show properties resembling each other so much as the spectrum in question and that of hydrogen. After the theory was put forward, it became clear, however, that the observed lines must belong to a spectrum of helium, but that they were not like the ordinary helium spectrum emitted from the neutral atom. They came from an ionized helium atom which consists of a single electron moving about a nucleus with double charge. In this way there was brought to light a new feature of the relationship between the elements, which corresponds exactly with our present ideas of atomic structure, according to which the physical and chemical properties of an element depend in the first instance only on the electric charge of the atomic nucleus.

Soon after this question was settled the existence of a similar general relationship between the properties of the elements was brought to light by Moseley's well-known investigations on the characteristic X-ray spectra of the elements, which was made possible by Laue's discovery of the interference of X-rays in crystals and the investigations of W.H. and W.L. Bragg on this subject. It appeared, in fact, that the X-ray spectra of the different elements possessed a much simpler structure and a much greater mutual resemblance than their optical spectra. In particular, it appeared that the spectra changed from element to element in a manner that corresponded closely to the formula given above for the spectrum emitted during the binding of an electron to a nucleus, provided N was put equal to the atomic number of the element concerned. This formula was even capable of expressing, with an approximation that could not be without significance, the frequencies of the strongest X-ray lines, if small whole numbers were substituted for n' and n''.

This discovery was of great importance in several respects. In the first place, the relationship between the X-ray spectra of different elements proved so simple that it became possible to fix without ambiguity the atomic number for all known substances, and in this way to predict with certainty the atomic number of all such hitherto unknown elements for which there is a place in the natural system. Figure 29.1 shows how the square root of the frequency for two characteristic X-ray lines depends on the atomic number. These lines belong to the group of so-called K-lines, which are the most penetrating of the characteristic rays. With very close approximation the points lie on straight lines, and the fact that they do so is conditioned not only by our taking account of known elements, but also by our leaving an open place between molybdenum (42) and ruthenium (44), just as in Mendeleev's original scheme of the natural system of the elements.

Further, the laws of X-ray spectra provide a confirmation of the general theoretical conceptions, both with regard to the constitution of the atom and the ideas that have served as a basis for the interpretation of spectra. Thus the similarity between X-ray spectra and the spectra emitted during the binding of a single electron to a

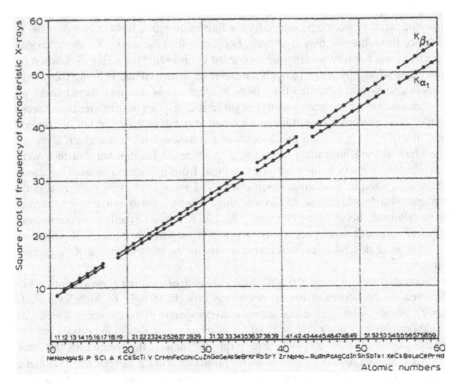

Fig. 29.1 The (square root of the) frequency of the K_α and K_β X-ray lines emitted from known elements scales linearly with atomic number; gaps clearly announce missing elements.—[K.K.]

nucleus may be simply interpreted from the fact that the transitions between stationary states with which we are concerned in X-ray spectra are accompanied by changes in the motion of an electron in the inner part of the atom, where the influence of the attraction of the nucleus is very great compared with the repulsive forces of the other electrons.

The relations between other properties of the elements are of a much more complicated character, which originates in the fact that we have to do with processes concerning the motion of the electrons in the outer part of the atom, where the forces that the electrons exert on one another are of the same order of magnitude as the attraction towards the nucleus, and where, therefore, the details of the interaction of the electrons play an important part. A characteristic example of such a case is afforded by the spatial extension of the atoms of the elements. Lothar Meyer himself directed attention to the characteristic periodic change exhibited by the ratio of the atomic weight to the density, the so-called atomic volume, of the elements in the natural system. An idea of these facts is given by Fig. 29.2, in which the atomic volume is represented as a function of the atomic number. A greater difference between this and the previous figure could scarcely be imagined. While the X-ray spectra vary

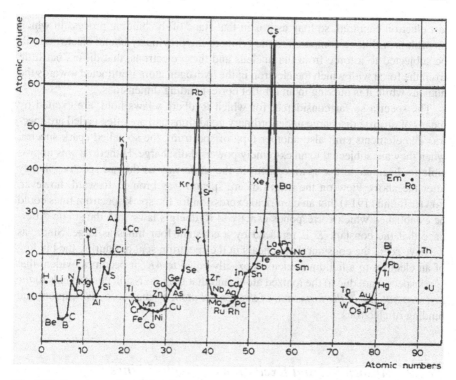

Fig. 29.2 Some properties of the elements—such as atomic volume—exhibit a periodic (rather than a linear) dependence on atomic number.—[K.K.]

uniformly with the atomic number, the atomic volumes show a characteristic periodic change which corresponds exactly to the change in the chemical properties of the elements.

Ordinary optical spectra behave in an analogous way. In spite of the dissimilarity between these spectra, Rydberg succeeded in tracing a certain general relationship between the hydrogen spectrum and other spectra. Even though the spectral lines of the elements with higher atomic number appear as combinations of a more complicated manifold of spectral terms which is not so simply co-ordinated with a series of whole numbers, still the spectral terms can be arranged in series each of which shows a strong similarity to the series of terms in the hydrogen spectrum. This similarity appears in the fact that the terms in each series can, as Rydberg pointed out, be very accurately represented by the formula $K/(n + \alpha)^2$, where K is the same constant that occurs in the hydrogen spectrum, often called the Rydberg constant, while n is the term number, and α a constant which is different for the different series.

This relationship with the hydrogen spectrum leads us immediately to regard these spectra as the *last step of a process whereby the neutral atom is built up by the capture and binding of electrons to the nucleus*, one by one. In fact, it is clear that the last electron captured, so long as it is in that stage of the binding process in which its orbit is still large compared to the orbits of the previously bound electrons, will be subjected to a force from the nucleus and these electrons, that differs but little from the force with which the electron in the hydrogen atom is attracted towards the nucleus while it is moving in an orbit of corresponding dimensions.

The spectra so far considered, for which Rydberg's laws hold, are excited by means of electric discharge under ordinary conditions and are often called arc spectra. The elements emit also another type of spectrum, the so-called spark spectra, when they are subjected to an extremely powerful discharge. Hitherto it was impossible to disentangle the spark spectra in the same way as the arc spectra. Shortly after the above view on the origin of arc spectra was brought forward, however, Fowler found (1914) that an empirical expression for the spark spectrum lines could be established which corresponds exactly to Rydberg's laws with the single difference that the constant K is replaced by a constant four times as large. Since, as we have seen, the constant that appears in the spectrum sent out during the binding of an electron to a helium nucleus is exactly equal to $4K$, it becomes evident that spark spectra are due to the ionized atom, and that their emission corresponds to *the last step but one in the formation of the neutral atom* by the successive capture and binding of electrons.

29.2.2 Absorption and Excitation of Spectral Lines

The interpretation of the origin of the spectra was also able to explain the characteristic laws that govern absorption spectra. As Kirchhoff and Bunsen had already shown, there is a close relation between the selective absorption of substances for radiation and their emission spectra, and it is on this that the application of spectrum analysis to the heavenly bodies essentially rests. Yet on the basis of the classical electromagnetic theory, it is impossible to understand why substances in the form of vapour show absorption for certain lines in their emission spectrum and not for others.

On the basis of the postulates given above we are, however, led to assume that the absorption of radiation corresponding to a spectral line emitted by a transition from one stationary state of the atom to a state of less energy is brought about by the return of the atom from the last-named state to the first. We thus understand immediately that in ordinary circumstances a gas or vapour can only show selective absorption for spectral lines that are produced by a transition from a state corresponding to an earlier stage in the binding process to the normal state. Only at higher temperatures or under the influence of electric discharges whereby an appreciable number of atoms are being constantly disrupted from the normal state, can we expect absorption for other lines in the emission spectrum in agreement with the experiments.

A most direct confirmation for the general interpretation of spectra on the basis of the postulates has also been obtained by investigations on the excitation of spectral lines and ionization of atoms by means of impact of free electrons with given velocities. A decided advance in this direction was marked by the well-known investigations of Franck and Hertz (1914). It appeared from their results that by means of electron impacts it was impossible to impart to an atom an arbitrary amount of energy, but only such amounts as corresponded to a transfer of the atom from its normal state to another stationary state of the existence of which the spectra assure us, and the energy of which can be inferred from the magnitude of the spectral term. Further, striking evidence was afforded of the independence that, according to the postulates, must be attributed to the processes which give rise to the emission of the different spectral lines of an element. Thus it could be shown directly that atoms that were transferred in this manner to a stationary state of greater energy were able to return to the normal state with emission of radiation corresponding to a single spectral line.

Continued investigations on electron impacts, in which a large number of physicists have shared, have also produced a detailed confirmation of the theory concerning the excitation of series spectra. Especially it has been possible to show that for the ionization of an atom by electron impact an amount of energy is necessary that is exactly equal to the work required, according to the theory, to remove the last electron captured from the atom. This work can be determined directly as the product of Planck's constant and the spectral term corresponding to the normal state, which, as mentioned above, is equal to the limiting value of the frequencies of the spectral series connected with selective absorption.

29.2.3   *The Quantum Theory of Multiply-Periodic Systems*

While it was thus possible by means of the fundamental postulates of the quantum theory to account directly for certain general features of the properties of the elements, a closer development of the ideas of the quantum theory was necessary in order to account for these properties in further detail. In the course of the last few years a more general theoretical basis has been attained through the development of formal methods that permit the fixation of the stationary states for electron motions of a more general type than those we have hitherto considered. For a simply periodic motion such as we meet in the pure harmonic oscillator, and at least to a first approximation, in the motion of an electron about a positive nucleus, the manifold of stationary states can be simply co-ordinated to a series of whole numbers. For motions of the more general class mentioned above, the so-called multiply periodic motions, however, the stationary states compose a more complex manifold, in which, according to these formal methods, each state is characterized by several whole numbers, the so-called "quantum numbers".

In the development of the theory a large number of physicists have taken part, and the introduction of several quantum numbers can be traced back to the work of Planck himself. But the definite step which gave the impetus to further work was made by Sommerfeld (1915) in his explanation of the fine structure shown by the hydrogen lines when the spectrum is observed with a spectroscope of high resolving

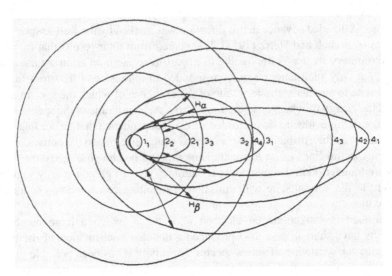

Fig. 29.3 Transitions corresponding to the *red* (H_α) and *green* (H_β) lines of atomic hydrogen. The precise frequencies of the lines depend on both the principle quantum number, n, and the subordinate quantum number, k, of the stationary states between which transitions occur.—[*K.K.*]

power. The occurrence of this fine structure must be ascribed to the circumstance that we have to deal, even in hydrogen, with a motion which is not exactly simply periodic. In fact, as a consequence of the change in the electron's mass with velocity that is claimed by the theory of relativity, the electron orbit will undergo a very slow precession in the orbital plane. The motion will therefore be doubly periodic, and besides a number characterizing the term in the Balmer formula, which we shall call the principal quantum number because it determines in the main the energy of the atom, the fixation of the stationary states demands another quantum number which we shall call the subordinate quantum number.

A survey of the motion in the stationary states thus fixed is given in the diagram (Fig. 29.3), which reproduces the relative size and form of the electron orbits. Each orbit is designated by a symbol n_k, where n is the principal quantum number and k the subordinate quantum number. All orbits with the same principal quantum number have, to a first approximation, the same major axis, while orbits with the same value of k have the same parameter, *i.e.* the same value for the shortest chord through the focus. Since the energy values for different states with the same value of n but different values of k differ a little from each other, we get for each hydrogen line corresponding to definite values of n' and n'' in the Balmer formula a number of different transition processes, for which the frequencies of the emitted radiation as calculated by the second postulate are not exactly the same. As Sommerfeld was able to show, the components this gives for each hydrogen line agree with the observations on the fine structure of hydrogen lines to within the limits of experimental error. In the figure the arrows designate the processes that give rise to the components of the red and green lines in the hydrogen spectrum, the frequencies of which are obtained by putting $n'' = 2$ and $n' = 3$ or 4 respectively in the Balmer formula.

In considering the figure it must not be forgotten that the description of the orbit is there incomplete, in so much as with the scale used the slow precession does not show at all. In fact, this precession is so slow that even for the orbits that rotate most rapidly the electron performs about 40,000 revolutions before the perihelion has gone round once. Nevertheless, it is this precession alone that is responsible for the multiplicity of the stationary states characterized by the subordinate quantum number. If, for example, the hydrogen atom is subjected to a small disturbing force which perturbs the regular precession, the electron orbit in the stationary states will have a form altogether different from that given in the figure. This implies that the fine structure will change its character completely, but the hydrogen spectrum will continue to consist of lines that are given to a close approximation by the Balmer formula, due to the fact that the approximately periodic character of the motion will be retained. Only when the disturbing forces become so large that even during a single revolution of the electron the orbit is appreciably disturbed, will the spectrum undergo essential changes. The statement often advanced that the introduction of two quantum numbers should be a necessary condition for the explanation of the Balmer formula must therefore be considered as a misconception of the theory.

Sommerfeld's theory has proved itself able to account not only for the fine structure of the hydrogen lines, but also for that of the lines in the helium spark spectrum. Owing to the greater velocity of the electron, the intervals between the components into which a line is split up are here much greater and can be measured with much greater accuracy. The theory was also able to account for certain features in the fine structure of X-ray spectra, where we meet frequency differences that may even reach a value more than a million times as great as those of the frequency differences for the components of the hydrogen lines.

Shortly after this result had been attained, Schwarzschild and Epstein (1916) simultaneously succeeded, by means of similar considerations, in accounting for the characteristic changes that the hydrogen lines undergo in an electric field, which had been discovered by Stark in the year 1914. Next, an explanation of the essential features of the Zeeman effect for the hydrogen lines was worked out at the same time by Sommerfeld and Debye (1917). In this instance the application of the postulates involved the consequence that only certain orientations of the atom relative to the magnetic field were allowable, and this characteristic consequence of the quantum theory has quite recently received a most direct confirmation in the beautiful researches of Stern and Gerlach on the deflexion of swiftly moving silver atoms in a nonhomogenous magnetic field.

29.2.4 The Correspondence Principle

While this development of the theory of spectra was based on the working out of formal methods for the fixation of stationary states, the present lecturer succeeded shortly afterwards in throwing light on the theory from a new viewpoint, by pursuing further the characteristic connexion between the quantum theory and classical electrodynamics already traced out in the hydrogen spectrum. In connexion with the important work of Ehrenfest and Einstein these efforts led to the formulation of the so-called correspondence principle, according to which the occurrence of transitions between the stationary states accompanied by emission of radiation is traced back to the harmonic components into which the motion of the atom may be resolved and which, according to the classical theory, determine the properties of the radiation to which the motion of the particles gives rise.

According to the correspondence principle, it is assumed that every transition process between two stationary states can be co-ordinated with a corresponding harmonic vibration component in such a way that the probability of the occurrence of the transition is dependent on the amplitude of the vibration. The state of polarization of the radiation emitted during the transition depends on the further characteristics of the vibration, in a manner analogous to that in which on the classical theory the intensity and state of polarization in the wave system emitted by the atom as a consequence of the presence of this vibration component would be determined respectively by the amplitude and further characteristics of the vibration.

With the aid of the correspondence principle it has been possible to confirm and to extend the above-mentioned results. Thus it was possible to develop a complete quantum theory explanation of the Zeeman effect for the hydrogen lines, which, in spite of the essentially different character of the assumptions that underlie the two theories, is very similar throughout to Lorentz's original explanation based on the classical theory. In the case of the Stark effect, where, on the other hand, the classical theory was completely at a loss, the quantum theory explanation could be so extended with the help of the correspondence principle as to account for the polarization of the different components into which the lines are split, and also for the characteristic intensity distribution exhibited by the components. This last question has been more closely investigated by Kramers, and the accompanying figure will give some impression of how completely it is possible to account for the phenomenon under consideration.

Figure 29.4 reproduces one of Stark's well-known photographs of the splitting up of the hydrogen lines. The picture displays very well the varied nature of the phenomenon, and shows in how peculiar a fashion the intensity varies from component to component. The components below are polarized perpendicular to the field, while those above are polarized parallel to the field.

Figure 29.5 gives a diagrammatic representation of the experimental and theoretical results for the line $H\gamma$, the frequency of which is given by the Balmer formula with $n'' = 2$ and $n' = 5$. The vertical lines denote the components into which the line is split up, of which the picture on the right gives the components which are polarized parallel to the field and that on the left those that are polarized perpendicular to it. The experimental results are represented in the upper half of the diagram, the distances from the dotted line representing the measured displacements of the components, and the lengths of the lines being proportional to the relative intensity as estimated by Stark from the blackening of the photographic plate. In the

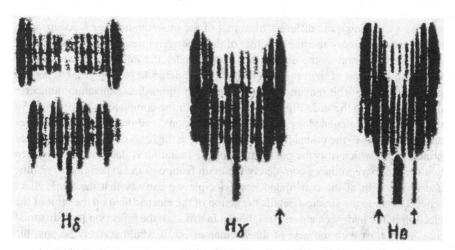

Fig. 29.4 Stark's observation of the splitting of the H_δ (*left*), H_γ (*center*) or H_β (*right*) hydrogen emission line by an applied electric field. The emission lines exhibit polarizations both parallel (*upper*) and perpendicular (*lower*) to the applied field.—[*K.K.*]

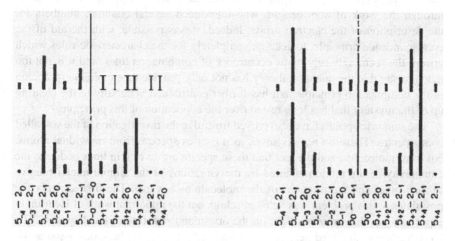

Fig. 29.5 A comparison of the experimental (*top*) and theoretical (*bottom*) splitting of the H_γ line by an electric field for both parallel (*right*) and perpendicular (*left*) polarizations. The lines are labelled according to the principle quantum number, n, and the subordinate quantum number, s, of the initial and final states.—[*K.K.*]

lower half is given for comparison a representation of the theoretical results from a drawing in Kramers' paper.

The symbol $(n'_{s'} - n''_{s''})$ attached to the lines gives the transitions between the stationary states of the atom in the electric field by which the components are emitted. Besides the principal quantum integer n, the stationary states are further characterized by a subordinate quantum integer s, which can be negative as well as positive and has a meaning quite different from that of the quantum number k occurring in the relativity theory of the fine structure of the hydrogen lines, which fixed the form of the electron orbit in the undisturbed atom. Under the influence of the electric field both the form of the orbit and its position undergo large changes, but certain properties of the orbit remain unchanged, and the surbordinate quantum number s is connected with these. In Fig. 29.5 the position of the components corresponds to the frequencies calculated for the different transitions, and the lengths of the lines are proportional to the probabilities as calculated on the basis of the correspondence principle, by which also the polarization of the radiation is determined. It is seen that the theory reproduces completely the main feature of the experimental results, and in the light of the correspondence principle we can say that the Stark effect reflects down to the smallest details the action of the electric field on the orbit of the electron in the hydrogen atom, even though in this case the reflection is so distorted that, in contrast with the case of the Zeeman effect, it would scarcely be possible directly to recognize the motion on the basis of the classical ideas of the origin of electromagnetic radiation.

Results of interest were also obtained for the spectra of elements of higher atomic number, the explanation of which in the meantime had made important progress through the work of Sommerfeld, who introduced several quantum numbers for the description of the electron orbits. Indeed, it was possible, with the aid of the correspondence principle, to account completely for the characteristic rules which govern the seemingly capricious occurrence of combination lines, and it is not too much to say that the quantum theory has not only provided a simple interpretation of the combination principle, but has further contributed materially to the clearing up of the mystery that has long rested over the application of this principle.

The same viewpoints have also proved fruitful in the investigation of the so-called band spectra. These do not originate, as do series spectra, from individual atoms, but from molecules; and the fact that these spectra are so rich in lines is due to the complexity of the motion entailed by the vibrations of the atomic nuclei relative to each other and the rotations of the molecule as a whole. The first to apply the postulates to this problem was Schwarzschild, but the important work of Heurhnger especially has thrown much light on the origin and structure of band spectra. The considerations employed here can be traced back directly to those discussed at the beginning of this lecture in connexion with Bjerrum's theory of the influence of molecular rotation on the infrared absorption lines of gases. It is true we no longer think that the rotation is reflected in the spectra in the way claimed by classical electrodynamics, but rather that the line components are due to transitions between stationary states which differ as regards rotational motion. That the phenomenon retains its essential feature, however, is a typical consequence of the correspondence principle.

29.3 Study Questions

QUES. 29.1. Do the properties of an element depend on the charge of its atomic nucleus?

a) In what way(s) are the spectra of hydrogen and singly ionized helium different? What is the cause of this difference? And why might the size of an electron's orbit depend on its nuclear charge?
b) Consider subsequent elements of increasing atomic number. Why are properties such as the atomic volume and the optical spectrum *periodic*, whereas the X-ray spectrum of is not?
c) Why are the emission spectra of very complicated elements (*i.e.,* those having many electrons and a highly charged nucleus) still quite similar to that of the hydrogen atom?
d) Is the absorption spectrum of a gas different than its emission spectrum? If so, why?
e) What is meant by the *ionization energy* of an atom? And what did Frank and Hertz demonstrate?

QUES. 29.2. What are quantum numbers, and what is their relationship to atomic spectra?

a) Was Bohr's theory able to account for the so-called *fine structure* of the atomic spectrum? How did Sommerfeld modify Bohr's theory?
b) How many quantum numbers are required to describe the fine structure of hydrogen? How does the orbital shape depend on the values of these quantum numbers? And what happens to an electron orbit when special relativity is taken into account?
c) Why does Bohr refer to systems characterized by more than one quantum number as *multiply periodic*? Do the electrons in an atom, which are characterized by particular quantum numbers, in fact *move*, or are they truly stationary?
d) What, according to the correspondence principle, determines the probability of a particular transition between stationary states?
e) Can Bohr's quantum theory account for the Stark effect? the Zeeman effect?

29.4 Exercises

EX. 29.1 (HELIUM RECOIL PROBLEM). What is the energy, momentum, and wavelength of a light quantum emitted by a singly-ionized helium atom when it makes a direct transition from an excited state with $n = 10$ to the ground state? Find the recoil speed of the helium atom in this process.

EX. 29.2 (QUANTUM NUMBERS). According to Bohr's original model of the atom, the energy of an orbiting electron was determined by its *principle quantum number, n*. Later, Sommerfeld generalized Bohr's model by introducing *subordinate quantum numbers* so as to account for the observed fine structure of atomic spectra. Accord to the Bohr-Sommerfeld theory, the electron orbits were ellipses; the principle quantum number, n, described the major axis and the azimuthal quantum number, k, described the minor axis. According to modern quantum theory—based on the work of Schrödinger and Heisenberg—atomic electron orbitals are now characterized by four quantum numbers. I simply list them here with minimal explanation:

n   The *principle quantum number* is sometimes called the *energy level* of the orbital. It can take any integer value beginning with unity: $n = 1, 2, 3, \ldots$

l   The *azimuthal quantum number* determines the orbital angular momentum—and hence the shape—of the orbital. Having specified n, the azimuthal quantum number can can take any integer value from $l = 0$ up to $l = n - 1$. The $l = 0$ orbital is called an s orbital; $l = 1$ a p orbital; $l = 2$ a d orbital, and $l = 3$ an f orbital. Subsequent values are referred to as $g, h, i, j \ldots$ orbitals (omitting l and s for obvious reasons).

m$_l$ The *magnetic quantum number* determines the projection the orbital angular momentum along a particular $(x, y,$ or $z)$ axis. Having specified l, the magnetic quantum number can take any integer value from $m_l = -l$ up to $m_l = +l$.

m$_s$ The *spin quantum number* describes the projection of the electron's intrinsic angular momentum—its so-called *spin*—along a particular axis. Depending on the orientation (so to speak) of the electron with respect to the axis, m_s can take values of either $m_s = -\frac{1}{2}$ or $m_s = +\frac{1}{2}$.

Notice that according to the old Bohr-Sommerfeld model, the azimuthal quantum number k had a minimum value of 1. From a common-sense perspective, it was thought that an orbiting electron could not have zero orbital angular momentum. According to modern quantum theory, it can. This illustrates one of the striking differences between common-sense (classical) and quantum behavior.[3] Now, as an exercise: how many distinct states can an electron in the $n = 2$ energy level have? Write out the four quantum numbers corresponding to each state. Repeat this for an electron having $n = 3$ and for an electron having $n = 4$.

29.5 Vocabulary

1. Atomic volume
2. Empirical
3. Frank-Hertz experiment
4. Ionization

---

[3] The relationship between the old and new azimuthal quantum numbers is given by $l = k - 1$; see Chaps. I.2 and I.4 of Herzberg, G., *Atomic Spectra and Atomic Structure*, 2nd ed., Dover Publications, New York, 1944.

5. Fine structure
6. Principle quantum number
7. Subordinate quantum number
8. Spectroscope
9. Perihelion
10. Zeeman effect
11. Stark effect
12. Stern–Gerlach experiment
13. Correspondence principle
14. Polarization
15. Capricious
16. Band spectra

Chapter 30
The Periodic Table of the Elements

We are therefore obliged to be modest in our demands and content ourselves with concepts which are formal in the sense that they do not provide a visual picture of the sort one is accustomed to require of the explanations with which natural philosophy deals.

—Niels Bohr

30.1 Introduction

Modern atomic theory grew largely out of an attempt to reconcile older "common-sense" concepts—such as the idea of electrons orbiting a nucleus like little moons around a planet—with the data emerging from new high-precision experiments on the emission and absorption of light by various substances. To what extent does Bohr's new atomic theory provide an adequate theoretical explanation of spectroscopic data? And more generally, what counts as an appropriate theoretical explanation? Must a theory explain *why* a particular phenomenon occurs? Must it be able to *predict* new phenomena? Must it provide an organizing principle which connects seemingly disparate phenomena? In the following reading selection, which brings us to the end of Bohr's 1922 Nobel lecture, Bohr attempts to explain how his new atomic model provides an appropriate theoretical explanation of the observed properties of the known elements. Do you find his arguments convincing? Or is there perhaps a better theoretical explanation for the periodic table of the elements?

30.2 Reading: Bohr, *The Structure of the Atom*

Bohr, N., The Structure of the Atom, in *Nobel Lectures, Physics 1922–1941*, Elsevier Publishing Company, 1965. Lecture delivered by Niels Bohr on December 11, 1922.

© Springer International Publishing Switzerland 2016
K. Kuehn, *A Student's Guide Through the Great Physics Texts*,
Undergraduate Lecture Notes in Physics, DOI 10.1007/978-3-319-21828-1_30

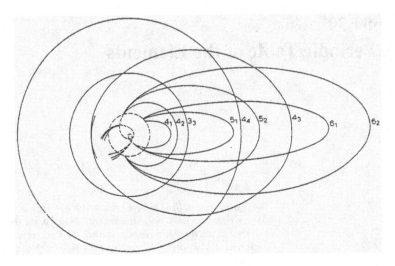

Fig. 30.1 Some stationary states having small subordinate quantum numbers (*e.g.* $k = 1$ and $k = 2$) have orbits which penetrate the region very near the nucleus. —[*K.K.*]

30.2.1   The Natural System of the Elements

The ideas of the origin of spectra outlined in the preceding have furnished the basis for a theory of the structure of the atoms of the elements which has shown itself suitable for a general interpretation of the main features of the properties of the elements, as exhibited in the natural system. This theory is based primarily on considerations of the manner in which the atom can be imagined to be built up by the capture and binding of electrons to the nucleus, one by one. As we have seen, the optical spectra of elements provide us with evidence on the progress of the last steps in this building-up process.

   An insight into the kind of information that the closer investigation of the spectra has provided in this respect may be obtained from Fig. 30.1, which gives a diagrammatic representation of the orbital motion in the stationary states corresponding to the emission of the arc-spectrum of potassium. The curves show the form of the orbits described in the stationary states by the last electron captured in the potassium atom, and they can be considered as stages in the process whereby the 19th electron is bound after the 18 previous electrons have already been bound in their normal orbits. In order not to complicate the figure, no attempt has been made to draw any of the orbits of these inner electrons, but the region in which they move is enclosed by a dotted circle. In an atom with several electrons the orbits will, in general, have a complicated character. Because of the symmetrical nature of the field of force about the nucleus, however, the motion of each single electron can be approximately described as a plane periodic motion on which is superimposed a uniform rotation in the plane of the orbit. The orbit of each electron will therefore be to a first approximation doubly periodic, and will be fixed by two quantum numbers, as are the stationary states in a hydrogen atom when the relativity precession is taken into account.

In Fig. 30.1, as in Fig. 29.3, the electron orbits are marked with the symbol n_k, where n is the principal quantum number and k the subordinate quantum number. While for the initial states of the binding process, where the quantum numbers are large, the orbit of the last electron captured lies completely outside of those of the previously bound electrons, this is not the case for the last stages. Thus, in the potassium atom, the electron orbits with subordinate quantum numbers 2 and 1 will, as indicated in the figure, penetrate partly into the inner region. Because of this circumstance, the orbits will deviate very greatly from a simple Kepler motion, since they will consist of a series of successive outer loops that have the same size and form, but each of which is turned through an appreciable angle relative to the preceding one. Of these outer loops only one is shown in the figure. Each of them coincides very nearly with a piece of a Kepler ellipse, and they are connected, as indicated, by a series of inner loops of a complicated character in which the electron approaches the nucleus closely. This holds especially for the orbit with subordinate quantum number 1, which, as a closer investigation shows, will approach nearer to the nucleus than any of the previously bound electrons.

On account of this penetration into the inner region, the strength with which an electron in such an orbit is bound to the atom will—in spite of the fact that for the most part it moves in a field of force of the same character as that surrounding the hydrogen nucleus—be much greater than for an electron in a hydrogen atom that moves in an orbit with the same principal quantum number, the maximum distance of the electron from the nucleus at the same time being considerably less than in such a hydrogen orbit. As we shall see, this feature of the binding process in atoms with many electrons is of essential importance in order to understand the characteristic periodic way in which many properties of the elements as displayed in the natural system vary with the atomic number.

In the accompanying table (Fig. 30.2) is given a summary of the results concerning the structure of the atoms of the elements to which the author has been led by a consideration of successive capture and binding of electrons to the atomic nucleus. The figures before the different elements are the atomic numbers, which give the total number of electrons in the neutral atom. The figures in the different columns give the number of electrons in orbits corresponding to the values of the principal and subordinate quantum numbers standing at the top. In accordance with ordinary usage we will, for the sake of brevity, designate an orbit with principal quantum number n as an n-quantum orbit. The first electron bound in each atom moves in an orbit that corresponds to the normal state of the hydrogen atom with quantum symbol 1_1. In the hydrogen atom there is of course only one electron; but we must assume that in the atoms of other elements the next electron also will be bound in such a 1-quantum orbit of type 1_1. As the table shows, the following electrons are bound in 2-quantum orbits. To begin with, the binding will result in a 2_1 orbit, but later electrons will be bound in 2_2 orbits, until, after binding the first 10 electrons in the atom, we reach a closed configuration of the 2-quantum orbits in which we assume there are four orbits of each type. This configuration is met for the first time

	1_1	$2_1\ 2_2$	$3_1\ 3_2\ 3_3$	$4_1\ 4_2\ 4_3\ 4_4$	$5_1\ 5_2\ 5_3\ 5_4\ 5_5$	$6_1\ 6_2\ 6_3\ 6_4\ 6_5\ 6_6$	$7_1\ 7_2$
1 H	1						
2 He	2						
3 Li	2	1					
4 Be	2	2					
5 B	2	2 (1)					
– –							
10 Ne	2	4 4					
11 Na	2	4 4	1				
12 Mg	2	4 4	2				
13 Al	2	4 4	2 1				
– –							
18 A	2	4 4	4 4				
19 K	2	4 4	4 4	1			
20 Ca	2	4 4	4 4	2			
21 Sc	2	4 4	4 4 1	(2)			
22 Ti	2	4 4	4 4 2	(2)			
– –							
29 Cu	2	4 4	6 6 6	1			
30 Zn	2	4 4	6 6 6	2			
31 Ga	2	4 4	6 6 6	2 1			
– –							
36 Kr	2	4 4	6 6 6	4 4			
37 Rb	2	4 4	6 6 6	4 4	1		
38 Sr	2	4 4	6 6 6	4 4	2		
39 Y	2	4 4	6 6 6	4 4 1	(2)		
40 Zr	2	4 4	6 6 6	4 4 2	(2)		
– –							
47 Ag	2	4 4	6 6 6	6 6 6	1		
48 Cd	2	4 4	6 6 6	6 6 6	2		
49 In	2	4 4	6 6 6	6 6 6	2 1		
– –							
54 X	2	4 4	6 6 6	6 6 6	4 4		
55 Cs	2	4 4	6 6 6	6 6 6	4 4	1	
56 Ba	2	4 4	6 6 6	6 6 6	4 4	2	
57 La	2	4 4	6 6 6	6 6 6	4 4 1	(2)	
58 Ce	2	4 4	6 6 6	6 6 6 1	4 4 1	(2)	
59 Pr	2	4 4	6 6 6	6 6 6 2	4 4 1	(2)	
– –							
71 Cp	2	4 4	6 6 6	8 8 8 8	4 4 1	(2)	
72 –	2	4 4	6 6 6	8 8 8 8	4 4 2	(2)	
– –							
79 Au	2	4 4	6 6 6	8 8 8 8	6 6 6		
80 Hg	2	4 4	6 6 6	8 8 8 8	6 6 6		
81 Tl	2	4 4	6 6 6	8 8 8 8	6 6 6	2 1	
– –							
86 Em	2	4 4	6 6 6	8 8 8 8	6 6 6	4 4	
87 –	2	4 4	6 6 6	8 8 8 8	6 6 6	4 4	1
88 Ra	2	4 4	6 6 6	8 8 8 8	6 6 6	4 4	2
89 Ac	2	4 4	6 6 6	8 8 8 8	6 6 6	4 4 1	(2)
90 Th	2	4 4	6 6 6	8 8 8 8	6 6 6	4 4 2	(2)
– –							
118 ?	2	4 4	6 6 6	8 8 8 8	8 8 8 8	6 6 6	4 4

Fig. 30.2 The building up of atomic orbitals by capture and binding of electrons. —[K.K.]

in the neutral neon atom, which forms the conclusion of the second period in the system of the elements. When we proceed in this system, the following electrons are bound in 3-quantum orbits, until, after the conclusion of the third period of the system, we encounter for the first time, in elements of the fourth period, electrons in 4-quantum orbits, and so on.

This picture of atomic structure contains many features that were brought forward by the work of earlier investigators. Thus the attempt to interpret the relations between the elements in the natural system by the assumption of a division of the electrons into groups goes as far back as the work of J.J. Thomson in 1904. Later, this viewpoint was developed chiefly by Kossel (1916), who, moreover, has connected such a grouping with the laws that investigations of X-ray spectra have brought to light.

Also G.R. Lewis and I. Langmuir have sought to account for the relations between the properties of the elements on the basis of a grouping inside the atom. These investigators, however, assumed that the electrons do not move about the nucleus, but occupy positions of equilibrium. In this way, though, no closer relation can be reached between the properties of the elements and the experimental results concerning the constituents of the atoms. Statical positions of equilibrium for the electrons are in fact not possible in cases in which the forces between the electrons and the nucleus even approximately obey the laws that hold for the attractions and repulsions between electrical charges.

The possibility of an interpretation of the properties of the elements on the basis of these latter laws is quite characteristic for the picture of atomic structure developed by means of the quantum theory. As regards this picture, the idea of connecting the grouping with a classification of electron orbits according to increasing quantum numbers was suggested by Moseley's discovery of the laws of X-ray spectra, and by Sommerfeld's work on the fine structure of these spectra. This has been principally emphasized by Vegard, who some years ago in connexion with investigations of X-ray spectra proposed a grouping of electrons in the atoms of the elements, which in many ways shows a likeness to that which is given in the above table.

A satisfactory basis for the further development of this picture of atomic structure has, however, only recently been created by the study of the binding processes of the electrons in the atom, of which we have experimental evidence in optical spectra, and the characteristic features of which have been elucidated principally by the correspondence principle. It is here an essential circumstance that the restriction on the course of the binding process, which is expressed by the presence of electron orbits with higher quantum numbers in the normal state of the atom, can be naturally connected with the general condition for the occurrence of transitions between stationary states, formulated in that principle.

Another essential feature of the theory is the influence, on the strength of binding and the dimensions of the orbits, of the penetration of the later bound electrons into the region of the earlier bound ones, of which we have seen an example in the discussion of the origin of the potassium spectrum. Indeed, this circumstance may be regarded as the essential cause of the pronounced periodicity in the properties of the elements, in that it implies that the atomic dimensions and chemical properties of homologous substances in the different periods, as, for example, the alkali-metals, show a much greater similarity than that which might be expected from a direct comparison of the orbit of the last electron bound with an orbit of the same quantum number in the hydrogen atom.

The increase of the principal quantum number which we meet when we proceed in the series of the elements, affords also an immediate explanation of the characteristic deviations from simple periodicity which are exhibited by the natural system and are expressed in Fig. 28.1 by the bracketing of certain series of elements in the later periods. The first time such a deviation is met with is in the 4th period, and the reason for it can be simply illustrated by means of our figure of the orbits of the last electron bound in the atom of potassium, which is the first element in this period. Indeed, in potassium we encounter for the first time in the sequence of the elements a case in which the principal quantum number of the orbit of the last electron bound is, in the normal state of the atom, larger than in one of the earlier stages of the binding process. The normal state corresponds here to a 4_1 orbit, which, because of the penetration into the inner region, corresponds to a much stronger binding of the electron than a 4-quantum orbit in the hydrogen atom. The binding in question is indeed even stronger than for a 2-quantum orbit in the hydrogen atom, and is therefore more than twice as strong as in the circular 3_3 orbit which is situated completely outside the inner region, and for which the strength of the binding differs but little from that for a 3-quantum orbit in hydrogen.

This will not continue to be true, however, when we consider the binding of the 19th electron in substances of higher atomic number, because of the much smaller relative difference between the field of force outside and inside the region of the first eighteen electrons bound. As is shown by the investigation of the spark spectrum of calcium, the binding of the 19th electron in the 4_1 orbit is here but little stronger than in 3_3 orbits, and as soon as we reach scandium, we must assume that the 3_3 orbit will represent the orbit of the 19th electron in the normal state, since this type of orbit will correspond to a stronger binding than a 4_1 orbit. While the group of electrons in 2-quantum orbits has been entirely completed at the end of the 2nd period, the development that the group of 3-quantum orbits undergoes in the course of the 3rd period can therefore only be described as a provisional completion, and, as shown in the table, this electron group will, in the bracketed elements of the 4th period, undergo a stage of further development in which electrons are added to it in 3-quantum orbits.

This development brings in new features, in that the development of the electron group with 4-quantum orbits comes to a standstill, so to speak, until the 3-quantum group has reached its final closed form. Although we are not yet in a position to account in all details for the steps in the gradual development of the 3-quantum electron group, still we can say that with the help of the quantum theory we see at once why it is in the 4th period of the system of the elements that there occur for the first time successive elements with properties that resemble each other as much as the properties of the iron group; indeed, we can even understand why these elements show their well known paramagnetic properties. Without further reference to the quantum theory, Eadenburg had on a previous occasion already suggested the idea of relating the chemical and magnetic properties of these elements with the development of an inner electron group in the atom.

I will not enter into many more details, but only mention that the peculiarities we meet with in the 5th period are explained in much the same way as those in the 4th period. Thus the properties of the bracketed elements in the 5th period as it appears in the table, depend on a stage in the development of the 4-quantum electron group that is initiated by the entrance in the normal state of electrons in 4_3 orbits. In the 6th period, however, we meet new features. In this period we encounter not only a stage of the development of the electron groups with 5- and 6-quantum orbits, but also the final completion of the development of the 4-quantum electron group, which is initiated by the entrance for the first time of electron orbits of the 4_4 type in the normal state of the atom. This development finds its characteristic expression in the occurrence of the peculiar family of elements in the 6th period, known as the rare-earths. These show, as we know, a still greater mutual similarity in their chemical properties than the elements of the iron family. This must be ascribed to the fact that we have here to do with the development of an electron group that lies deeper in the atom. It is of interest to note that the theory can also naturally account for the fact that these elements, which resemble each other in so many ways, still show great differences in their magnetic properties.

The idea that the occurrence of the rare-earths depends on the development of an inner electron group has been put forward from different sides. Thus it is found in the work of Vegard, and at the same time as my own work, it was proposed by Bury in connexion with considerations of the systematic relation between the chemical properties and the grouping of the electrons inside the atom from the point of view of Langmuir's static atomic model. While until now it has not been possible, however, to give any theoretical basis for such a development of an inner group, we see that our extension of the quantum theory provides us with an unforced explanation. Indeed, it is scarcely an exaggeration to say that if the existence of the rare earths had not been established by direct chemical investigation, the occurrence of a family of elements of this character within the 6th period of the natural system of the elements might have been theoretically predicted.

When we proceed to the 7th period of the system, we meet for the first time with 7-quantum orbits, and we shall expect to find within this period features that are essentially similar to those in the 6th period, in that besides the first stage in the development of the 7-quantum orbits, we must expect to encounter further stages in the development of the group with 6- or 5-quantum orbits. However, it has not been possible directly to confirm this expectation, because only a few elements are known in the beginning of the 7th period. The latter circumstance may be supposed to be intimately connected with the instability of atomic nuclei with large charges, which is expressed in the prevalent radioactivity among elements with high atomic number.

30.2.2   X-Ray Spectra and Atomic Constitution

In the discussion of the conceptions of atomic structure we have hitherto placed the emphasis on the formation of the atom by successive capture of electrons. Our picture would, however, be incomplete without some reference to the confirmation of the theory afforded by the study of X-ray spectra. Since the interruption of Moseley's fundamental researches by his untimely death, the study of these spectra has been continued in a most admirable way by Prof. Siegbahn in Lund. On the basis of the large amount of experimental evidence adduced by him and his collaborators, it has been possible recently to give a classification of X-ray spectra that allows an immediate interpretation on the quantum theory. In the first place it has been possible, just as in the case of the optical spectra, to represent the frequency of each of the X-ray lines as the difference between two out of a manifold of spectral terms characteristic of the element in question. Next, a direct connexion with the atomic theory is obtained by the assumption that each of these spectral terms multiplied by Planck's constant is equal to the work which must be done on the atom to remove one of its inner electrons. In fact, the removal of one of the inner electrons from the completed atom may, in accordance with the above considerations on the formation of atoms by capture of electrons, give rise to transition processes by which the place of the electron removed is taken by an electron belonging to one of the more loosely bound electron groups of the atom, with the result that after the transition an electron will be lacking in this latter group.

The X-ray lines may thus be considered as giving evidence of stages in a process by which the atom undergoes a reorganization after a disturbance in its interior. According to our views on the stability of the electronic configuration such a disturbance must consist in the removal of electrons from the atom, or at any rate in their transference from normal orbits to orbits of higher quantum numbers than those belonging to completed groups; a circumstance which is clearly illustrated in the characteristic difference between selective absorption in the X-ray region, and that exhibited in the optical region.

The classification of the X-ray spectra, to the achievement of which the above-mentioned work of Sommerfeld and Kossel has contributed materially, has recently made it possible, by means of a closer examination of the manner in which the terms occurring in the X-ray spectra vary with the atomic number, to obtain a very direct test of a number of the theoretical conclusions as regards the structure of the atom. In Fig. 30.3 the abscissæ are the atomic numbers and the ordinates are proportional to the square roots of the spectral terms, while the symbols K, L, M, N, O, for the individual terms refer to the characteristic discontinuities in the selective absorption of the elements for X-rays; these were originally found by Barkla before the discovery of the interference of X-rays in crystals had provided a means for the closer investigation of X-ray spectra. Although the curves generally run very uniformly, they exhibit a number of deviations from uniformity which have been especially brought to light by the recent investigation of Coster, who has for some years worked in Siegbahn's laboratory.

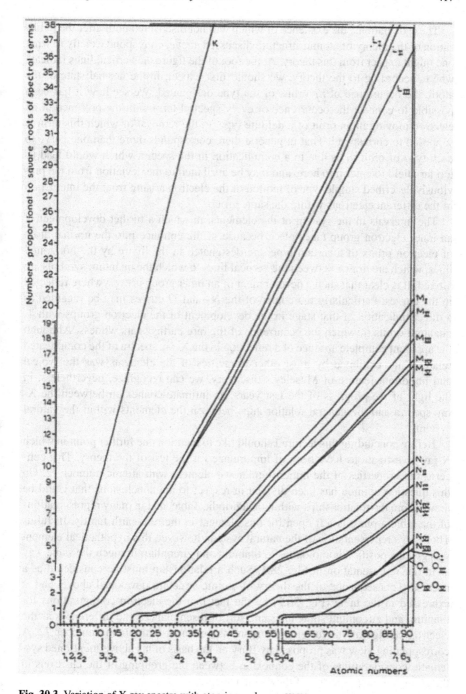

Fig. 30.3 Variation of X-ray spectra with atomic number. —[*K.K.*]

These deviations, the existence of which was not discovered until after the publication of the theory of atomic structure discussed above, correspond exactly to what one might expect from this theory. At the foot of the figure the vertical lines indicate where, according to the theory, we should first expect, in the normal state of the atom, the occurrence of n_k orbits of the type designated. We see how it has been possible to connect the occurrence of every spectral term with the presence of an electron moving in an orbit of a definite type, to the removal of which this term is supposed to correspond. That in general there corresponds more than one curve to each type of orbit n_k is due to a complication in the spectra which would lead us too far afield to enter into here, and may be attributed to the deviation from the previously described simple type of motion of the electron arising from the interaction of the different electrons within the same group.

The intervals in the system of the elements, in which a further development of an inner electron group takes place because of the entrance into the normal atom of electron orbits of a certain type, are designated in the figure by the horizontal lines, which are drawn between the vertical lines to which the quantum symbols are affixed. It is clear that such a development of an inner group is everywhere reflected in the curves. Particularly the course of the N- and O-curves may be regarded as a direct indication of that stage in the development of the electron groups with 4-quantum orbits of which the occurrence of the rare-earths bears witness. Although the apparent complete absence of a reflection in the X-ray spectra of the complicated relationships exhibited by most other properties of the elements was the typical and important feature of Moseley's discovery, we can recognize, nevertheless, in the light of the progress of the last years, an intimate connexion between the X-ray spectra and the general relationships between the elements within the natural system.

Before concluding this lecture I should like to mention one further point in which X-ray investigations have been of importance for the test of the theory. This concerns the properties of the hitherto unknown element with atomic number 72. On this question opinion has been divided in respect to the conclusions that could be drawn from the relationships within the Periodic Table, and in many representations of the table a place is left open for this element in the rare-earth family. In Julius Thomsen's representation of the natural system, however, this hypothetical element was given a position homologous to titanium and zirconium in much the same way as in our representation in Fig. 28.1. Such a relationship must be considered as a necessary consequence of the theory of atomic structure developed above, and is expressed in the table (Fig. 30.2) by the fact that the electron configurations for titanium and zirconium show the same sort of resemblances and differences as the electron configurations for zirconium and the element with atomic number 72. A corresponding view was proposed by Bury on the basis of his above-mentioned systematic considerations of the connexion between the grouping of the electrons in the atom and the properties of the elements.

Recently, however, a communication was published by Dauvillier announcing the observation of some weak lines in the X-ray spectrum of a preparation containing rare-earths. These were ascribed to an element with atomic number 72 assumed to be identical with an element of the rare-earth family, the existence of which in the preparation used had been presumed by Urbain many years ago. This conclusion would, however, if it could be maintained, place extraordinarily great, if not unsurmountable, difficulties in the way of the theory, since it would claim a change in the strength of the binding of the electrons with the atomic number which seems incompatible with the conditions of the quantum theory. In these circumstances Dr. Coster and Prof. Hevesy, who are both for the time working in Copenhagen, took up a short time ago the problem of testing a preparation of zircon-bearing minerals by X-ray spectroscopic analysis. These investigators have been able to establish the existence in the minerals investigated of appreciable quantities of an element with atomic number 72, the chemical properties of which show a great similarity to those of zirconium and a decided difference from those of the rare-earths.[1]

I hope that I have succeeded in giving a summary of some of the most important results that have been attained in recent years in the field of atomic theory, and I should like, in concluding, to add a few general remarks concerning the viewpoint from which these results may be judged, and particularly concerning the question of how far, with these results, it is possible to speak of an explanation, in the ordinary sense of the word. By a theoretical explanation of natural phenomena we understand in general a classification of the observations of a certain domain with the help of analogies pertaining to other domains of observation, where one presumably has to do with simpler phenomena. The most that one can demand of a theory is that this classification can be pushed so far that it can contribute to the development of the field of observation by the prediction of new phenomena.

When we consider the atomic theory, we are, however, in the peculiar position that there can be no question of an explanation in this last sense, since here we have to do with phenomena which from the very nature of the case are simpler than in any other field of observation, where the phenomena are always conditioned by the combined action of a large number of atoms. We are therefore obliged to be modest in our demands and content ourselves with concepts which are formal in the sense that they do not provide a visual picture of the sort one is accustomed to require of the explanations with which natural philosophy deals. Bearing this in mind I have sought to convey the impression that the results, on the other hand, fulfill, at least in some degree, the expectations that are entertained of any theory; in fact, I have attempted to show how the development of atomic theory has contributed to the classification of extensive fields of observation, and by its predictions has pointed out the way to the completion of this classification. It is scarcely necessary, however, to emphasize that the theory is yet in a very preliminary stage, and many fundamental questions still await solution.

---

[1] For the result of the continued work of Coster and Hevesy with the new element, for which they have proposed the name hafnium, the reader may be referred to their letters in *Nature* of January 20, February 10 and 24, and April 7.

30.3 Study Questions

QUES. 30.1. Does Bohr's model of the atom provide an appropriate theoretical explanation for the periodic table of the elements?

a) What is the maximum number of electrons which can be bound to an element having principle quantum number $n = 1$? $n = 2$?

b) Do electron orbits having large principle quantum numbers always lie outside orbits having smaller principle quantum numbers?

c) What are the quantum numbers of the electrons orbiting the nucleus in potassium? Generally speaking, when an atom's orbits are successively filled with captured electrons, are the n-quantum orbits always populated before the $(n + 1)$-quantum orbits?

d) What feature dictates the chemical and magnetic properties of an element? In particular, how does the classification of electron orbits according to quantum numbers allow one to group the elements?

e) What is unique about the rare-earth elements? In which period do they appear?

f) Why are high-atomic number elements comparatively rare? What physical process limits their stability?

g) How is the X-ray spectrum of an element connected to its atomic structure?

h) What constitutes an appropriate theoretical explanation of the periodic table? Can Bohr's quantum theory explain the classification of the elements? Did it make any testable predictions? Have these predictions been verified?

30.4 Exercises

EX. 30.1 (THE AUFBAU PRINCIPLE AND ELECTRON CONFIGURATIONS). The hypothetical scheme according to which an atomic nucleus captures successive electrons and places them into atomic orbitals is referred to as the *aufbau principle* (German for "building-up principle"). According to the building-up principle, an electron will tend to occupy the lowest energy state which is available, subject to a constraint imposed by the *Pauli exclusion principle*: no electron may have the same four quantum numbers as another electron. For example, the two bound electrons in a neutral (and unexcited) helium atom have quantum numbers $(n, l, m_l, m_s) = (1, 0, 0, +\frac{1}{2})$ and $(1, 0, 0, -\frac{1}{2})$.

A standard notation has been adopted for denoting the electron configuration of an atom. This consists of a string of numbers and letters denoting the occupancy of each orbital. For example, the ground state electron configuration of a hydrogen atom is denoted by $1s^1$ (pronounced "one ess one"). The 1s denotes the orbital having $n = 1$ and $l = 0$; the superscript indicates that the 1s orbital is occupied by a single electron. The ground state of lithium is $1s^2 2s^1$ ("one ess two, two ess one"). This can also be written more compactly as [He] $2s^1$, since lithium is like helium with an additional electron in the 2s orbital.

As an exercise, write down the ground state electron configurations of your favorite element from each period of the periodic table. In so doing, you should keep in mind the following general rule: orbitals having the smallest principle quantum number, n, are filled before ones having larger principle quantum numbers. This rule is true only insofar as the azimuthal quantum number, l, has no effect on the energy of the orbital. Recall that according to Sommerfeld, orbitals having small orbital angular momenta have very elliptical orbits (classically speaking) and thus penetrate very near to the atomic nucleus. This gives rise to a strong (negative) binding energy. The building-up principle is then governed by the *Madelung rule*: orbitals with lower $n + l$ values tend to be filled before ones with higher $n + l$ values. All of this is based on the (somewhat dubious) assumption that an electron's energy is independent of the locations of neighboring electrons; a more comprehensive treatment of the aufbau principle necessitate the careful consideration of electron-electron interactions. By applying the Madelung rule, what is the first element in which a higher n orbital is occupied before a lower n orbital? Is this consistent with the ground-state electron configuration reported by Bohr? What is the first element that does *not* obey Madelung's rule? (HINT: You may wish to double-check your answers by looking at a modern reference which reports the ground-state electron configurations of various elements.)

EX. 30.2 (X-RAY EMISSION SPECTRUM OF COPPER). Suppose that a beam of electrons is accelerated towards a copper target. One of the incoming electrons has sufficient energy to penetrate the outer shells of a copper atom and eject one of the innermost electrons from its orbital. Subsequently, one of the electrons having principal quantum number $n = 2$ makes a transition to the newly vacated $n = 1$ orbital. This produces the K_α spectral emission line.

a) What is the wavelength of the photon emitted as a result of this $n = 2$ to $n = 1$ transition in copper? How does this compare to the wavelength of the photon emitted during a similar $n = 2$ to $n = 1$ transition for a helium atom?

b) Suppose, instead, that one of the $n = 3$ electrons is ejected from copper, and that an $n = 4$ electron falls into the vacated $n = 3$ orbital? Can you approximate the wavelength of the emitted photon? (HINT: The remaining inner-orbital electrons act to shield the positively charged copper nucleus, making its Coulomb force on the outer-orbital electron appear much weaker.)

EX. 30.3 (THEORY AND EXPLANATION ESSAY). Does the Bohr model provide a theoretical explanation of the periodic table of the elements? According to Bohr, what must any theoretical explanation accomplish? Does Bohr's quantum theory fulfill this criteria? Do you generally agree with Bohr's views on what constitutes a theoretical explanation? If not, how do your views differ?

30.5   Vocabulary

1. Prevalent
2. Elucidate
3. Homologous
4. Paramagnetic
5. Rare-earth
6. Manifold
7. Discontinuity
8. Abscissæ
9. Ordinate
10. Configuration
11. Analogy
12. Pertain

Chapter 31
Wave Mechanics

*The atom in reality is merely the diffraction phenomenon of an
electron wave captured as it were by the nucleus of the atom.*
—Erwin Schrödinger

31.1 Introduction

Erwin Schrödinger (1887–1961) was born in Vienna, Austria. He was home-schooled by his parents and tutors until the age of 11, when he attended the academic Gymnasium in Vienna. His early interests included Latin and Greek grammar, German poetry, and science. In 1906, he enrolled at the University of Vienna, where he began to attend lectures on theoretical physics given by Friedrich Hasenöhrl, the successor of Ludwig Boltzmann. After receiving his doctoral degree in 1910, he became an assistant to Franz Exner and supervised physics laboratory courses. After serving as an artillery officer in World War I, he pursued an academic career. In 1920, he took a position as assistant to Max Wien, then served as an extraordinary professor at Stuttgart and an ordinary professor at Breslau before replacing von Laue at the University of Zurich. In 1927, he moved to Berlin to succeed Max Planck at the University until 1933. During the War years, he spent time in Oxford, at the University of Graz and at the University of Ghent before moving to the new Institute for Advanced Studies in Dublin to serve as the Director of the School of Theoretical Physics until his retirement in 1955.

While at Zurich, Schrödinger published much of his most famous work, including theoretical papers on the specific heat of solids, thermodynamics, atomic spectra, and the foundation of quantum mechanics. Indeed Schrödinger is known as one of the primary architects of modern quantum theory. The reading selection that follows is the lecture delivered by Schrödinger in 1933 upon receiving the Nobel Prize in physics.

© Springer International Publishing Switzerland 2016
K. Kuehn, *A Student's Guide Through the Great Physics Texts*,
Undergraduate Lecture Notes in Physics, DOI 10.1007/978-3-319-21828-1_31

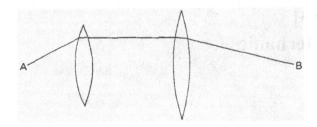

Fig. 31.1 Snell's law governs the refraction of a light ray at the boundary between two media.—
[*K.K.*]

31.2   Reading: Schrödinger, *The Fundamental Idea of Wave Mechanics*

Schrödinger, E., Wave Mechanics, in *Nobel Lectures, Physics 1922–1941*, Elsevier
Publishing Company, 1965. Lecture delivered by Erwin Schrödinger in 1933.

On passing through an optical instrument, such as a telescope or a camera lens, a
ray of light is subjected to a change in direction at each refracting or reflecting surface. The path of the rays can be constructed if we know the two simple laws which
govern the changes in direction: the law of refraction which was discovered by Snellius a few hundred years ago, and the law of reflection with which Archimedes was
familiar more than 2000 years ago. As a simple example, Fig. 31.1 shows a ray
$A - B$ which is subjected to refraction at each of the four boundary surfaces of two
lenses in accordance with the law of Snellius.

Fermat defined the total path of a ray of light from a much more general point of
view. In different media, light propagates with different velocities, and the radiation
path gives the appearance as if the light must arrive at its destination as quickly
as possible. (Incidentally, it is permissible here to consider any two points along
the ray as the starting- and end-points.) The least deviation from the path actually
taken would mean a delay. This is the famous Fermat principle of the shortest light
time, which in a marvellous manner determines the entire fate of a ray of light
by a single statement and also includes the more general case, when the nature of
the medium varies not suddenly at individual surfaces, but gradually from place to
place. The atmosphere of the earth provides an example. The more deeply a ray of
light penetrates into it from outside, the more slowly it progresses in an increasingly
denser air. Although the differences in the speed of propagation are infinitesimal,
Fermat's principle in these circumstances demands that the light ray should curve
earthward (see Fig. 31.2), so that it remains a little longer in the higher "faster"
layers and reaches its destination more quickly than by the shorter straight path
(broken line in the figure; disregard the square, $W W W' W'$ for the time being). I

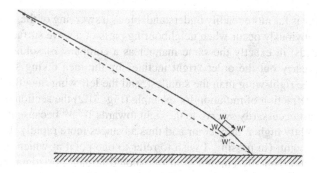

Fig. 31.2 The curvature of a light ray in Earth's atmosphere is a consequence of Fermat's principle, which states that a light ray will follow the path of least time (rather than the path of least distance) between two points.—[*K.K.*]

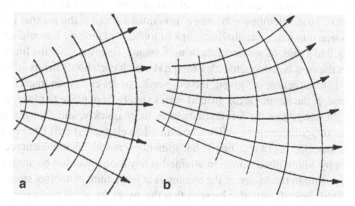

Fig. 31.3 According to the wave theory, light rays are fictitious entities constructed orthogonal to wave surfaces (**a**); the curvature of a light ray is associated with the swerving of a wave surface as it advances through an inhomogeneous medium (**b**).—[*K.K.*]

think, hardly any of you will have failed to observe that the sun when it is deep on the horizon appears to be not circular but flattened: its vertical diameter looks to be shortened. This is a result of the curvature of the rays.

According to the wave theory of light, the light rays, strictly speaking, have only fictitious significance. They are not the physical paths of some particles of light, but are a mathematical device, the so-called orthogonal trajectories of wave surfaces, imaginary guide lines as it were, which point in the direction normal to the wave surface in which the latter advances (*cf.* Fig. 31.3a which shows the simplest case of concentric spherical wave surfaces and accordingly rectilinear rays, whereas Fig. 31.3b illustrates the case of curved rays).

It is surprising that a general principle as important as Fermat's relates directly to these mathematical guide lines, and not to the wave surfaces, and one might be inclined for this reason to consider it a mere mathematical curiosity. Far from it. It becomes properly understandable only from the point of view of wave theory and ceases to be a divine miracle. From the wave point of view, the so-called curvature of the light ray is far more readily understandable as a swerving of the wave surface, which must obviously occur when neighbouring parts of a wave surface advance at different speeds; in exactly the same manner as a company of soldiers marching forward will carry out the order "right incline" by the men taking steps of varying lengths, the right-wing man the smallest, and the left-wing man the longest. In atmospheric refraction of radiation for example (Fig. 31.2) the section of wave surface WW must necessarily swerve to the right towards $W'W'$ because its left half is located in slightly higher, thinner air and thus advances more rapidly than the right part at lower point. (In passing, I wish to refer to one point at which the Snellius' view fails. A horizontally emitted light ray should remain horizontal because the refraction index does not vary in the horizontal direction. In truth, a horizontal ray curves more strongly than any other, which is an obvious consequence of the theory of a swerving wave front.) On detailed examination the Fermat principle is found to be completely tantamount to the trivial and obvious statement that—given local distribution of light velocities—the wave front must swerve in the manner indicated. I cannot prove this here, but shall attempt to make it plausible. I would again ask you to visualize a rank of soldiers marching forward. To ensure that the line remains dressed, let the men be connected by a long rod which each holds firmly in his hand. No orders as to direction are given; the only order is: let each man march or run as fast as he can. If the nature of the ground varies slowly from place to place, it will be now the right wing, now the left that advances more quickly, and changes in direction will occur spontaneously. After some time has elapsed, it will be seen that the entire path travelled is not rectilinear, but somehow curved. That this curved path is exactly that by which the destination attained at any moment could be attained most rapidly according to the nature of the terrain, is at least quite plausible, since each of the men did his best. It will also be seen that the swerving also occurs invariably in the direction in which the terrain is worse, so that it will come to look in the end as if the men had intentionally "bypassed" a place where they would advance slowly.

The Fermat principle thus appears to be the trivial quintessence of the wave theory. It was therefore a memorable occasion when Hamilton made the discovery that the true movement of mass points in a field of forces (*e.g.* of a planet on its orbit around the sun or of a stone thrown in the gravitational field of the earth) is also governed by a very similar general principle, which carries and has made famous the name of its discoverer since then. Admittedly, the Hamilton principle does not say exactly that the mass point chooses the quickest way, but it does say something so similar—the analogy with the principle of the shortest travelling time of light is so close, that one was faced with a puzzle. It seemed as if Nature had realized one and the same law twice by entirely different means: first in the case of light, by means of a fairly obvious play of rays; and again in the case of the mass points, which was anything but obvious, unless somehow wave nature were to be attributed to them also. And this, it seemed impossible to do. Because the "mass points" on which the laws of mechanics had really been confirmed experimentally at that time were only the large, visible, sometimes very large bodies, the planets, for which a thing like "wave nature" appeared to be out of the question.

The smallest, elementary components of matter which we today, much more specifically, call "mass points", were purely hypothetical at the time. It was only after the discovery of radioactivity that constant refinements of methods of measurement permitted the properties of these particles to be studied in detail, and now permit the paths of such particles to be photographed and to be measured very exactly (stereophotogrammetrically) by the brilliant method of C.T.R. Wilson. As far as the measurements extend they confirm that the same mechanical laws are valid for particles as for large bodies, planets, *etc.* However, it was found that neither the molecule nor the individual atom can be considered as the "ultimate component": but even the atom is a system of highly complex structure. Images are formed in our minds of the structure of atoms consisting of particles, images which seem to have a certain similarity with the planetary system. It was only natural that the attempt should at first be made to consider as valid the same laws of motion that had proved themselves so amazingly satisfactory on a large scale. In other words, Hamilton's mechanics, which, as I said above, culminates in the Hamilton principle, were applied also to the "inner life" of the atom. That there is a very close analogy between Hamilton's principle and Fermat's optical principle had meanwhile become all but forgotten. If it was remembered, it was considered to be nothing more than a curious trait of the mathematical theory.

Now, it is very difficult, without further going into details, to convey a proper conception of the success or failure of these classical-mechanical images of the atom. On the one hand, Hamilton's principle in particular proved to be the most faithful and reliable guide, which was simply indispensable; on the other hand one had to suffer, to do justice to the facts, the rough interference of entirely new incomprehensible postulates, of the so-called quantum conditions and quantum postulates. Strident disharmony in the symphony of classical mechanics—yet strangely familiar—played as it were on the same instrument. In mathematical terms we can formulate this as follows: whereas the Hamilton principle merely postulates that a given integral must be a minimum, without the numerical value of the minimum being established by this postulate, it is now demanded that the numerical value of the minimum should be restricted to integral multiples of a universal natural constant, Planck's quantum of action. This incidentally. The situation was fairly desperate. Had the old mechanics failed completely, it would not have been so bad. The way would then have been free to the development of a new system of mechanics. As it was, one was faced with the difficult task of saving the soul of the old system, whose inspiration clearly held sway in this microcosm, while at the same time flattering it as it were into accepting the quantum conditions not as gross interference but as issuing from its own innermost essence.

The way out lay just in the possibility, already indicated above, of attributing to the Hamilton principle, also, the operation of a wave mechanism on which the point-mechanical processes are essentially based, just as one had long become accustomed to doing in the case of phenomena relating to light and of the Fermat principle which governs them. Admittedly, the individual path of a mass point loses its proper physical significance and becomes as fictitious as the individual isolated ray of light. The essence of the theory, the minimum principle, however, remains not only intact, but reveals its true and simple meaning only under the wave-like aspect, as already explained. Strictly speaking, the new theory is in fact not new, it is a completely organic development, one might almost be tempted to say a more elaborate exposition, of the old theory.

How was it then that this new more "elaborate" exposition led to notably different results; what enabled it, when applied to the atom, to obviate difficulties which the old theory could not solve? What enabled it to render gross interference acceptable or even to make it its own?

Again, these matters can best be illustrated by analogy with optics. Quite properly, indeed, I previously called the Fermat principle the quintessence of the wave theory of light: nevertheless, it cannot render dispensible a more exact study of the wave process itself. The so-called refraction and interference phenomena of light can only be understood if we trace the wave process in detail because what matters is not only the eventual destination of the wave, but also whether at a given moment it arrives there with a wave peak or a wave trough. In the older, coarser experimental arrangements, these phenomena occurred as small details only and escaped observation. Once they were noticed and were interpreted correctly, by means of waves, it was easy to devise experiments in which the wave nature of light finds expression not only in small details, but on a very large scale in the entire character of the phenomenon.

Allow me to illustrate this by two examples, first, the example of an optical instrument, such as telescope, microscope, *etc.* The object is to obtain a sharp image, *i.e.* it is desired that all rays issuing from a point should be reunited in a point, the so-called focus (*cf.* Fig. 31.4 a). It was at first believed that it was only geometrical-optical difficulties which prevented this: they are indeed considerable. Later it was found that even in the best designed instruments focussing of the rays was considerably inferior than would be expected if each ray exactly obeyed the Fermat principle independently of the neighbouring rays. The light which issues from a point and is received by the instrument is reunited behind the instrument not in a single point any more, but is distributed over a small circular area, a so-called diffraction disc, which, otherwise, is in most cases a circle only because the apertures and lens contours are generally circular. For, the cause of the phenomenon which we call diffraction is that not all the spherical waves issuing from the object point can be accommodated by the instrument. The lens edges and any apertures merely cut out a part of the wave surfaces (*cf.* Fig. 31.4b) and—if you will permit me to use a more suggestive expression—the injured margins resist rigid unification in a point and produce the somewhat blurred or vague image. The degree of blurring is closely associated with the wavelength of the light and is completely inevitable because of this deep-seated theoretical relationship. Hardly noticed at first, it governs and restricts the performance of the modern microscope which has mastered all other errors of reproduction. The images obtained of structures not much coarser or even still finer than the wavelengths of light are only remotely or not at all similar to the original.

A second, even simpler example is the shadow of an opaque object cast on a screen by a small point light source. In order to construct the shape of the shadow, each light ray must be traced and it must be established whether or not the opaque

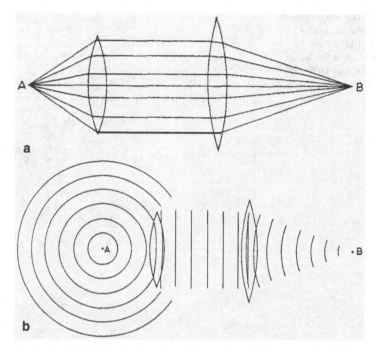

Fig. 31.4 Light from an object (*A*) is focused by lenses into an image (*B*) using a ray picture (*above*) and a wave picture (*below*). Only the latter picture provides a complete explanation of the inevitable blurring of the image. —[*K.K.*]

object prevents it from reaching the screen. The margin of the shadow is formed by those light rays which only just brush past the edge of the body. Experience has shown that the shadow margin is not absolutely sharp even with a point-shaped light source and a sharply defined shadow-casting object. The reason for this is the same as in the first example. The wave front is as it were bisected by the body (*cf.* Fig. 31.5) and the traces of this injury result in blurring of the margin of the shadow which would be incomprehensible if the individual light rays were independent entities advancing independently of one another without reference to their neighbours.

This phenomenon—which is also called diffraction—is not as a rule very noticeable with large bodies. But if the shadow-casting body is very small at least in one dimension, diffraction finds expression firstly in that no proper shadow is formed at all, and secondly—much more strikingly—in that the small body itself becomes as it were its own source of light and radiates light in all directions (preferentially to be sure, at small angles relative to the incident light). All of you are undoubtedly familiar with the so-called "motes of dust" in a light beam falling into a dark room. Fine blades of grass and spiders' webs on the crest of a hill with the sun behind it, or the errant locks of hair of a man standing with the sun behind often light up mysteriously by diffracted light, and the visibility of smoke and mist is based on it. It comes not really from the body itself, but from its immediate surroundings, an

Fig. 31.5 The fringes which appear in the shadow of an illuminated object can be understood from the wave (and not the ray) picture of light.—[*K.K.*]

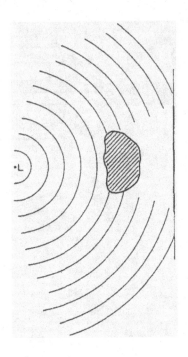

area in which it causes considerable interference with the incident wave fronts. It is interesting, and important for what follows, to observe that the area of interference always and in every direction has at least the extent of one or a few wavelengths, no matter how small the disturbing particle may be. Once again, therefore, we observe a close relationship between the phenomenon of diffraction and wavelength. This is perhaps best illustrated by reference to another wave process, *i.e.* sound. Because of the much greater wavelength, which is of the order of centimetres and metres, shadow formation recedes in the case of sound, and diffraction plays a major, and practically important, part: we can easily hear a man calling from behind a high wall or around the corner of a solid house, even if we cannot see him.

Let us return from optics to mechanics and explore the analogy to its fullest extent. In optics the old system of mechanics corresponds to intellectually operating with isolated mutually independent light rays. The new undulatory mechanics corresponds to the wave theory of light. What is gained by changing from the old view to the new is that the diffraction phenomena can be accommodated or, better expressed, what is gained is something that is strictly analogous to the diffraction phenomena of light and which on the whole must be very unimportant, otherwise the old view of mechanics would not have given full satisfaction so long. It is, however, easy to surmise that the neglected phenomenon may in some circumstances make itself very much felt, will entirely dominate the mechanical process, and will face the old system with insoluble riddles, if the entire mechanical system is comparable in extent with the wavelengths of the "waves of matter" which play the same part in mechanical processes as that played by the light waves in optical processes.

This is the reason why in these minute systems, the atoms, the old view was bound to fail, which though remaining intact as a close approximation for gross mechanical processes, but is no longer adequate for the delicate interplay in areas of the order of magnitude of one or a few wavelengths. It was astounding to observe the manner in which all those strange additional requirements developed spontaneously from the new undulatory view, whereas they had to be forced upon the old view to adapt them to the inner life of the atom and to provide some explanation of the observed facts.

Thus, the salient point of the whole matter is that the diameters of the atoms and the wavelength of the hypothetical material waves are of approximately the same order of magnitude. And now you are bound to ask whether it must be considered mere chance that in our continued analysis of the structure of matter we should come upon the order of magnitude of the wavelength at this of all points, or whether this is to some extent comprehensible. Further, you may ask, how we know that this is so, since the material waves are an entirely new requirement of this theory, unknown anywhere else. Or is it simply that this is an assumption which had to be made?

The agreement between the orders of magnitude is no mere chance, nor is any special assumption about it necessary; it follows automatically from the theory in the following remarkable manner. That the heavy nucleus of the atom is very much smaller than the atom and may therefore be considered as a point centre of attraction in the argument which follows may be considered as experimentally established by the experiments on the scattering of alpha rays done by Rutherford and Chadwick. Instead of the electrons we introduce hypothetical waves, whose wavelengths are left entirely open, because we know nothing about them yet. This leaves a letter, say a, indicating a still unknown figure, in our calculation. We are, however, used to this in such calculations and it does not prevent us from calculating that the nucleus of the atom must produce a kind of diffraction phenomenon in these waves, similarly as a minute dust particle does in light waves. Analogously, it follows that there is a close relationship between the extent of the area of interference with which the nucleus surrounds itself and the wavelength, and that the two are of the same order of magnitude. What this is, we have had to leave open; but the most important step now follows: we identify the area of interference, the diffraction halo, with the atom; we assert that the atom in reality is merely the diffraction phenomenon of an electron wave captured as it were by the nucleus of the atom. It is no longer a matter of chance that the size of the atom and the wavelength are of the same order of magnitude: it is a matter of course. We know the numerical value of neither, because we still have in our calculation the one unknown constant, which we called a. There are two possible ways of determining it, which provide a mutual check on one another. First, we can so select it that the manifestations of life of the atom, above all the spectrum lines emitted, come out correctly quantitatively; these can after all be measured very accurately. Secondly, we can select a in a manner such that the diffraction halo acquires the size required for the atom. These two determinations of a (of which the second is admittedly far more imprecise because "size of the atom" is no clearly defined term) are in complete agreement with one another. Thirdly, and lastly, we can remark that the constant remaining unknown, physically speaking, does not in

Fig. 31.6 From a wave
picture, which of these
equally-possible parallel rays
corresponds to the true path
of a point particle?—[K.K.]

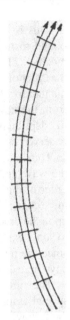

fact have the dimension of a length, but of an action, *i.e.* energy × time. It is then an obvious step to substitute for it the numerical value of Planck's universal quantum of action, which is accurately known from the laws of heat radiation. It will be seen that we return, with the full, now considerable accuracy, to the first (most accurate) determination.

Quantitatively speaking, the theory therefore manages with a minimum of new assumptions. It contains a single available constant, to which a numerical value familiar from the older quantum theory must be given, first to attribute to the diffraction halos the right size so that they can be reasonably identified with the atoms, and secondly, to evaluate quantitatively and correctly all the manifestations of life of the atom, the light radiated by it, the ionization energy, *etc.* I have tried to place before you the fundamental idea of the wave theory of matter in the simplest possible form. I must admit now that in my desire not to tangle the ideas from the very beginning, I have painted the lily. Not as regards the high degree to which all sufficiently, carefully drawn conclusions are confirmed by experience, but with regard to the conceptual ease and simplicity with which the conclusions are reached. I am not speaking here of the mathematical difficulties, which always turn out to be trivial in the end, but of the conceptual difficulties. It is, of course, easy to say that we turn from the concept of a curved path to a system of wave surfaces normal to it. The wave surfaces, however, even if we consider only small parts of them include at least a narrow bundle of possible curved paths (Fig. 31.6) to all of which they stand in the same relationship. According to the old view, but not according to the new, one of them in each concrete individual case is distinguished from all the others which are "only possible", as that "really travelled". We are faced here with the full force of the logical opposition between an

either—or (point mechanics)

and a

both—and (wave mechanics)

This would not matter much, if the old system were to be dropped entirely and to be replaced by the new. Unfortunately, this is not the case. From the point of view of wave mechanics, the infinite array of possible point paths would be merely fictitious, none of them would have the prerogative over the others of being that really travelled in an individual case. I have, however, already mentioned that we have yet really observed such individual particle paths in some cases. The wave theory can represent this, either not at all or only very imperfectly. We find it confoundedly difficult to interpret the traces we see as nothing more than narrow bundles of equally possible paths between which the wave surfaces establish cross-connections. Yet, these cross-connections are necessary for an understanding of the diffraction and interference phenomena which can be demonstrated for the same particle with the same plausibility—and that on a large scale, not just as a consequence of the theoretical ideas about the interior of the atom, which we mentioned earlier. Conditions are admittedly such that we can always manage to make do in each concrete individual case without the two different aspects leading to different expectations as to the result of certain experiments. We cannot, however, manage to make do with such old, familiar, and seemingly indispensible terms as "real" or "only possible"; we are never in a position to say what really is or what really happens, but we can only say what will be observed in any concrete individual case. Will we have to be permanently satisfied with this...? On principle, yes. On principle, there is nothing new in the postulate that in the end exact science should aim at nothing more than the description of what can really be observed. The question is only whether from now on we shall have to refrain from tying description to a clear hypothesis about the real nature of the world. There are many who wish to pronounce such abdication even today. But I believe that this means making things a little too easy for oneself.

I would define the present state of our knowledge as follows. The ray or the particle path corresponds to a longitudinal relationship of the propagation process (*i.e.* in the direction of propagation), the wave surface on the other hand to a transversal relationship (*i.e.* normal to it). Both relationships are without doubt real; one is proved by photographed particle paths, the other by interference experiments. To combine both in a uniform system has proved impossible so far. Only in extreme cases does either the transversal, shell-shaped or the radial, longitudinal relationship predominate to such an extent that we think we can make do with the wave theory alone or with the particle theory alone.

31.3   Study Questions

QUES. 31.1. What is Fermat's principle?

a) What happens to a ray of light when passing from one medium into another?
b) What two laws govern the reflection and refraction of light rays? How are these related to Fermat's principle?
c) What is the relationship between the law of refraction and the wave theory of light?
d) In what sense is Fermat's principle the "quintessence of the wave theory"?

QUES. 31.2. How is Hamilton's principle similar to Fermat's principle?

a) Do Hamilton's and Fermat's principles deal with the same physical entities?
b) What puzzle does the similarity of the two principles then present?
c) How is a mass point similar to an isolated ray of light? What does this imply about its physical significance?

QUES. 31.3. Under what conditions is the wave-like character of light most clearly expressed?

a) How do the physical dimensions of a body affect the degree to which the wave-like character of light is noticeable?
b) What everyday examples does Schrödinger provide in which the wave-like character of light is observed?
c) Why can one hear, but not see, a person calling from behind a wall?

QUES. 31.4. Is the wave-like character of matter always observable? Is it ever observable?

a) Why had the wave-like character of matter been overlooked until now? What systems most clearly illustrate this wave-like character?
b) Is it mere chance that the characteristic length-scale of matter-waves is the same as the size of an atom? What does Schrödinger mean when he asserts that "the atom in reality is merely the diffraction phenomenon of an electron wave captured as it were by the nucleus of the atom"?
c) What significant role does Planck's constant play in the theory of wave mechanics?

QUES. 31.5. What is the fundamental idea of wave mechanics?

a.) Schrödinger states that it is "difficult to interpret the traces we see as nothing more than narrow bundles of equally possible paths between which the wave surfaces establish cross-connections." Why, then then does he suggest this interpretation?
b.) Schrödinger states that it is an old idea that "exact science should aim at nothing more than the description of what can really be observed." Do you agree with this postulate? Can you name one or more scientists who would disagree with this postulate?

c.) Does Schrödinger himself wish to "refrain from tying description to a clear hypothesis about the real nature of the world"?

d.) How does Schrödinger summarize the present state of knowledge?

31.4 Exercises

EX. 31.1 (THE TIME-DEPENDENT SCHRÖDINGER EQUATION). In 1926, Schrödinger published a famous paper in which he set forth his theory of wave mechanics. The heart of this theory is the so-called *time-dependent Schrödinger equation*. This (linear, homogeneous, second-order, partial) differential equation may be expressed (in cartesian coordinates) as follows:

$$i\hbar \frac{\partial}{\partial t} \Psi = -\frac{\hbar^2}{2m} \left(\frac{\partial^2}{\partial x^2} + \frac{\partial^2}{\partial y^2} + \frac{\partial^2}{\partial z^2} \right) \Psi + V\Psi. \tag{31.1}$$

Here, $\hbar$ is Planck's constant (divided by 2π), m is the mass of the particle under consideration, V is the potential energy function to which mass m is subjected, and $i = \sqrt{-1}$. The solutions, Ψ, of this differential equation are the matter-waves to which Schrödinger refers in his nobel lecture. Generally speaking, they are time- and space- dependent functions whose form depends on the potential energy function $V(x, y, z)$. For example, for an electron placed in a spherically symmetric inverse-square potential (such as the potential of a hydrogen nucleus), the solutions, Ψ, yield the various stationary state orbitals of a neutral hydrogen atom.[1] In Ex. 31.4 and 31.5, we will explore these solutions for a simple particle-in-a-box potential. In the present problem, we will motivate the development of the Schrödinger equation itself.

The time-dependent Schrödinger equation may be understood as the $F = ma$ (so to speak) of wave mechanics, in that it serves as the axiom[2] upon which any analysis of matter-waves is based.[3] Considered as an axiom, Eq. 31.1 cannot be derived. Nonetheless, one can begin to understand why the fundamental equation of wave mechanics must have this particular form by considering another partial differential equation—the classical wave equation:

$$\frac{\partial^2}{\partial t^2} \Psi = c^2 \frac{\partial^2}{\partial x^2} \Psi \tag{31.2}$$

---

[1] See Bohr's discussion of stationary states of they hydrogen atom in Chaps. 28–30 of the present volume.

[2] In his *Principia*, Newton described his second law of motion not as an empirical law but as an *Axiom*; see Chap. 21 of volume II.

[3] Strictly speaking, Schrödinger's equation is only applicable to non-relativistic quantum systems; when treating relativistic particles the Dirac equation must be used.

The one-dimensional wave equation, as written above, governs the propagation of waves in diverse classical systems, from longitudinal sound waves to transverse electromagnetic waves.

a) Prove that the plane-wave $\Psi(x, t) = \Psi_0 \cos(kx - \omega t)$ is a solution to the one-dimensional wave equation, provided that the wave speed, c, the wave-number, k, and the angular frequency, ω, obey the *dispersion relation*:

$$\omega^2 = c^2 k^2 \qquad (31.3)$$

How are k and ω related to the wavelength, λ, and the period, T, of the traveling wave? In which direction is this wave propagating?

b) If we take Einstein's theory of light seriously, then electromagnetic waves may be understood as photons whose energy and momentum are given by $E = \hbar\omega$ and $p = h/\lambda$, respectively. Prove that the energy and momentum of a quantum of light are then related by

$$E = cp. \qquad (31.4)$$

c) Unlike electrons, photons have zero mass. So presumably, Eq. 31.4 does not work for electrons. In fact, from classical mechanics, we know that the energy and momentum of a particle of mass m moving in a potential V are related by

$$E = \frac{p^2}{2m} + V. \qquad (31.5)$$

The question now is: can we begin with Eq. 31.5 and work backwards to construct a wave-equation which governs the behavior of massive particles? To begin, let us assume that the energy and momentum of a massive particle are still given by the Einstein and de Broglie relations. Show that Eq. 31.5 may then be written as

$$\hbar\omega = \frac{\hbar^2 k^2}{2m} + V.$$

d) Notice that in the dispersion relation for a massive particle, ω and k are no longer linearly related, as they were in the dispersion relation for a photon. We might therefore anticipate that the new partial differential equation we seek has a first derivative of Ψ with respect to t on the left-hand-side and a second derivative of Ψ with respect to x on the right-hand-side. Let us try an equation of the form

$$A \frac{\partial}{\partial t} \Psi = B \frac{\partial^2}{\partial x^2} \Psi + V\Psi.$$

Here, A and B are (soon to-be-determined) constants. Is $\Psi(x, t) = \Psi_0 \cos(kx - \omega t)$ a solution to this equation? What problem do you run into?

e) Show that this problem can be fixed by (i) using a *complex* plane-wave solution of the form $\Psi(x, t) = \Psi_0 e^{i(kx - \omega t)}$, and (ii) appropriately choosing the values of A and B. This should allow you to obtain a one-dimensional form of the time-dependent Schrödinger equation, Eq. 31.1.

Ex. 31.2 (THE SUPEPOSITION PRINCIPLE AND SCHRÖDINGER'S CAT). For a free particle which obeys the Schrödinger equation, the wave-number (and frequency) can have any values whatsoever. For a bound particle, on the other hand, we will soon find that they can only take on certain, discrete values.[4] Anticipating this development, let us introduce an integer, n, and denote the allowed wave-numbers and frequencies by k_n and ω_n, respectively. The nth allowed solution to the Schrödinger equation is then written as

$$\Psi_n(x, t) = \Psi_{0,n} e^{i(k_n x - \omega_n t)} \tag{31.6}$$

Now it is a general feature of linear differential equations (such as the Schrödinger equation), that any linear combination of solutions is itself a solution. To illustrate this, use Eq. 31.6 to prove that if Ψ_1 and Ψ_2 are solutions of the time-dependent Schrödinger equation, then so is the linear combination

$$\Psi(x, t) = a_1 \Psi_1(x, t) + a_2 \Psi_2(x, t). \tag{31.7}$$

Equation 31.7 expresses the *superposition principle*. This principle curiously implies that the state of an electron can (theoretically) be represented by a wave-function, Ψ, which has more than one wavelength and more than one frequency. An electron in such a *superposition state* will typically not have a unique momentum or a unique energy—at least until it is actually observed by a measuring apparatus. When observed, however, the electron will be found to have the momentum (or energy) associated with one of the particular states which comprise the superposition state. For example, for an electron in the state given by Eq. 31.7, whose constituent states Ψ_1 and Ψ_2 have energy values $E_1 = \hbar\omega_1$ and $E_2 = \hbar\omega_2$, the probability of finding the energy of the electron to be E_1 will be $|a_1|^2$, and the probability of finding the energy of the electron to be E_2 will be $|a_2|^2$. After this measurement process, the electron will persist in whichever state happened to be measured; the superposition state, Ψ, is said to have "collapsed" during the measurement process into one of the possible "energy eigenstates" Ψ_1 or Ψ_2.

It is interesting to note that Schrödinger himself considered the existence of such superposition states to be rather dubious, for the following reason. Perhaps a microscopic body (such as an electron) might exist in a superposition state of different energies. But when this concept is extended to macroscopic bodies, the superposition principle implies that a large organism (such as a cat) might exist in a superposition state in which it is both alive and dead. Schrödinger thought this situation to be absurd. Is it?[5]

Ex. 31.3 (THE TIME-INDEPENDENT SCHRÖDINGER EQUATION). In Ex. 31.1, we developed a partial differential equation which yields the correct dispersion relation

---

[4] See Ex. 31.4, below.

[5] Other paradoxes associated with the so-called Copenhagen interpretation of quantum theory will be presented by Werner Heisenberg in the subsequent chapter.

for a massive particle whose energy and momentum are given by $E = \hbar\omega$ and $p = h/\lambda$. This partial differential equation is the time-dependent Schrödinger equation. In many cases, the solutions to this equation can be factored into the product of a space-dependent function and a time-dependent function. For example, in one spatial dimension, Ψ may be written as

$$\Psi(x, t) = \psi(x)\phi(t). \tag{31.8}$$

a) Show that when this is the case, we may re-write the time-dependent Schrödinger equation as

$$\frac{1}{\psi(x)}\left(\frac{-\hbar^2}{2m}\right)\frac{\partial^2}{\partial x^2}\psi(x) + V = \frac{i\hbar}{\phi(t)}\frac{\partial}{\partial t}\phi(t) \tag{31.9}$$

b) Notice that if the potential function is time-independent, then the left-hand-side of Eq. 31.9 is only a function of x, and the right-hand-side is only a function of t. We have separated the variables. Moreover, since x and t are *independent* variables, the only way for solutions of Eq. 31.9 to exist is if the left and right sides are separately equal to the same (space- and time-independent) constant. Using this technique of separation of variables, show that $\phi(t)$ takes the form

$$\phi(t) = e^{-iEt/\hbar}, \tag{31.10}$$

and that $\psi(x)$ obeys the (one-dimensional) *time-independent Schrödinger equation*:

$$\frac{-\hbar^2}{2m}\frac{\partial^2}{\partial x^2}\psi(x) + V(x)\psi(x) = E\psi(x) \tag{31.11}$$

c) Finally, show that in three spatial dimensions the time-independent Schrödinger equation becomes

$$\frac{-\hbar^2}{2m}\left(\frac{\partial^2}{\partial x^2} + \frac{\partial^2}{\partial y^2} + \frac{\partial^2}{\partial z^2}\right)\psi(x, y, z) + V(x, y, z)\psi(x, y, z) = E\psi(x, y, z) \tag{31.12}$$

Ex. 31.4 (PARTICLE IN A 1-D BOX). In this exercise, we will determine the energy of an electron confined in a hypothetical one-dimensional box of length a which has absolutely impenetrable walls. This box may be described by the following potential function.

$$V(x) = \begin{cases} 0 & \text{if } 0 < x < a \\ \infty & \text{if } x \leq 0 \text{ or } x \geq a \end{cases} \tag{31.13}$$

Notice that $V(x)$ doesn't depend on time. Thus, we can use the time-independent Schrödinger equation to determine $\psi(x)$.

a) First, demonstrate that the function

$$\psi(x) = A\sin(kx) \tag{31.14}$$

is a viable solution to the time-independent Schrödinger equation *inside the box.* (HINT Plug it into Eq. 31.11 and check!) *A* is here a constant which we will determine shortly.

b) Does our solution work outside the box? Notice that $\psi(x)$ must vanish (become zero) in the regions where $V = \infty$; otherwise the left-hand side of Eq. 31.11 would be infinite, and this would require an infinite energy, E, on the right-hand side. So now show that in order to satisfy the boundary conditions, the wave-number can only have certain, discrete values:

$$k_l = \frac{l\pi}{a}, \tag{31.15}$$

where $l = 0, 1, 2, \ldots$ (HINT: where are the nodes of $\sin(kx)$?)

c) In order to solve the Schrödinger equation for the particle in a one-dimensional box, we had to introduce an integer, l; this is called a *quantum number*. Prove that the electron's energy is given by

$$E_l = \frac{h^2 l^2}{8ma^2}. \tag{31.16}$$

The energies E_l are said to be the *eigenvalues* (special values) which correspond to each of the *eigenfunctions* (special functions) ψ_l.

d) Thus far, we have been focusing on the space-dependent part of the wave-function, $\psi(x)$; recall that the solutions to Schrödinger's equation also contain a time-dependent factor, $\phi(t)$. So the wave-functions of the electron in a one-dimensional box should be written as

$$\Psi_l(x, t) = A e^{-i E_l t/\hbar} \sin(l\pi x/a), \tag{31.17}$$

where E_l is given by Eq. 31.16. This is a complex space- and time-dependent function. Show that $|\Psi_l(x, t)|^2$, which is the product of $\Psi_l(x, t)$ and its complex conjugate, is a time-independent and real function of x.

e) Schrödinger initially suggested that the aforementioned quantity, $|\Psi_l(x)|^2$, represented the charge density of the electron. In other words, he thought the electron charge was somehow smeared out over a region of space. Later, Max Born interpreted this same quantity as a probability density function. That is, he understood

$$P(x)\,dx = |\Psi_l(x)|^2\,dx \tag{31.18}$$

as the probability that an electron might be found in a small region of space between x and $x + dx$ (if it was measured, for example, with an ionization chamber or a scintillation counter). In order to treat $P(x)$ as a probability density function, there must be a 100% likelihood of finding the electron *somewhere* inside the box. Carry out the integral

$$\int_{x=0}^{x=a} P(x)\,dx = 1 \tag{31.19}$$

and thereby determine the appropriate value of A which *normalizes* the wave-function $\Psi_l(x, t)$.

f) Make a sketch of the probability density function, $P(x)$, for an electron in (i) the ground state ($l = 0$) and (ii) the first excited state ($l = 1$). Where is the electron most likely to be located when in each of these two states? Where is it least likely to be located?

g) What is the wavelength of the photon which is emitted when the electron makes a transition from the first excited state to the ground state? How small would the box need to be for this photon to lie in the visible range of the spectrum?

Ex. 31.5 (PARTICLE IN A 3-D BOX). In this exercise, you will attempt to generalize the analysis of Ex. 31.4 to the case of a three-dimensional box having sides of length a, b, and c.

a) First, you should construct a wave-function $\psi_{l,m,n}(x, y, z)$ which satisfies the three-dimensional time-independent Schrödinger equation. Remember that these wave-functions must vanish outside of the box. This will require the introduction of three quantum numbers, l, m, and n.

b) Next, you should find a mathematical expression for the energy eigenvalues, $E_{l,m,n}$, which correspond to the eigenfunctions $\psi_{l,m,n}$. Assuming the dimensions of the box to be $(a, b, c) = (10, 12, 15)$ nanometers, what are the quantum numbers, (l, m, n), which correspond to the lowest five energy states? Calculate the wavelength of a photon emitted when the confined electron makes a transition from the first excited state to the ground state.

c) Finally, make a sketch of the probability density function for an electron in the ground state and in the first excited state. Where is the electron most likely to be found when in each of these two states? Where is it least likely to be found?

31.5   Vocabulary

1. Refraction
2. Reflection
3. Infinitesimal
4. Tantamount
5. Postulate
6. Strident
7. Quantum
8. Microcosm
9. Fictitious
10. Exposition
11. Obviate
12. Render
13. Quintessence
14. Diffraction

15. Surmise
16. Gross
17. Salient
18. Trivial
19. Prerogative
20. Confound
21. Longitudinal
22. Transverse

Chapter 32
The Quantum Paradox

> One would get into hopeless difficulties if one tried to describe
> what happens between two consecutive observations.
>
> —Werner Heisenberg

32.1 Introduction

Werner Heisenberg (1901–1976) was born in the city of Würzburg, part of the German Empire. In 1920, he enrolled at the University of Munich where his father was a professor of Byzantine studies. Although Heisenberg shared his father's love of classical languages, he chose to study physics under Arnold Sommerfeld while at the University. Sommerfeld was one of the principle architects of the "old quantum theory" which used as its starting point Rutherford's planetary model of the atom.[1] After struggling through a laboratory course on experimental physics (led by Wilhelm Wien), Heisenberg earned his Ph.D. in 1923. His dissertation was "On the Stability and Turbulence of Fluid Flow." Immediately thereafter, he traveled to Göttingen to begin working as an assistant to Max Born, who would later be awarded the Nobel Prize in physics for his statistical interpretation of the wave function which appeared in Schrödinger's wave-mechanical quantum theory.[2] Heisenberg had already spent some time with Born in Göttingen while his doctoral advisor, Sommerfeld, was lecturing at the University of Wisconsin during the winter of 1922. While working as an assistant to Born, Heisenberg briefly traveled to the University of Copenhagen to collaborate with Niels Bohr and his assistant Hendrik Kramers. By the time he completed his habilitation, which qualified him to teach at the university level, Heisenberg's attention was devoted almost entirely to developing a new quantum theory of the atom. His breakthrough came in 1925, when he and Born published a paper entitled "Quantum-Theoretical Re-interpretation of Kinematic and Mechanical Relations."

---

[1] See Chaps. 28–30 of the present volume.

[2] Max Born was awarded the Nobel prize in physics in 1954 for work which he had published in 1926. For a brief explanation of the probability interpretation of $|\Psi^*\Psi|$, see Ex. 31.4 in the previous chapter of the present volume.

© Springer International Publishing Switzerland 2016
K. Kuehn, *A Student's Guide Through the Great Physics Texts*,
Undergraduate Lecture Notes in Physics, DOI 10.1007/978-3-319-21828-1_32

Unlike the old quantum theory of Bohr and Sommerfeld which incorporated classical ideas (such as planet-like electron orbits), Heisenberg's new quantum mechanics was expressed entirely in terms of quantities which are observable—at least in principle—such as the position and the momentum of an electron. One of the most striking and significant features of the new quantum mechanics is the formulation of *Heisenberg's uncertainty (or indeterminacy) principle*, which appeared in his 1927 publication "On the Perceptual Content of Quantum Theoretical Kinematics and Mechanics." Heisenberg's *indeterminacy principle* places a numerical limit on the precision with which one may simultaneously determine the values of certain pairs of observable quantities. For example, the position, q, and the momentum, p, of a particle obey the uncertainty relation

$$\Delta q \, \Delta p \geq \frac{h}{4\pi}. \tag{32.1}$$

Here, Δq and Δp refer to the uncertainties of the particle's position and momentum, respectively, and h is Planck's constant.[3] In other words, for a subatomic particle, "the more precisely the position is determined, the less precisely the momentum is known in this instant, and *vice versa*."[4]

Shortly after his development of the new quantum mechanics, Heisenberg was appointed Lecturer in Theoretical Physics at the University of Copenhagen. Then in 1927 he was appointed Professor of Theoretical Physics at the University of Leipzig. He was awarded the Nobel Prize in Physics in 1932. After the discovery of the neutron in 1932 by James Chadwick, Heisenberg helped to develop a theory of proton-neutron interactions within the atomic nucleus (now called the strong nuclear force). In 1941, he was appointed Professor of Physics at the University of Berlin and then the Director of the Kaiser Wilhelm Institute for Physics. Along with several other nuclear scientists who remained and worked for Germany during the war, Heisenberg was taken prisoner by Allied troops and relocated to England. Upon his return to Germany in 1946, he assisted in reorganizing his former "Kaiser Wilhlem" Institute into the (politically uncontroversial) "Max Planck" Institute. Later, he would serve as the President of the Alexander von Humboldt Foundation and as Professor of Theoretical Physics at the University of Munich, where he had begun his scientific studies under Arnold Sommerfeld so many years before.

The reading selection that follows was one of Heisenberg's Gifford Lectures delivered at the University of Saint Andrews in the winter of 1955–1956. Herein, he describes the *Copenhagen Interpretation* of quantum theory. This interpretation takes the indeterminacy principle as the cornerstone, so to speak, of quantum theory. In other words, it assumes that our inability to simultaneously know the position and momentum of a sub-atomic particle with unlimited precision is *not* simply an artifact of one particular method of measurement. Rather, it is an essential feature of nature

---

[3] See Heisenberg, W., *The Physical Principles of Quantum Theory*, University of Chicago Press, Chicago, 1930.
[4] This is a translated quote from Heisenberg's 1927 publication.

itself, and hence cannot be circumvented under any conceivable circumstances. As you explore the following text, you might consider the following questions: what (if any) are the philosophical and scientific implications of accepting the indeterminacy principle? In particular, does the Copenhagen interpretation of quantum theory retain the notion of causality? Do you think that Heisenberg's conclusions are correct?

32.2   Reading: Heisenberg, *The Copenhagen Interpretation of Quantum Theory*

Heisenberg, W., *Physics and Philosophy*, World Perspectives, George Allen & Unwin, London, 1958. Chap. 3.

The Copenhagen interpretation of quantum theory starts from a paradox. Any experiment in physics, whether it refers to the phenomena of daily life or to atomic events, is to be described in the terms of classical physics. The concepts of classical physics form the language by which we describe the arrangement of our experiments and state the results. We cannot and should not replace these concepts by any others. Still the application of these concepts is limited by the relations of uncertainty. We must keep in mind this limited range of applicability of the classical concepts while using them, but we cannot and should not try to improve them.

For a better understanding of this paradox it is useful to compare the procedure for the theoretical interpretation of an experiment in classical physics and in quantum theory. In Newton's mechanics, for instance, we may start by measuring the position and the velocity of the planet whose motion we are going to study. The result of the observation is translated into mathematics by deriving numbers for the co-ordinates and the momenta of the planet from the observation. Then the equations of motion are used to derive from these values of the co-ordinates and momenta at a given time the values of these co-ordinates or any other properties of the system at a later time, and in this way the astronomer can predict the properties of the system at a later time. He can, for instance, predict the exact time for an eclipse of the moon.

In quantum theory the procedure is slightly different. We could for instance be interested in the motion of an electron through a cloud chamber and could determine by some kind of observation the initial position and velocity of the electron. But this determination will not be accurate; it will at least contain the inaccuracies following from the uncertainty relations and will probably contain still larger errors due to the difficulty of the experiment. It is the first of these inaccuracies which allows us to translate the result of the observation into the mathematical scheme of quantum theory. A probability function is written down which represents the experimental situation at the time of the measurement, including even the possible errors of the measurement.

This probability function represents a mixture of two things, partly a fact and partly our knowledge of a fact. It represents a fact in so far as it assigns at the initial time the probability unity (*i.e.*, complete certainty) to the initial situation: the electron moving with the observed velocity at the observed position; "observed" means observed within the accuracy of the experiment. It represents our knowledge in so far as another observer could perhaps know the position of the electron more accurately. The error in the experiment does—at least to some extent—not represent a property of the electron but a deficiency in our knowledge of the electron. Also this deficiency of knowledge is expressed in the probability function.

In classical physics one should in a careful investigation also consider the error of the observation. As a result one would get a probability distribution for the initial values of the co-ordinates and velocities and.therefore something very similar to the probability function in quantum mechanics. Only the necessary uncertainty due to the uncertainty relations is lacking in classical physics.

When the probability function in quantum theory has been determined at the initial time from the observation, one can from the laws of quantum theory calculate the probability function at any later time and can thereby determine the probability for a measurement giving a specified value of the measured quantity. We can, for instance, predict the probability for finding the electron at a later time at a given point in the cloud chamber. It should be emphasized, however, that the probability function does not in itself represent a course of events in the course of time. It represents a tendency for events and our knowledge of events. The probability function can be connected with reality only if one essential condition is fulfilled: if a new measurement is made to determine a certain property of the system. Only then does the probability function allow us to calculate the probable result of the new measurement. The result of the measurement again will be stated in terms of classical physics.

Therefore, the theoretical interpretation of an experiment requires three distinct steps: (1) the translation of the initial experimental situation into a probability function; (2) the following up of this function in the course of time; (3) the statement of a new measurement to be made of the system, the result of which can then be calculated from the probability function. For the first step the fulfillment of the uncertainty relations is a necessary condition. The second step cannot be described in terms of the classical concepts; there is no description of what happens to the system between the initial observation and the next measurement. It is only in the third step that we change over again from the "possible" to the "actual."

Let us illustrate these three steps in a simple ideal experiment. It has been said that the atom consists of a nucleus and electrons moving around the nucleus; it has also been stated that the concept of an electronic orbit is doubtful. One could argue that it should at least in principle be possible to observe the electron in its orbit. One should simply look at the atom through a microscope of a very high resolving power, then one would see the electron moving in its orbit. Such a high resolving power could to be sure not be obtained by a microscope using ordinary light, since the inaccuracy of the measurement of the position can never be smaller than the wave length of the light. But a microscope using γ-rays with a wave length smaller than the size of the atom would do. Such a microscope has not yet been constructed but that should not prevent us from discussing the ideal experiment.

Is the first step, the translation of the result of the observation into a probability function, possible? It is possible only if the uncertainty relation is fulfilled after the observation. The position of the electron will be known with an accuracy given by the wave length of the γ-ray. The electron may have been practically at rest before the observation. But in the act of observation at least one light quantum of the γ-ray must have passed the microscope and must first have been deflected by the electron. Therefore, the electron has been pushed by the light quantum, it has changed its momentum and its velocity, and one can show that the uncertainty of this change is just big enough to guarantee the validity of the uncertainty relations. Therefore, there is no difficulty with the first step.

At the same time one can easily see that there is no way of observing the orbit of the electron around the nucleus. The second step shows a wave pocket moving not around the nucleus but away from the atom, because the first light quantum will have knocked the electron out from the atom. The momentum of light quantum of the γ-ray is much bigger than the original momentum of the electron if the wave length of the γ-ray is much smaller than the size of the atom. Therefore, the first light quantum is sufficient to knock the electron out of the atom and one can never observe more than one point in the orbit of the electron; therefore, there is no orbit in the ordinary sense. The next observation—the third step—will show the electron on its path from the atom. Quite generally there is no way of describing what happens between two consecutive observations. It is of course tempting to say that the electron must have been somewhere between the two observations and that therefore the electron must have described some kind of path or orbit even if it may be impossible to know which path. This would be a reasonable argument in classical physics. But in quantum theory it would be a misuse of the language which, as we will see later, cannot be justified. We can leave it open for the moment, whether this warning is a statement about the way in which we should talk about atomic events or a statement about the events themselves, whether it refers to epistemology or to ontology. In any case we have to be very cautious about the wording of any statement concerning the behavior of atomic particles.

Actually we need not speak of particles at all. For many experiments, it is more convenient to speak of matter waves; for instance, of stationary matter waves around the atomic nucleus. Such a description would directly contradict the other description if one does not pay attention to the limitations given by the uncertainty relations. Through the limitations the contradiction is avoided. The use of "matter waves" is convenient, for example, when dealing with the radiation emitted by the atom. By means of its frequencies and intensities the radiation gives information about the oscillating charge distribution in the atom, and there the wave picture comes much nearer to the truth than the particle picture. Therefore, Bohr advocated the use of both pictures, which he called "complementary" to each other. The two pictures are of course mutually exclusive, because a certain thing cannot at the same time be a particle (*i.e.*, substance confined to a very small volume) and a wave (*i.e.*, a field spread out over a large space), but the two complement each other. By playing with both pictures, by going from the one picture to the other and back again, we finally get the right impression of the strange kind of reality behind our atomic experiments,

Bohr uses the concept of "complementarity" at several places in the interpretation of quantum theory. The knowledge of the position of a particle is complementary to the knowledge of its velocity or momentum. If we know the one with high accuracy we cannot know the other with high accuracy; still we must know both for determining the behavior of the system. The space-time description of the atomic events is complementary to their deterministic description. The probability function obeys an equation of motion as the co-ordinates did in Newtonian mechanics; its change in the course of time is completely determined by the quantum mechanical equation, but it does not allow a description in space and time. The observation, on the other hand, enforces the description in space and time but breaks the determined continuity of the probability function by changing our knowledge of the system.

Generally the dualism between two different descriptions of the same reality is no longer a difficulty since we know from the mathematical formulation of the theory that contradictions cannot arise. The dualism between the two complementary pictures—waves and particles—is also clearly brought out in the flexibility of the mathematical scheme. The formalism is normally written to resemble Newtonian mechanics, with equations of motion for the co-ordinates and the momenta of the particles. But by a simple transformation it can be rewritten to resemble a wave equation for an ordinary three-dimensional matter wave. Therefore, this possibility of playing with different complementary pictures has its analogy in the different transformations of the mathematical scheme; it does not lead to any difficulties in the Copenhagen interpretation of quantum theory.

A real difficulty in the understanding of this interpretation arises, however, when one asks the famous question: But what happens "really" in an atomic event? It has been said before that the mechanism and the results of an observation can always be stated in terms of the classical concepts. But what one deduces from an observation is a probability function, a mathematical expression that combines statements about possibilities or tendencies with statements about our knowledge of facts. So we cannot completely objectify the result of an observation, we cannot describe what "happens" between this observation and the next. This looks as if we had introduced an element of subjectivism into the theory, as if we meant to say: what happens depends on our way of observing it or on the fact that we observe it. Before discussing this problem of subjectivism it is necessary to explain quite clearly why one would get into hopeless difficulties if one tried to describe what happens between two consecutive observations.

For this purpose it is convenient to discuss the following ideal experiment: We assume that a small source of monochromatic light radiates toward a black screen with two small holes in it. The diameter of the holes may be not much bigger than the wave length of the light, but their distance will be very much bigger. At some distance behind the screen a photographic plate registers the incident light. If one describes this experiment in terms of the wave picture, one says that the primary wave penetrates through the two holes; there will be secondary spherical waves starting from the holes that interfere with one another, and the interference will produce a pattern of varying intensity on the photographic plate.

The blackening of the photographic plate is a quantum process, a chemical reaction produced by single light quanta. Therefore, it must also be possible to describe the experiment in terms of light quanta. If it would be permissible to say what happens to the single light quantum between its emission from the light source and its absorption in the photographic plate, one could argue as follows: The single light quantum can come through the first hole or through the second one. If it goes through the first hole and is scattered there, its probability for being absorbed at a certain point of the photographic plate cannot depend upon whether the second hole is closed or open. The probability distribution on the plate will be the same as if only the first hole was open. If the experiment is repeated many times and one takes together all cases in which the light quantum has gone through the first hole, the blackening of the plate due to these cases will correspond to this probability distribution. If one considers only those light quanta that go through the second hole, the blackening should correspond to a probability distribution derived from the assumption that only the second hole is open. The total blackening, therefore, should just be the sum of the blackenings in the two cases; in other words; there should be no interference pattern. But we know this is not correct, and the experiment will show the interference pattern. Therefore, the statement that any light quantum must have gone either through the first *or* through the second hole is problematic and leads to contradictions. This example shows clearly that the concept of the probability function does not allow a description of what happens between two observations. Any attempt to find such a description would lead to contradictions; this must mean that the term "happens" is restricted to the observation.

Now, this is a very strange result, since it seems to indicate that the observation plays a decisive role in the event and that the reality varies, depending upon whether we observe it or not. To make this point clearer we have to analyze the process of observation more closely.

To begin with, it is important to remember that in natural science we are not interested in the universe as a whole, including ourselves, but we direct our attention to some part of the universe and make that the object of our studies. In atomic physics this part is usually a very small object, an atomic particle or a group of such particles, sometimes much larger—the size does not matter; but it is important that a large part of the universe, including ourselves, does *not* belong to the object. Now, the theoretical interpretation of an experiment starts with the two steps that have been discussed. In the first step we have to describe the arrangement of the experiment, eventually combined with a first observation, in terms of classical physics and translate this description into a probability function. This probability function follows the laws of quantum theory, and its change in the course of time, which is continuous, can be calculated from the initial conditions; this is the second step. The probability function combines objective and subjective elements. It contains statements about possibilities or better tendencies ("potentia" in Aristotelian philosophy), and these statements are completely objective, they do not depend on any observer; and it contains statements about our knowledge of the system, which of course are subjective in so far as they may be different for different observers. In ideal cases the subjective element in the probability function may be practically negligible as compared with the objective one. The physicists then speak of a "pure case."

When we now come to the next observation, the result of which should be predicted from the theory, it is very important to realize that our object has to be in contact with the other part of the world, namely, the experimental arrangement, the measuring rod, *etc.*, before or at least at the moment of observation. This means that the equation of motion for the probability function does now contain the influence of the interaction with the measuring device. This influence introduces a new element of uncertainty, since the measuring device is necessarily described in the terms of classical physics; such a description contains all the uncertainties concerning the microscopic structure of the device which we know from thermodynamics, and since the device is connected with the rest of the world, it contains in fact the uncertainties of the microscopic structure of the whole world. These uncertainties may be called objective in so far as they are simply a consequence of the description in the terms of classical physics and do not depend on any observer. They may be called subjective in so far as they refer to our incomplete knowledge of the world.

After this interaction has taken place, the probability function contains the objective element of tendency and the subjective element of incomplete knowledge, even if it has been a "pure case" before. It is for this reason that the result of the observation cannot generally be predicted with certainty; what can be predicted is the probability of a certain result of the observation, and this statement about the probability can be checked by repeating the experiment many times. The probability function does—unlike the common procedure in Newtonian mechanics—not describe a certain event but, at least during the process of observation, a whole ensemble of possible events.

The observation itself changes the probability function discontinuously; it selects of all possible events the actual one that has taken place. Since through the observation our knowledge of the system has changed discontinuously, its mathematical representation also has undergone the discontinuous change and we speak of a "quantum jump." When the old adage "Natura non facit saltus" is used as a basis for criticism of quantum theory, we can reply that certainly our knowledge can change suddenly and that this fact justifies the use of the term "quantum jump."

Therefore, the transition from the "possible" to the "actual" takes place during the act of observation. If we want to describe what happens in an atomic event, we have to realize that the word "happens" can apply only to the observation, not to the state of affairs between two observations. It applies to the physical, not the psychical act of observation, and we may say that the transition from the "possible" to the "actual" takes place as soon as the interaction of the object with the measuring device, and thereby with the rest of the world, has come into play; it is not connected with the act of registration of the result by the mind of the observer. The discontinuous change in the probability function, however, takes place with the act of registration, because it is the discontinuous change of our knowledge in the instant of registration that has its image in the discontinuous change of the probability function.

To what extent, then, have we finally come to an objective description of the world, especially of the atomic world? In classical physics science started from the belief—or should one say from the illusion?—that we could describe the world or at least parts of the world without any reference to ourselves. This is actually possible to a large extent. We know that the city of London exists whether we see it or not. It may be said that classical physics is just that idealization in which we can speak about parts of the world without any reference to ourselves. Its success has led to the general ideal of an objective description of the world. Objectivity has become the first criterion for the value of any scientific result. Does the Copenhagen interpretation of quantum theory still comply with this ideal? One may perhaps say that quantum theory corresponds to this ideal as far as possible. Certainly quantum theory does not contain genuine subjective features, it does not introduce the mind of the physicist as a part of the atomic event. But it starts from the division of the world into the "object" and the rest of the world, and from the fact that at least for the rest of the world we use the classical concepts in our description. This division is arbitrary and historically a direct consequence of our scientific method; the use of the classical concepts is finally a consequence of the general human way of thinking. But this is already a reference to ourselves and in so far our description is not completely objective.

It has been stated in the beginning that the Copenhagen interpretation of quantum theory starts with a paradox. It starts from the fact that we describe our experiments in the terms of classical physics and at the same time from the knowledge that these concepts do not fit nature accurately. The tension between these two starting points is the root of the statistical character of quantum theory. Therefore, it has sometimes been suggested that one should depart from the classical concepts altogether and that a radical change in the concepts used for describing the experiments might possibly lead back to a nonstatistical,[5] completely objective description of nature.

This suggestion, however, rests upon a misunderstanding. The concepts of classical physics are just a refinement of the concepts of daily life and are an essential part of the language which forms the basis of all natural science. Our actual situation in science is such that we *do* use the classical concepts for the description of the experiments, and it was the problem of quantum theory to find theoretical interpretation of the experiments on this basis. There is no use in discussing what could be done if we were other beings than we are. At this point we have to realize, as von Weizsäcker has put it, that "Nature is earlier than man, but man is earlier than natural science." The first part of the sentence justifies classical physics, with its ideal of complete objectivity. The second part tells us why we cannot escape the paradox of quantum theory, namely, the necessity of using the classical concepts.

We have to add some comments on the actual procedure in the quantum-theoretical interpretation of atomic events. It has been said that we always start with a division of the world into an object, which we are going to study, and the rest of the world, and that this division is to some extent arbitrary. It should indeed not make any difference in the final result if we, *e.g.*, add some part of the measuring device or the whole device to the object and apply the laws of quantum theory to

---

[5] I have taken the liberty to correct this from "nonstatical"—[*K.K.*].

this more complicated object. It can be shown that such an alteration of the theoretical treatment would not alter the predictions concerning a given experiment. This follows mathematically from the fact that the laws of quantum theory are for the phenomena in which Planck's constant can be considered as a very small quantity, approximately identical with the classical laws. But it would be a mistake to believe that this application of the quantum-theoretical laws to the measuring device could help to avoid the fundamental paradox of quantum theory.

The measuring device deserves this name only if it is in close contact with the rest of the world, if there is an interaction between the device and the observer. Therefore, the uncertainty with respect to the microscopic behavior of the world will enter into the quantum-theoretical system here just as well as in the first interpretation. If the measuring device would be isolated from the rest of the world, it would be neither a measuring device nor could it be described in the terms of classical physics at all.

With regard to this situation Bohr has emphasized that it is more realistic to state that the division into the object and the rest of the world is not arbitrary. Our actual situation in research work in atomic physics is usually this: we wish to understand a certain phenomenon, we wish to recognize how this phenomenon follows from the general laws of nature. Therefore, that part of matter or radiation which takes part in the phenomenon is the natural "object" in the theoretical treatment and should be separated in this respect from the tools used to study the phenomenon. This again emphasizes a subjective element in the description of atomic events, since the measuring device has been constructed by the observer, and we have to remember that what we observe is not nature in itself but nature exposed to our method of questioning. Our scientific work in physics consists in asking questions about nature in the language that we possess and trying to get an answer from experiment by the means that are at our disposal. In this way quantum theory reminds us, as Bohr has put it, of the old wisdom that when searching for harmony in life one must never forget that in the drama of existence we are ourselves both players and spectators. It is understandable that in our scientific relation to nature our own activity becomes very important when we have to deal with parts of nature into which we can penetrate only by using the most elaborate tools.

32.3   Study Questions

QUES. 32.1. Does quantum theory allow for the prediction of future events?

a) How is the future position of the moon predicted in the context of classical physics? What observations are required? What laws must be invoked? And what limits the precision with which such predictions can be made?

b) Is the prediction procedure identical when using quantum physics? In particular, are the sources of uncertainty the same? Are the laws and equations which govern motion the same? And what is the significance of the measurement process itself?

c) What (perhaps surprising) assertion does Heisenberg make regarding the description of a system between individual acts of measurement?

QUES. 32.2. Do electrons actually orbit around the nucleus of an atom?

a) How might a microscope be used to observe the location of an orbiting electron? Can ordinary light be used? What limits the precision of any such measurement?

b) What alternative type of radiation might be instead employed? What difficulty does this raise? And how does this limit the ability of observing subsequent points of the electron's orbit?

c) Are these experimental difficulties associated with the particular type of measurement employed? Could they perhaps be circumvented by some other measurement strategy? If not, then what does this imply about the existence of an atomic orbit?

d) Is the question of whether an atomic orbit exists a problem of ontology or a problem of epistemology?

QUES. 32.3. Can an object be both a particle and a wave at the same time?

a) What is the difference between a particle and a wave? In which context is the particle picture of an electron more appropriate? In which context is the wave picture more appropriate?

b) Which did Bohr advocate, the wave picture or the particle picture? What does it mean that the position and momentum of a particle are *complementary*? How is the space-time description (of atomic events) complementary to a deterministic description (of such events)?

QUES. 32.4. What happens between two consecutive observations of a particle?

a) Describe the setup of Heisenberg's idealized two-slit experiment. Where is the source of light? The screen? The photographic plate?

b) Sketch the illumination of the photographic plate, as you would expect from applying the classical wave theory of light.

c) Similarly, sketch the illumination of the photographic plate, as you would expect from a single photon passing through the left slit in the barrier (if the right slit were blocked).

d) Now sketch the illumination of the photographic plate, as you would expect from a large number of photons passing through the left slit in the barrier (the right slit is still blocked).

e) Next sketch the illumination of the photographic plate, as you would expect from a large number of photons passing through the right slit in the barrier (the left hole is now blocked).

f) Finally, sketch the illumination of the photographic plate, as you would expect from a large number of photons passing through the barrier while both slits are left open.

g) Is it true that each individual photon passes through either the left slit or the right slit? If so, what problem arises? If not, what can we say about the trajectory of each individual photon?

QUES. 32.5. Does London exist when nobody is looking at it?

a) How is the wave function introduced in quantum theory? In particular, how is the shape of the wave function of an electron determined at the moment of measurement—for instance, when registering its position using the click of a geiger counter or a condensation trail in a cloud chamber?

b) What happens to the wave function of the electron during the time interval between measurements? Does the wave function evolve in a predictable way? What equation governs its time evolution? Is quantum theory then a deterministic theory?

c) How would you describe the ontological status of the wave function. Does it exist? If so, in what sense? How does Heisenberg's quantum theory resurrect (so to speak) the Aristotelian concept of *potentia*? At what moment is there a transition from the *possible* to the *actual*?

d) What, then, is the ontological status of the electron itself? Does the electron describe a real trajectory through space between acts of measurement? Does the electron exist between acts of measurement? What about larger objects, such as bugs, persons and cities?

QUES. 32.6. Does quantum theory provide an objective description of the world?

a) What does Heisenberg mean by the term "classical concepts"? Provide an example. Can such concepts adequately describe atomic events? If not, should they be employed? What is the alternative to employing classical concepts?

b) In what sense, then, does the Copenhagen interpretation of quantum theory start from a paradox? Are there any alternatives—such as using a completely new set of concepts?

c) What is the significance of von Weizsäcker's statement that "Nature is earlier than man, but man is earlier than natural science"?

32.4   Exercises

EX. 32.1 (THE PRINCIPLE OF COMPLEMENTARITY ESSAY). What do you think: does the principle of complementarity advocated by Niels Bohr undermine science by giving license to sloppy thinking, or perhaps even to outright contradiction?

EX. 32.2 (THE HEISENBERG UNCERTAINTY PRINCIPLE—SINGLE SLIT DIFFRACTION). In this exercise, we will demonstrate how *Heisenberg's uncertainty relation,*

$$\Delta x \Delta p_x \gtrsim h, \tag{32.2}$$

follows from de Broglie's concept of matter waves. As mentioned in the introduction to this chapter, this principle refers to how precisely one can simultaneously know—under any conceivable circumstances—the values of certain pairs of measurable quantities.[6] In the case of Eq. 32.2, these two quantities are the position (x) and

---

[6] See, for example, Heisenberg, W., *The Physical Principles of Quantum Theory*, University of Chicago Press, Chicago, 1930.

the x-directed momentum (p_x) of a particle; the uncertainties in these quantities are written as Δx and Δp_x, respectively.

a) To begin, suppose that an electron having momentum p_y is aimed directly at a screen in which is cut a small vertical slit of width d. Supposing the electron to be a matter-wave, what is the wavelength of the electron approaching the slit?[7]
b) Write an expression for the angular width, α, of the central diffraction peak after the electron passes through the slit. According to the probability interpretation of the wave-function, the electron has a non-zero probability of being found at any location within this diffraction peak.[8]
c) The width of this peak may thus be interpreted as an indication of the range of possible transverse momenta (which we may call Δp_x) acquired by the electron upon passage through the slit. Write down an expression for Δp_x in terms of α, λ and h.
d) From the previous considerations, obtain an expression for the product $\Delta x \Delta p_x$. What happens to Δp_x as the slit width is narrowed? Do your results agree with Eq. 32.1? What do your results imply?
e) How general are your conclusions? For example, is your uncertainty relation limited to considerations of electrons? Or would it apply to any particle? And what minimal set of assumptions underly the Heisenberg uncertainty relation?

Ex. 32.3 (THE HEISENBERG UNCERTAINTY PRINCIPLE—ELECTRON-IN-A-BOX). Consider the ground-state solution to the time-independent Schrödinger equation for an electron confined in a one-dimensional impenetrable box of width a (see Ex. 31.4). Demonstrate that the uncertainty in the position, Δx, and momentum, Δp_x, of the electron are consistent with the Heisenberg uncertainty relation. (HINT: Notice that when in the ground state, the momentum of the electron has a definite magnitude, but *not* a definite direction. Use the (inverse of the) Euler formula to re-write the standing wave solution as a sum of rightward and leftward traveling waves, and thereby find the uncertainty in the electron's momentum.)

Ex. 32.4 (HUMAN TWO-SLIT EXPERIMENT). Suppose that you visit a peculiar world in which Planck's constant is 100 J-s, rather than the customary value of 6.626×10^{-34} J-s. While visiting this world, you run directly towards a wall with two open doorways into a hallway, as depicted cartoonishly in Fig. 32.1. Suppose that your mass is 50 kg, your speed is 2.0 m/s, the doorways are each 1.0 m wide and are separated by 2.0 m. The width of the hallway is 3.0 m.

a) What is your momentum and your de Broglie wavelength?

---

[7] For an experimental justification of the concept of matter waves, refer back to Davisson and Germer's work on electron scattering from crystals; this is described in Chap. 26 of the present volume.

[8] The probability density function, $|\Psi|^2$, was introduced by Max Born; see Ex. 31.4 of the previous chapter.

Fig. 32.1 A matter-
wave/person running towards
two open doors

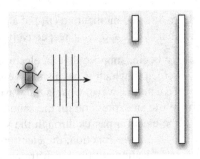

b) Sketch your two-slit interference pattern on the wall opposite the doorways.
   What is the separation between your interference maxima?
c) Provide an interpretation of this interference pattern. In particular, what do the
   maxima and minima of the interference pattern imply? Where do you actually
   hit the wall?
d) Do you believe that the concept of matter waves is equally valid for electrons,
   sand grains and people? If not, why?

EX. 32.5 (SCIENCE AND OBJECTIVITY ESSAY). Should scientific theories provide an
objective description of the world? Does classical theory? Does (the Copenhagen
interpretation of) quantum theory?

32.5   Vocabulary

1. Paradox
2. Gamma ray
3. Matter wave
4. Complementarity
5. Dualism
6. Light quantum
7. Epistemology
8. Ontology
9. Objective
10. Subjective
11. Potentia
12. Quantum jump

Bibliography

Arons, A., and M. B. Peppard, Einstein's Proposal of the Photon Concept—a Translation of the
Annalen der Physik Paper of 1905, *American Journal of Physics, 33*(5), 367–374, 1965.

Bart, D., and J. Bart, Sir William Thomson, on the 150th Anniversary of the Atlantic Cable, *Antique
Wireless Association Review, 21,* 2008.

Bohr, N., *Atomic Theory and the Description of Nature*, Cambridge University Press, Cambridge,
1934.

Bohr, N., The Structure of the Atom, in *Nobel Lectures, Physics 1922–1941*, Elsevier Publishing
Company, 1965.

Boscovich, R. J., *A Theory of Natural Philosophy*, Open Court Publishing Co., Chicago and
London, 1922.

Brown, R., A brief account of microsopical observations made in the months of June, July and
August 1827, on the particles contained in the pollen of plants; and on the general existence
of active molecules in organic and inorganic bodies, *Philosophical Magazine Series 2, 4*(21),
1828.

Campbell, J., *Rutherford: Scientist Supreme*, AAS Publications, Christchurch, New Zealand, 1999.

Carnot, S., *Reflections on the Motive Power of Heat*, second ed., John Wiley & Sons and Chapman
& Hall, New York and London, 1897.

Chadwick, J., The Existence of a Neutron, *Proceedings of the Royal Society of London. Series A,
Containing Papers of a Mathematical and Physical Character, 136*(830), 692–708, 1932.

Chase, G., S. Rituper, and J. Sulcoski, *Experiments in Nuclear Science*, Burgess Pub. Co., 1971.

Clausius, R., *Mechanical Theory of Heat, with its Applications to the Steam-Engine and to the
Physical Properties of Bodies*, John Van Voorst, London, 1867.

Compton, A. H., X-Rays as a Branch of Optics, *Journal of the Optical Society of America and
Review of Scientific Instruments, 16*(2), 71–86, 1928.

Davisson, C., The Diffraction of Electrons by a Crystal of Nickel, *Bell System Technical Journal,
7*(1), 90–105, 1928.

Eliot, C. W. (Ed.), *Scientific Papers: Physics, Chemistry, Astronomy, Geology*, vol. 30, P.F. Collier
& Son, New York, 1910.

Fourier, J., *The Analytical Theory of Heat*, Cambridge University Press, London, 1878.

Gastineau, J. E., *Nuclear Radiation with Computers and Calculators*, 3rd ed., Vernier Software &
Technology, 2003.

Grattan-Guinness, I., *Joseph Fourier 1768–1830*, The Massachusettes Institute of Technology,
1972.

Gray, A., *Lord Kelvin: An Account of His Scientific Life and Work*, J.M. Dent & Co. and E.P. Dutton
& Co., London and New York, 1908.

Haynes, W. M. (Ed.), *The CRC Handbook of Chemistry and Physics*, 95 ed., The Chemical Rubber
Company, 2014.

Heisenberg, W., *The Physical Principles of Quantum Theory*, University of Chicago Press,
Chicago, 1930.

© Springer International Publishing Switzerland 2016

K. Kuehn, *A Student's Guide Through the Great Physics Texts*,
Undergraduate Lecture Notes in Physics, DOI 10.1007/978-3-319-21828-1

Heisenberg, W., *Physics and Philosophy*, World Perspectives, George Allen & Unwin, London, 1958.

Herzberg, G., *Atomic Spectra and Atomic Structure*, 2nd ed., Dover Publications, New York, 1944.

Hoppe, H.-H., In Defense of Extreme Rationalism, *Rev. Austrian Econ.*, *3*(1), 179–214, 1989.

Laue, M., Concerning the detection of X-ray interferences, in *Nobel Lectures, Physics 1901–1921*, Elsevier Publishing Company, 1914.

Lawrence, B. W., The diffraction of X-rays by crystals, in *Nobel Lectures, Physics 1901–1921*, Elsevier Publishing Company, 1915.

Leff, H. S., and A. F. Rex (Eds.), *Maxwell's Demon: Entropy, Information, Computing*, Princeton Series in Physics, Princeton University Press, 1990.

Maxwell, J. C., *Theory of Heat*, tenth ed., Longmans, Green, and Co., London and New York, 1891.

Melsen, A. G., *From Atomos to Atom*, Duquesne University Press, Pittsburg, 1952.

Millikan, R. A., A direct photoelectric determination of Planck's "h", *Physical Review*, *7*(3), 355, 1916.

Newton, I., *Opticks: or A Treatise of the Reflections, Refractions, Inflections & Colours of Light*, 4th ed., William Innes at the West-End of St. Pauls, London, 1730.

Planck, M., *Eight Lectures on Theoretical Physics*, Columbia University Press, New York, 1915.

Reif, F., *Fundamentals of statistical and thermal physics*, McGraw-Hill, 1965.

Röntgen, W. C., On a New Kind of Rays, *Nature*, *3*(59), 277–231, 1896.

Rutherford, E., XLI. The Mass and Velocity of the α particles expelled from Radium and Actinium., *Philosophical Magazine Series 6*, *12*(70), 348–371, 1906.

Rutherford, E., Bakerian Lecture: Nuclear Constitution of Atoms, *Proceedings of the Royal Society of London. Series A, Containing Papers of a Mathematical and Physical Character*, *97*(686), 374–400, 1920.

Schrödinger, E., Wave Mechanics, in *Nobel Lectures, Physics 1922–1941*, Elsevier Publishing Company, 1965.

Thomson, J. J., *The Corpuscular Theory of Matter*, Charles Scribner's Sons, New York, 1907.

Thomson, S. W., Kinetic Theory of the Dissipation of Energy, *Nature*, pp. 441–444, 1874.

Thomson, S. W., The Sorting Demon of Maxwell, *Proceedings of the Royal Institution*, *ix*, 113, 1879.

Thomson, S. W., *Mathematical and Physical Papers*, Cambridge University Press, 1882.

Thomson, W., On Vortex Atoms, *Proceedings of the Royal Society of Edinburgh*, *6*, 1867.

Wallace, W. A., The Problem of Causality in Galileo's Science, *The Review of Metaphysics*, *36*(3), 607–632, 1983.

Index

© Springer International Publishing Switzerland 2016   459
K. Kuehn, *A Student's Guide Through the Great Physics Texts*,
Undergraduate Lecture Notes in Physics, DOI 10.1007/978-3-319-21828-1

Printed in the United States
By Bookmasters